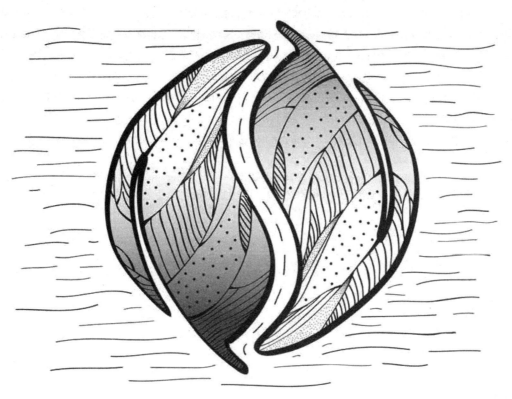

Spring 5
企业级开发实战

周冠亚 黄文毅 著

清华大学出版社
北京

内 容 简 介

Spring 框架是为了降低解决企业系统开发的复杂度而产生的，掌握并学会使用 Spring 框架进行项目开发，是 Java 开发人员必备技能之一，本书从企业应用开发的角度出发，深入浅出地讲解了 Spring 5 的新特性和 Spring 集成开发技术。全书共 19 章，第 1 章~第 3 章主要讲解如何搭建 Spring 开发环境以及 Spring IoC 和 AOP 容器的原理及代码分析。第 4 章和第 5 章概述 Spring 5 和 Java 8 的新特性。第 6 章和第 7 章讲解 Spring 5 新特性——WebFlux 响应式编程、开发和调试。第 8 章和第 9 章主要讲解 Spring 5 集成 Kotlin 语言以及更多 Spring 5 新特性的细节。第 10 章~第 19 章主要介绍 Spring 集成其他热门技术，例如，Log4j2 日志框架、Spring MVC、MyBatis、Redis 缓存、ZooKeeper、Kafka 消息中间件、Mycat 分库分表中间件、Sharding-JDBC 和 Dubbo 服务治理框架等。附录部分介绍本书涉及的以及在面试中常见的设计模式。

本书适用于所有 Java 编程语言开发人员、分布式系统开发爱好者以及计算机专业的学生等。

本书封面贴有清华大学出版社防伪标签，无标签者不得销售。
版权所有，侵权必究。侵权举报电话：010-62782989　13701121933

图书在版编目（CIP）数据

Spring 5 企业级开发实战/周冠亚，黄文毅著. —北京：清华大学出版社，2019
ISBN 978-7-302-53102-9

Ⅰ．①S… Ⅱ．①周… ②黄… Ⅲ．①JAVA 语言－程序设计 Ⅳ．①TP312.8

中国版本图书馆 CIP 数据核字（2019）第 101115 号

责任编辑：王金柱
封面设计：王　翔
责任校对：闫秀华
责任印制：丛怀宇

出版发行：清华大学出版社
网　　址：http://www.tup.com.cn，http://www.wqbook.com
地　　址：北京清华大学学研大厦 A 座
邮　　编：100084
社 总 机：010-62770175
邮　　购：010-62786544
投稿与读者服务：010-62776969，c-service@tup.tsinghua.edu.cn
质 量 反 馈：010-62772015，zhiliang@tup.tsinghua.edu.cn

印 装 者：三河市龙大印装有限公司
经　　销：全国新华书店
开　　本：190mm×260mm
印　　张：34
字　　数：870 千字
版　　次：2019 年 7 月第 1 版
印　　次：2019 年 7 月第 1 次印刷
定　　价：118.00 元

产品编号：082126-01

推 荐 语

从基础再到深入浅出，用极其简单的例子详解了 Spring 的每个知识点，更重要的是每一个知识点都有极其详细生动的例子搭配讲解，特别是 Spring AOP 业务和系统功能分离的思想，看到之后原来都可以这么简单；所以非常推荐此书给大家。

前苏宁易购系统架构师·JackLiu

这是一本获取 Spring 5 知识和经验的必备图书。本书通过理论和实际应用相结合的方式对 Spring 的核心知识点进行深入剖析，同时也介绍了 Spring 5 的新特性。在阅读完本书后，可以让读者更好地理解 Spring 的实现原理和底层架构，能够使用 Spring 的强大功能至上而下地构建复杂的 Spring 应用程序。感谢作者花了大量时间和精力创作了一本 Spring 领域的百科全书。

驴妈妈旅游网资深研发工程师·邓贤文

本书很好地讲述了 Spirng 5 在实际开发的各种重要核心技术和最新实用技术，深入浅出地论述了每个技术的应用场景，解释深入，通俗易懂。不仅适合入门者系统地学习 Spring 技术，也适合有一定工作经验的人来加强和深入对 Spring 的理解，是一本质量很高的 Spring 技术图书。

中泰证券股份有限公司科技研发部技术经理·王祥来

仅从书的目录来看，本书实战型比较强，通过具体地、常用的实战例子，引导、激发大家学习 Spring 的热情，相信这将会是一本不错的 Spring 5 参考书。

美图技术专家，导师·阮龙生

本书由浅入深地讲解Spring 5，作者成功地将复杂的理论以很容易理解的方式解释出来。同时本书指导读者如何在实际工作中运用这些方法，有助于读者结合实践去阅读理解源码。虽然关于Spring 5的图书很多，但是本书是难得一见的佳作。

中国电信号百商旅电子商务有限公司项目经理 • 刘俊

如果你想在项目中熟练使用Spring或者想深入了解Spring的工作原理，这本书就是你想要的！本书从基础出发，由浅入深、循序渐进地阐述Spring的重点（IoC/AOP），并且扩展整合了很多在实际应用场景中常用的技术，构成了一套完整的项目框架体系，是一本非常实用的Spring著作。

上海卓赞教育科技有限公司（DaDa英语）资深研发工程师 • 宋庭勇

从书的目录来看，内容丰富，由浅及深，相信这会是一本不错的Spring 5参考书。

瑞幸咖啡测试经理 • 陈茂川

凌晨已至，脑子却还在飞速运转，距离翻开此书已不知不觉过了三个小时。书中内容由浅入深，简明扼要，如果你是第一次接触Spring，这本书势必成为你的启蒙老师。

美团高级前端研发工程师 • 张奇雄

自从Rod在2003年创建Spring框架开始，一路借助于完整的生态体系建设和与时俱进的自我革新，Spring已经成为Java应用研发框架的事实标准，多年来在各个行业信息化建设中表现优异。本书完整地衔接了理论与工程实践，不单对Spring最新相关特性做了全面阐述，同时也覆盖了Spring与各种主流中间件及框架结合的最佳实践。对于一线研发人员而言，相信本书可以帮助你做出睿智的决策。

同程艺龙智慧交通技术负责人 • 杨继龙

Spring 框架作为企业级别常用的成熟框架，已作为主流在市场上应用和实践多年，本书不单详解了 Spring 相关的基础内容，更是在 Spring 5 上有非常系统的讲解。本书不仅对初学者有很好的指导作用，对于有相关开发经验的工程师来说，本书也是不可多得的 Spring 佳作，能为有经验的工程师的技术决策起到积极作用，值得推荐给大家！

陆金所服务器端高级研发工程师·周雅君

周冠亚老师对技术有着异常执着的热情，多年的一线互联网大厂工作经历，也让周老师练就了一身不凡的本领。本书是周老师的得意之作，是对 Spring 相关技术钻研的个人心得和成果，也是对多年 Spring 项目实战经验的总结和分享。该书从 Spring 项目实战到源码分析，再到原理讲解，深入浅出地从多个角度解读 Spring，能够帮助技术人员快速了解、掌握甚至深入 Spring，是一本不可多得的佳作。

云析学院创始人，Java 架构师，金牌讲师·赵新

作者把自己多年的开发经验总结付梓，从实际应用的角度出发系统地将 Spring 的核心概念、高级特性、系统集成整合到一起，引领读者轻松踏上 Spring 企业开发的旅途，易懂易学，用处很大。

《从 Lucene 到 Elasticsearch 全文检索实战》一书作者·姚攀

Spring 作为一个互联网公司的必备框架，由 Rod Johnson 创建。它是为了解决企业应用开发的复杂性而创建的，为应用提供一站式（one-stopshop）的解决方案。Spring 的发展日新月异，已经进化到了 5.0 的阶段，本书除了透彻地介绍了 Spring 标准的模块之外，把 5.0 的新特征很翔实地展示给了读者，实例也很精炼，此外，Spring 和其他模块集成的快速体验也实战化，给读者快速地实战落地提供了良好的指导。

网易资深开发工程师·震升

前　　言

　　Spring 在如今的 Java 企业开发中占据十分重要的地位。一路走来，作者经历过的上百个项目无一例外都是使用 Spring 开发的。2017 年 9 月 Spring 5 发布了通用版本（GA），标志着自 2013 年 12 月以来第一个主要 Spring Framework 版本诞生。本书从企业实战角度出发，讲解最新版本的 Spring 5.0\5.1 的新特性，并将常见互联网技术与 Spring 集成，力争让读者通过本书能够又快又好地掌握 Spring 企业级开发技能，并能学以致用。

　　本书涵盖 Spring 基础知识讲解，Spring 5 新特性和 Spring 集成开发等知识。本书从结构上可以分三部分，第一部分是 Spring 基础篇，介绍 Spring 核心概念和原理，涉及第 1 章~第 3 章。第二部分是 Spring 5 高级特性篇，涉及第 4 章~第 9 章。第三部分是 Spring 系统集成篇，主要讲解 Spring 框架与互联网公司常用的技术集成开发，涉及第 10 章~第 19 章。附录部分还介绍了本书涉及的以及在面试中常见的设计模式。

本书结构

　　本书共 19 章和 1 个附录，各章内容概述如下：

　　第 1 章　介绍 Spring 开发所需的环境和工具。包括 JDK 的安装，Intellij IDEA 安装、Tomcat 安装和配置、Maven 安装。

　　第 2 章　对 Spring 框架核心概念 IoC 容器进行讲解，并通过代码分析的方式阐述 IoC 容器的实现原理。

　　第 3 章　对 Spring 框架核心概念 AOP 进行讲解，并说明如何通过不同的方式实现 AOP，最后通过代码解析的方式阐述 AOP 的实现原理。

　　第 4 章　概述 Spring 5 的新特性。

　　第 5 章　概述 Java 8 的一些新特性，这些特性在 Spring 5 中得到了支持。

　　第 6 章　讲解使用 Spring 5 的新特性 WebFlux 进行编程和 Reactor 编程。

　　第 7 章　讲解 Spring 5 提供的响应式客户端编程。

　　第 8 章　讲解 Spring 5 集成 Kotlin 进行编程。

　　第 9 章　讲解更多 Spring 5 的新特性及细节。

　　第 10 章　讲解 Spring 集成 Log4j2 进行日志控制。

　　第 11 章　讲解 Spring 如何集成 Spring MVC 模块进行 Web 开发，并分析 Spring MVC 底层代码实现。

　　第 12 章　讲解 Spring 如何集成 MyBatis 进行数据库持久层开发，并分析 MyBatis 框架底层的代码。

　　第 13 章　讲解 Spring 对事务的支持，并分析 Spring 事务管理的底层代码实现。

第 14 章　讲解 Spring 集成 Redis 开发，并分析 Redis 各种不同部署方式之间的区别，本章最后分享在高并发场景下使用 Redis 需要注意的一些要点。

第 15 章　讲解 ZooKeeper 如何进行开发，并分析 ZooKeeper 在特定场景下的一些高级用法。

第 16 章　讲解 Spring 如何集成 Kafka 进行开发，并分析 Kafka 的核心架构。

第 17 章　讲解 Spring 如何集成 Mycat 进行分库分表开发，及如何将 Spring、Mybatis 和 Mycat 集成进行数据库持久化层的开发。

第 18 章　讲解 Spring 如何集成 Sharding-JDBC 进行分库分表开发，并讲解一些 Sharding-JDBC 的高级特性。

第 19 章　讲解 Spring 如何集成 Dubbo 进行 RPC 服务开发，并分析 Dubbo 框架的底层代码。

附录 A　讲解本书代码分析过程中的设计模式和企业开发过程中常见的设计模式。

本书预备知识

Java 基础

需要读者掌握 Java SE 基础知识，这是最基本的也是最重要的。

Linux 基础

本书讲解的 Spring 集成中间件开发部分，中间件都是基于 Linux 服务器进行部署的，因此读者应当掌握常用的 Linux 命令。

数据库基础

本书会涉及 Spring 对事务的支持和 Spring 集成 Mycat 或 Sharding-JDBC 进行分库分表操作，因此读者对数据库基础知识应有较好的掌握。

分布式系统基础

本书 Spring 系统集成部分会涉及当前互联网公司比较主流的分布式技术，读者需要对分布式系统的基础知识有一定的了解。

本书使用的软件版本

本书使用到的开发环境如下：

- 操作系统 MacOS 10.14.3
- 开发工具 Intellij IDEA 2018.1
- JDK 版本 1.8
- Tomcat 9.0.10
- maven-3.5.0
- Spring 最新版 5.1.5.RELEASE

本书系统集成部分使用到的多种组件的具体版本请参考对应章节。

读者对象

本书适合所有 Java 编程语言开发人员，所有对 Spring 感兴趣的开发人员，对分布式系统感兴趣的开发人员以及对各类技术原理有求知欲的开发人员。

源代码下载

GitHub 源代码下载地址：https://github.com/online-demo/spring5projectdemo.git

勘误与交流

限于笔者水平和写作时间有限，欢迎大家通过电子邮件等方式批评指正。

笔者的邮箱：zhouguanya20@163.com　　huangwenyi10@163.com
笔者的博客：http://blog.csdn.net/huangwenyi1010
笔者的微信公众号：A_GallopingSnail

致谢

本书能够顺利出版，首先要感谢清华大学出版社王金柱编辑给笔者一次与各位读者分享技术、交流学习的机会，感谢王金柱编辑在本书出版过程的辛勤付出。感谢好友黄文毅，也是笔者的同事，对笔者在写作思路和排版上的帮助和支持。

感谢汉海信息技术（上海）有限公司（简称美团点评），书中很多的知识点和项目实战经验都来源于"美团点评"，感谢主管章成峰、导师吕波和贾钧翔以及同事叶雄和孙成飞，感谢"饿了么"资深架构师唐斌对笔者技术和学习上的支持，感谢行业前辈杨继龙对笔者职业生涯的提点，感谢笔者的好友黄子涵对笔者写作进度的监督和指导，感谢英语老师吴定山对笔者英语能力的培养。

谨以此书献给我敬爱的父母，愿他们健康长寿。

周冠亚
2019 年 2 月 24 日

目 录

第一篇 Spring 基础篇

第1章 环境搭建 ············· 3
- 1.1 Spring 介绍 ············· 3
  - 1.1.1 Spring 设计目标 ············· 3
  - 1.1.2 Spring 各个子模块 ············· 3
  - 1.1.3 Spring 使用场景 ············· 4
  - 1.1.4 Spring 与 Spring MVC 的关系 ······ 5
  - 1.1.5 Spring 5 高级特性 ············· 5
- 1.2 环境准备 ············· 5
  - 1.2.1 安装 JDK ············· 5
  - 1.2.2 安装 IntelliJ IDEA ············· 6
  - 1.2.3 安装 Apache Maven ············· 6
  - 1.2.4 安装 Apache Tomcat ············· 7
- 1.3 快速搭建 Spring 5 项目 ············· 8
  - 1.3.1 使用 IntelliJ IDEA 创建 Spring 5 + Spring MVC 项目 ············· 8
  - 1.3.2 测试部署 ············· 9
- 1.4 小结 ············· 9

第2章 Spring IoC 容器原理 ············· 10
- 2.1 IoC 容器揭秘 ············· 10
  - 2.1.1 IoC 的概念 ············· 10
  - 2.1.2 依赖倒置原则 ············· 11
  - 2.1.3 依赖注入 ············· 16
- 2.2 Spring IoC 的实现方式 ············· 17
  - 2.2.1 XML 方式实现 ············· 17
  - 2.2.2 通过注解方式实现 ············· 20
- 2.3 Spring IoC 实现原理解析 ············· 21
  - 2.3.1 BeanFactory 代码解析 ············· 21
  - 2.3.2 ApplicationContext 代码解析 ······ 22
  - 2.3.3 BeanDefinition 代码解析 ············· 23
  - 2.3.4 Spring IoC 代码分析 ············· 23
- 2.4 Spring IoC 容器中 Bean 的生命周期 ······ 44
- 2.5 小结 ············· 52

第3章 Spring AOP 揭秘 ············· 53
- 3.1 AOP 前置知识 ············· 53
  - 3.1.1 JDK 动态代理 ············· 53
  - 3.1.2 CGLIB 动态代理 ············· 56
  - 3.1.3 AOP 联盟 ············· 58
- 3.2 AOP 概述 ············· 58
  - 3.2.1 AOP 基本概念 ············· 58
  - 3.2.2 Spring AOP 相关概念 ············· 59
- 3.3 Spring AOP 实现 ············· 60
  - 3.3.1 基于 JDK 动态代理实现 ············· 60
  - 3.3.2 基于 CGLIB 动态代理实现 ············· 65
- 3.4 基于 Spring AOP 的实战 ············· 70
  - 3.4.1 增强类型 ············· 70
  - 3.4.2 前置增强 ············· 71
  - 3.4.3 后置增强 ············· 73
  - 3.4.4 环绕增强 ············· 74
  - 3.4.5 异常抛出增强 ············· 75
  - 3.4.6 引介增强 ············· 75
  - 3.4.7 切入点类型 ············· 77
- 3.5 Spring 集成 AspectJ 实战 ············· 78
  - 3.5.1 使用 AspectJ 方式配置 Spring AOP ············· 78

3.5.2 AspectJ 各种切点指示器 ………… 81
3.5.3 args()与"@args()" ………… 81
3.5.4 @annotation() ………… 87
3.5.5 execution ………… 89
3.5.6 target()与"@target()" ………… 90
3.5.7 this() ………… 92
3.5.8 within()与"@within()" ………… 95

3.6 Spring AOP 的实现原理 ………… 98
   3.6.1 设计原理 ………… 99
   3.6.2 JdkDynamicAopProxy ………… 106
   3.6.3 CglibAopProxy ………… 110
3.7 小结 ………… 115

第二篇　Spring 5 新特性篇

第 4 章　Spring 5 新特性概述 ………… 117
4.1 Spring 5.0 新特性 ………… 117
   4.1.1 运行环境 ………… 117
   4.1.2 删除的代码 ………… 118
   4.1.3 核心修改 ………… 118
   4.1.4 核心容器更新 ………… 118
   4.1.5 Spring Web MVC 更新 ………… 118
   4.1.6 Spring WebFlux ………… 119
   4.1.7 对 Kotlin 的支持 ………… 119
   4.1.8 测试改进 ………… 120
4.2 Spring 5.1 新特性 ………… 121
   4.2.1 核心修改 ………… 121
   4.2.2 核心容器更新 ………… 121
   4.2.3 Web 修改 ………… 121
   4.2.4 Spring Web MVC 更新 ………… 121
   4.2.5 Spring WebFlux 更新 ………… 122
   4.2.6 Spring Messaging 更新 ………… 122
   4.2.7 Spring ORM 更新 ………… 122
   4.2.8 测试更新 ………… 122

第 5 章　Java 8 新特性概述 ………… 123
5.1 Lambda 表达式 ………… 123
   5.1.1 Lambda 表达式初探 ………… 123
   5.1.2 Lambda 表达式作用域 ………… 125
   5.1.3 在线程中使用 Lambda 表达式 ………… 126
   5.1.4 在集合中使用 Lambda 表达式 ………… 127
   5.1.5 在 Stream 中使用 Lambda 表达式 ………… 128
5.2 接口默认方法 ………… 129
5.3 小结 ………… 132

第 6 章　Spring WebFlux 响应式编程 ………… 133
6.1 传统的编程模型 ………… 133
6.2 响应式编程模型 ………… 134
6.3 Reactor ………… 135
   6.3.1 Flux 与 Mono ………… 135
   6.3.2 subscribe() ………… 137
   6.3.3 操作符（Operator） ………… 139
   6.3.4 线程模型 ………… 144
6.4 Spring WebFlux ………… 146
   6.4.1 基于注解的 WebFlux 开发方式 ………… 146
   6.4.2 基于函数式的 WebFlux 开发方式 ………… 147
6.5 小结 ………… 152

第 7 章　WebClient 响应式客户端 ………… 153
7.1 RestTemplate 调试 Spring MVC ………… 153
7.2 WebClient 调试 Spring WebFlux ………… 156
7.3 小结 ………… 158

第 8 章　Spring 5 结合 Kotlin 编程……159

- 8.1　Kotlin 简介……159
  - 8.1.1　Kotlin 的特性……159
  - 8.1.2　Kotlin 基本数据类型……161
  - 8.1.3　Kotlin 开发环境搭建……161
  - 8.1.4　在 Kotlin 中定义常量与变量……162
  - 8.1.5　字符串模板……162
  - 8.1.6　NULL 检查机制……163
  - 8.1.7　For 循环和区间……163
  - 8.1.8　定义函数……166
  - 8.1.9　类和对象……167
  - 8.1.10　Kotlin 与 Java 互操作……169
- 8.2　Spring 5 集成 Kotlin……170
- 8.3　小结……172

第 9 章　Spring 5 更多新特性……173

- 9.1　Resource 接口……173
- 9.2　HTTP 2……174
  - 9.2.1　HTTP 的现状……174
  - 9.2.2　HTTP 2 的新特性……174
  - 9.2.3　多路复用与长连接的区别……175
- 9.3　JUnit 5……176
  - 9.3.1　JUnit 5 简介……176
  - 9.3.2　JUnit 5 快速体验……176
  - 9.3.3　JUnit 5 常用注解……178
- 9.4　小结……179

第三篇　Spring 系统集成篇

第 10 章　Spring 集成 Log4j2……181

- 10.1　Log4j2 配置详解……181
- 10.2　Log4j2 日志级别……184
- 10.3　Log4j2 实战演练……185
- 10.4　小结……188

第 11 章　Spring 集成 Spring MVC……189

- 11.1　Spring MVC 快速体验……189
  - 11.1.1　web.xml 配置……189
  - 11.1.2　创建 Spring MVC 的配置文件……190
  - 11.1.3　创建 Spring MVC 的视图文件……190
  - 11.1.4　创建控制器……191
  - 11.1.5　测试运行……191
- 11.2　Spring MVC 视图呈现……192
  - 11.2.1　FreeMarker 视图的实现……192
  - 11.2.2　XML 视图的实现……193
  - 11.2.3　JSON 视图的实现……195
- 11.3　Spring MVC 拦截器……196
- 11.4　Spring MVC 代码解析……198
- 11.5　小结……225

第 12 章　Spring 集成 MyBatis……226

- 12.1　Spring、Spring MVC 和 MyBatis 集成快速体验……226
- 12.2　MyBatis 代码解析……236
- 12.3　小结……243

第 13 章　Spring 事务管理……244

- 13.1　事务的特性……244
- 13.2　事务的隔离级别……244
  - 13.2.1　READ_UNCOMMITTED……245
  - 13.2.2　READ_COMMITTED……245
  - 13.2.3　REPEATABLE_READ……246
  - 13.2.4　SERIALIZABLE……246
- 13.3　JDBC 方式使用事务……247
- 13.4　Spring 事务管理快速体验……248

13.5　Spring 事务隔离级别⋯⋯⋯⋯⋯254
13.6　Spring 事务传播行为⋯⋯⋯⋯⋯255
13.7　Spring 事务代码分析⋯⋯⋯⋯⋯255
13.8　小结⋯⋯⋯⋯⋯⋯⋯⋯⋯⋯⋯263

第 14 章　Spring 集成 Redis⋯⋯⋯⋯264

14.1　Redis 单节点安装⋯⋯⋯⋯⋯⋯264
14.2　Redis 支持的数据类型⋯⋯⋯⋯266
　　14.2.1　Redis String 类型的使用方式⋯⋯⋯⋯⋯⋯⋯⋯266
　　14.2.2　Redis Hash 类型的使用方式⋯⋯⋯⋯⋯⋯⋯⋯271
　　14.2.3　Redis List 类型的使用方式⋯⋯⋯⋯⋯⋯⋯⋯274
　　14.2.4　Redis Set 类型的使用方式⋯⋯⋯⋯⋯⋯⋯⋯277
　　14.2.5　Redis SortedSet 类型的使用方式⋯⋯⋯⋯⋯⋯⋯⋯280
14.3　Redis 持久化策略⋯⋯⋯⋯⋯⋯284
　　14.3.1　Redis RDB 持久化⋯⋯⋯284
　　14.3.2　Redis AOF 持久化⋯⋯⋯285
14.4　Redis 主从复制模式⋯⋯⋯⋯⋯287
　　14.4.1　Redis 一主一从拓扑结构⋯⋯289
　　14.4.2　Redis 一主多从拓扑结构⋯⋯289
　　14.4.3　Redis 树形拓扑结构⋯⋯290
　　14.4.4　Redis 主从架构的缺点⋯⋯291
14.5　Redis 哨兵模式⋯⋯⋯⋯⋯⋯⋯291
　　14.5.1　Redis 哨兵模式简介⋯⋯291
　　14.5.2　Redis 哨兵定时监控任务⋯⋯292
　　14.5.3　主观下线和客观下线⋯⋯294
　　14.5.4　Redis 哨兵选举领导者⋯⋯295
　　14.5.5　故障转移⋯⋯⋯⋯⋯⋯295
　　14.5.6　Redis 哨兵模式安装部署⋯⋯296
14.6　Redis 集群模式⋯⋯⋯⋯⋯⋯⋯302
　　14.6.1　Redis 集群模式数据共享⋯⋯303
　　14.6.2　Redis 集群中的主从复制⋯⋯303
　　14.6.3　Redis 集群中的一致性问题⋯⋯⋯⋯⋯⋯⋯⋯303
　　14.6.4　Redis 集群架构⋯⋯⋯⋯304
　　14.6.5　Redis 集群容错⋯⋯⋯⋯304
　　14.6.6　Redis 集群环境搭建⋯⋯305
14.7　Spring、MyBatis 和 Redis 集成快速体验⋯⋯⋯⋯⋯⋯⋯⋯⋯313
14.8　Redis 缓存穿透和雪崩⋯⋯⋯⋯323
　　14.8.1　Redis 缓存穿透⋯⋯⋯⋯323
　　14.8.2　Redis 缓存雪崩⋯⋯⋯⋯326
14.9　小结⋯⋯⋯⋯⋯⋯⋯⋯⋯⋯⋯329

第 15 章　Spring 集成 ZooKeeper⋯⋯⋯330

15.1　ZooKeeper 集群安装⋯⋯⋯⋯⋯330
15.2　ZooKeeper 总体架构⋯⋯⋯⋯⋯336
　　15.2.1　ZooKeeper 选举机制⋯⋯336
　　15.2.2　ZooKeeper 数据模型⋯⋯338
15.3　Spring 集成 ZooKeeper 快速体验⋯⋯342
15.4　ZooKeeper 发布订阅⋯⋯⋯⋯⋯346
　　15.4.1　NodeCache⋯⋯⋯⋯⋯346
　　15.4.2　PathChildrenCache⋯⋯349
　　15.4.3　TreeCache⋯⋯⋯⋯⋯352
15.5　ZooKeeper 分布式锁⋯⋯⋯⋯⋯356
15.6　小结⋯⋯⋯⋯⋯⋯⋯⋯⋯⋯⋯361

第 16 章　Spring 集成 Kafka⋯⋯⋯⋯362

16.1　Kafka 集群安装⋯⋯⋯⋯⋯⋯⋯362
16.2　Kafka 总体架构⋯⋯⋯⋯⋯⋯⋯365
16.3　Spring 集成 Kafka 快速体验⋯⋯374
16.4　小结⋯⋯⋯⋯⋯⋯⋯⋯⋯⋯⋯377

第 17 章　Spring 集成 Mycat⋯⋯⋯⋯379

17.1　Mycat 分库分表⋯⋯⋯⋯⋯⋯⋯379
17.2　Mycat 分库分表实战⋯⋯⋯⋯⋯381
17.3　Spring+MyBatis+Mycat 快速体验⋯⋯386
17.4　小结⋯⋯⋯⋯⋯⋯⋯⋯⋯⋯⋯397

第 18 章 Spring 集成 Sharding-JDBC · 398

- 18.1 Spring 集成 Sharding-JDBC 快速体验 ············ 398
- 18.2 Sharding-JDBC 强制路由 ············ 407
- 18.3 Sharding-JDBC 分布式主键 ············ 408
- 18.4 小结 ············ 413

第 19 章 Spring 集成 Dubbo ············ 414

- 19.1 远程过程调用协议 ············ 414
- 19.2 Spring 集成 Dubbo 快速体验 ············ 415
- 19.3 Dubbo 代码分析 ············ 419
- 19.4 小结 ············ 452

附录 A 设计模式 ············ 453

- A.1 工厂模式 ············ 453
- A.2 抽象工厂模式 ············ 456
- A.3 单例模式 ············ 462
- A.4 建造者模式 ············ 463
- A.5 原型模式 ············ 468
- A.6 适配器模式 ············ 472
- A.7 桥接模式 ············ 476
- A.8 标准模式 ············ 477
- A.9 组合模式 ············ 481
- A.10 装饰器模式 ············ 483
- A.11 外观模式 ············ 486
- A.12 享元模式 ············ 489
- A.13 代理模式 ············ 491
- A.14 责任链模式 ············ 493
- A.15 命令模式 ············ 496
- A.16 解释器模式 ············ 499
- A.17 迭代器模式 ············ 502
- A.18 中介者模式 ············ 504
- A.19 备忘录模式 ············ 506
- A.20 观察者模式 ············ 509
- A.21 状态模式 ············ 511
- A.22 空对象模式 ············ 513
- A.23 策略模式 ············ 516
- A.24 模板模式 ············ 519
- A.25 拦截过滤器模式 ············ 522

参考文献 ············ 365

第一篇

Spring基础篇

第 1 章

环 境 搭 建

本章主要介绍学习 Spring 5 之前的环境准备，Spring 5 项目各个模块介绍，以及开发 Spring 5 项目需要用到的一些开发工具，并在本章末尾搭建一个可运行的 Spring MVC 项目。

1.1 Spring 介绍

1.1.1 Spring 设计目标

Spring 是一个于 2003 年兴起的轻量级的 Java 开源开发框架，由 Rod Johnson 在其著作 Expert One-On-One J2EE Development and Design 中阐述的部分理念和原型衍生而来。Spring 让开发人员有更多的精力投入到业务逻辑开发中，而不需要将其应用程序绑定到特定的部署环境，是为了降低企业应用开发的复杂性而创建的。Spring 不是创造轮子（技术、框架），而是使现有的轮子运转得更好的工具。可以把 Spring 理解成一个大容器，这个容器可以整合现有的各种技术框架。Spring 框架的主要优势是方便各种框架集成，降低了 Java EE 开发的难度。Spring 使用基本的 JavaBean 来完成以前只可能由 EJB 完成的事情。Spring 的用途不限于服务器端应用程序的开发。

Spring 框架提供了很多子模块供开发人员使用。

1.1.2 Spring 各个子模块

Spring 是一个大容器，可以集成各种技术。如图 1-1 所示是 Spring 支持的各种技术。

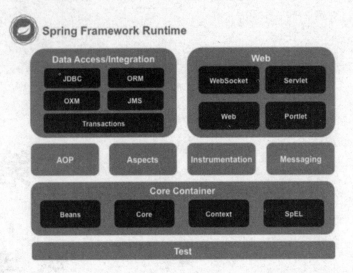

图1-1 Spring 支持的技术体系

下面依次介绍 Spring 支持的各种技术。

（1）Core Container

Spring 的核心容器由 Beans、Core、Context、SpEL 等模块组成。所有 Spring 的其他模块都是建立在 Core Container 基础模块上的。该模块规定了创建和维护 Bean 的方式，提供了控制反转（IoC）和依赖注入（DI）等特性。

（2）Data Access/Integration

数据访问/集成模块提供了对 JDBC、ORM、OXM、JMS 和 Transaction 等模块的集成。使主流的 ORM 框架，持久化框架和消息中间件可以很方便地集成到 Spring 中，降低开发人员对这些框架的维护成本，提升了开发效率。

（3）Web 模块

Web 模块提供了对 Web 开发相关技术的集成，对开发模型-视图-控制器（MVC）项目提供便利。

（4）AOP 模块

AOP 模块提供了 AOP 联盟提倡的面向切面编程的实现环境。AOP 将代码按功能进行分离，降低了模块间代码的耦合度。

（5）Test 模块

Test 模块支持 JUnit 和 TestNG 等单元测试模块的集成，还提供了 mock 对象，使开发人员可以更加独立的测试代码。

1.1.3 Spring 使用场景

（1）管理依赖的资源

在企业开发中，经常需要管理各种配置文件，如 JDBC 连接配置文件，ORM 配置文件等。可以通过 Spring 管理这些文件。如加载 JDBC 的配置文件 jdbc.properties 就可以使用如下代码方式配置 Spring，这样 Spring 启动时会在此路径下自动搜索名称为 jdbc.properties 的配置文件，并将其加载到内存中：

```
<context:property-placeholder location="classpath*:jdbc.properties"/>
```

（2）Bean 管理

一个企业项目中，会有很多 Bean，每次都手动创建和管理这些 Bean 的对象是很低效的。Spring 提供了管理 Bean 的 IoC 容器，并在需要用到相关 Bean 的时候，提供依赖注入（DI）将相关的 Bean 注入。

（3）事务管理

Spring 提供的事务管理，使开发人员在做数据库操作时，无须再手动执行对数据库的提交或回滚操作，并且 Spring 还提供了对事务传播的支持，可以实现更加复杂的事务嵌套的逻辑，对数据一致性提供了更好的支持。

Spring 的使用场景远不止这里提到的三点，更多 Spring 使用场景，请见 Spring 系统集成篇。

1.1.4　Spring 与 Spring MVC 的关系

Spring 和 Spring MVC 两者名字类似，但是两者却有着本质的不同。Spring 是一个巨大的容器，可以集成各种技术。Spring MVC 是一个 Web 技术，Spring MVC 可以集成到 Spring 中。用数学上集合的概念来解释，Spring MVC 是 Spring 的一个子集。

1.1.5　Spring 5 高级特性

截止本书出版，Spring 最新版本已经升级到了 5.1 版本。Spring 5 相比于 Spring 之前的历史版本，带来了以下新的特性：

- Spring 5 整个框架基于 Java 8
- 支持 HTTP/2
- Spring Web MVC 支持最新 API
- Spring WebFlux 响应式编程
- 支持 Kotlin 函数式编程

更多有关 Spring 5 的高级特性请见 Spring 5 高级特性篇。

1.2　环境准备

1.2.1　安装 JDK

JDK（Java SE Development Kit）建议使用 JDK 1.8 及以上的版本。JDK 官方下载路径：http://www.oracle.com/technetwork/java/javase/downloads/index.html，安装过程此处不过多描述，读者可以根据各自的电脑操作系统配置选择合适的 JDK 安装包进行安装，笔者电脑的操作系统是 MacOS 10.14.3。安装包下载完成之后，双击下载软件，按照提示安装即可。

安装好 JDK 后，在英文输入法状态下，按"command + 空格"组合键，调出 Spotlight 搜索界面，输入 terminal 选择【终端】然后按 Enter 键，便可以快速启动终端，具体如图 1-2 所示。

图 1-2　打开 terminal 终端

在【终端】输入"java -version"，如果看到具体的 JDK 版本，则说明 JDK 安装成功，如图 1-3 所示。

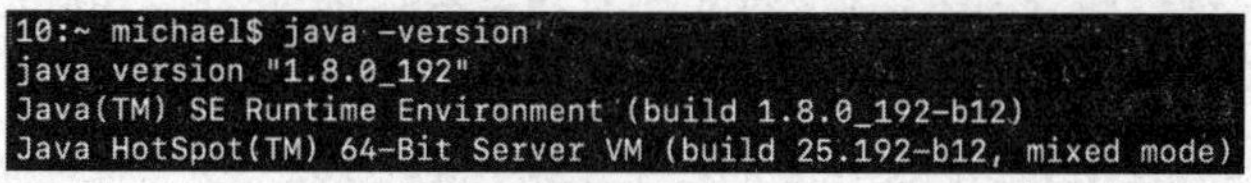

图 1-3　验证 JDK 安装成功

1.2.2　安装 IntelliJ IDEA

在 Intellij IDEA 的官方网站 http://www.jetbrains.com/idea/ 可以免费下载 IDEA。下载完 IDEA 后，运行安装程序，按提示安装即可。本书使用的 Intellij IDEA 的版本为 2018.1.2，也可以使用其他版本的 IDEA，版本只要不过低即可。安装成功之后，软件打开界面如图 1-4 所示。

图 1-4　Intellij IDEA 软件打开界面

1.2.3　安装 Apache Maven

Apache Maven 是目前流行的项目管理和构建自动化工具，可以通过 Maven 的官网 http://maven.apache.org/download.cgi 下载最新版的 Maven。本书的 Maven 版本为 apache-maven-3.5。

下载完后解压缩即可，例如解压到/usr/local/Cellar/maven@3.5/3.5.4/libexec。

接下来讲解 Maven 如何集成到 IntelliJ IDEA 中。

在 Intellij IDEA 界面中，选择 File→Settings，在出现的窗口中找到 Maven 选项，分别把 Maven home directory、User settings file、Local repository 设置成读者自己的 Maven 的相关目录，如图 1-5 所示。

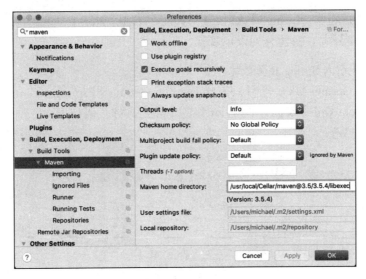

图 1-5　Intellij IDEA 集成 Apache Maven 窗口

1.2.4　安装 Apache Tomcat

Apache Tomcat 是目前主流的 Web 容器，可以在 Apache Tomcat 官网下载。下载地址：https://tomcat.apache.org/。本书的 Maven 版本为 apache-tomcat-9.0.10，下载后将压缩文件解压。在 Intellij IDEA 中集成 Apache Tomcat，如图 1-6 所示。

集成 Apache Tomcat 后，在 Intellij IDEA 中启动 Tomcat，Intellij IDEA 控制台输出如图 1-7 所示，此时 Apache Tomcat 已经启动。

图 1-6　Intellij IDEA 集成 Apache Tomcat 窗口

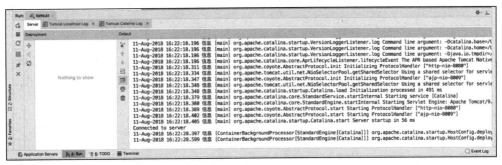

图 1-7　Intellij IDEA 集成 Apache Tomcat 启动窗口

1.3 快速搭建 Spring 5 项目

1.3.1 使用 IntelliJ IDEA 创建 Spring 5 + Spring MVC 项目

本节将通过 Spring 5 + Spring MVC 快速搭建一个 Hello World Web 程序。搭建本书项目全部是基于 Spring 5.1 开发。搭建分为以下步骤。

（1）在 pom 中引入 Spring 相关依赖。

一个最简单的 Spring MVC 项目只需要 javax.servlet-api 和 spring-webmvc 这两个 jar 包。在 maven 项目的 pom.xml 中加入这两个 jar 包的依赖，具体代码如下：

```xml
<dependency>
  <groupId>javax.servlet</groupId>
  <artifactId>javax.servlet-api</artifactId>
  <version>4.0.1</version>
</dependency>
<dependency>
  <groupId>org.springframework</groupId>
  <artifactId>spring-webmvc</artifactId>
  <version>${spring.framework.version}</version>
</dependency>
```

（2）创建 HelloWorldController，输出文字 Hello World。

Controller 是 Spring MVC 控制层模块，创建一个简单 Controller，提供"/hello"这个 HTTP 接口，然后使用浏览器访问该接口，即可输出文字"Hello World"。

```java
@RestController
@RequestMapping("/")
public class HelloWorldController {
    @RequestMapping("hello")
    public String sayHello() {
        return "Hello World";
    }
}
```

（3）配置文件 springmvc.xml。定义扫描包的路径和视图解析器。

```xml
<!-- 配置自动扫描的包 -->
<context:component-scan base-package="com.test"></context:component-scan>
<!-- 配置视图解析器 把 handler 方法返回值解析为实际的物理视图 -->
<bean class="org.springframework.web.servlet.view.InternalResourceViewResolver">
    <property name = "prefix" value="/pages/"></property>
    <property name = "suffix" value = ".jsp"></property>
</bean>
```

（4）配置 web.xml 文件。指定 Spring MVC 核心 Servlet 和相关的配置文件即可：

```xml
<web-app>
  <display-name>Archetype Created Web Application</display-name>
  <servlet>
    <servlet-name>springDispatcherServlet</servlet-name>
    <servlet-class>org.springframework.web.servlet.DispatcherServlet</servlet-class>
    <!-- 配置 Spring mvc 下的配置文件的位置和名称 -->
    <init-param>
      <param-name>contextConfigLocation</param-name>
      <param-value>classpath:spring*.xml</param-value>
    </init-param>
    <load-on-startup>1</load-on-startup>
  </servlet>
  <servlet-mapping>
    <servlet-name>springDispatcherServlet</servlet-name>
    <url-pattern>/</url-pattern>
  </servlet-mapping>
</web-app>
```

1.3.2 测试部署

使用 Intellij IDEA 集成的 Tomcat 发布整个 Spring MVC 应用，在浏览器中访问 http://localhost:8080/hello 接口，效果如图 1-8 所示。

图 1-8　Intellij IDEA 集成 Apache Tomcat 启动窗口

1.4　小　　结

本章主要介绍了 Spring 技术体系的构成，并初步讲解了构建 Spring 项目需要用到的一些开发工具的安装和使用。通过 Spring 构建一个简单的 Spring MVC 项目，再通过浏览器访问 Spring MVC 项目提供的 HTTP 接口，即可打印文字，例如 Hello World。下一章将讲解 Spring 框架的核心概念——IoC。

第 2 章

Spring IoC 容器原理

Spring 框架在企业开发中，是很常见很基础的技术。本章将讲解 Spring 框架的核心概念——IoC。本章从简单的案例入手，一步深入剖析 IoC 的核心思想。在本章最后，通过对 Spring 代码进行解析，揭秘 Spring IoC 实现的方式。

2.1 IoC 容器揭秘

2.1.1 IoC 的概念

IoC 是 Inversion of Control 的简写，翻译成汉语就是"控制反转"。IoC 并不是一门技术，而是一种设计思想。在没有 IoC 设计的场景下，开发人员在使用所需的对象时，需手动创建各种对象，如 new Student()。如图 2-1 所示是传统 Java 开发方式。

有了 IoC 这样的设计思想，在开发中，意味着将设计好的对象交给容器管理，而不再是像传统的编程方式中，在对象内部直接控制对象。如图 2-2 所示是使用 IoC 的 Java 开发方式。

如何理解 IoC 呢？从 IoC 的字面意思上，无非是要把握好两块：控制、反转。下面将从这两方面着手分析 IoC。

1. 控制：由谁控制，控制了什么？

在传统 Java 程序设计中，开发人员直接在某个对象内部通过 new 关键字创建另一个对象，是开发人员主动去创建依赖对象；而 IoC 的设计思想，是通过专门的对象容器来创建和维护对象。于是就可以解答这个问题了。对于谁控制这个问题的回答是，由 IoC 容器来控制；对于控制了什么这个问题的回答是：控制了对象。

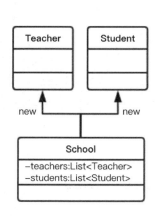 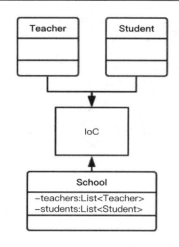

图 2-1　传统 Java 开发方式　　　　图 2-2　使用 IoC 的 Java 开发方式

2. 反转：什么是反转，反转了哪些方面？

在传统 Java 应用程序开发中，由开发人员在对象中主动控制其需要的依赖对象；而反转则是把对象依赖的过程颠倒了，即开发人员不再控制其依赖对象，而是由容器来帮助开发人员创建其需要的对象。于是就可以解答这个问题了。对于什么是反转这个问题的回答是，由容器帮开发人员创建依赖对象，对象只是被动地接受依赖对象，对象的控制权不再是开发人员，而是容器；对于反转了哪些方面这个问题的回答是，开发人员需要依赖的对象被反转。

现在再来回顾一下 IoC 的概念就好理解了，IoC——控制反转，即对象的控制权转移了，从开发人员转移到对象容器了。

2.1.2　依赖倒置原则

软件工程理论中的六大设计原则。

1. 单一职责原则

不存在多于一个的因素导致类的状态发生变更，即一个类只负责一项单一的职责。

2. 里氏替换原则

基类出现的地方都可以用其子类进行替换，而不会引起任何不适应的问题。

3. 接口隔离原则

客户端不应该依赖于其不需要的接口，类间的依赖关系应该建立在最小的接口之上。

4. 迪米特法则

一个对象对其他对象有最少的了解。

5. 开闭原则

软件设计对于扩展是开放的（Open for extension），即模块的行为是可以扩展的。
软件设计对于修改是关闭的（Closed for modification），即模块的行为是不可修改的。

6. 依赖倒置原则

高层次的模块不应该依赖于低层次的模块,都应该依赖于抽象。

如图 2-3 所示,CTO 是整个组织架构的高层模块,其他模块都是 CTO 的底层模块。同理,研发 1 部是业务研发 1 组和业务研发 2 组的高层模块,业务研发 1 组和业务研发 2 组是研发 1 部的底层模块,以此类推。

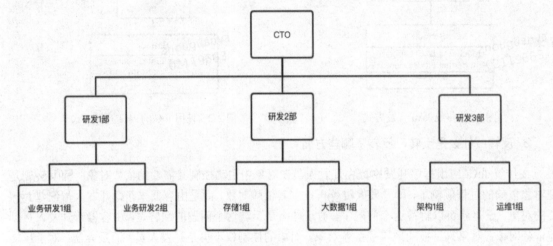

图 2-3 组织架构示意图

抽象不应该依赖于具体实现,具体实现应该依赖于抽象。例如,人就是一个抽象,具体到黄种人、白种人和黑种人就是具体的实现。具体对应到软件工程领域,抽象可以是抽象类或接口。具体实现就是继承或实现这些抽象类或接口的类。如下代码就是一个典型的抽象和具体实现:

```java
/**
 * 抽象接口
 */
public interface Eatable {
    /**
     * 吃方法
     */
    void eat();
}

/**
 * 具体实现类 Apple
 */
public class Apple implements Eatable {
    @Override
    public void eat() {
        System.out.println("吃苹果");
    }
```

```java
}

/**
 * 具体实现类Banana
 */
public class Banana implements Eatable {
    @Override
    public void eat() {
        System.out.println("吃香蕉");
    }
}
```

通过上述代码,分析了依赖倒置原则的定义后,下面将对比不遵循依赖倒置原则和遵循依赖倒置原则的两种不同软件设计风格。

假设有如下场景,一个人,需要通过某种交通工具去上班。在不遵循倒置原则的情况下,软件的设计风格如下。

当员工住的离公司比较近的时候,骑自行车上班即可:

```java
public class Person {
    /**
     * 人拥有一辆自行车
     */
    private Bike bike;

    /**
     * 创建对象,同时创建一辆自行车
     */
    public Person () {
        bike = new Bike();
    }

    /**
     * 上班
     */
    public void go () {
        System.out.println(bike.go("骑车去上班"));
    }
}
```

当员工住的离公司远,需要改乘公交上班时,代码需要修改成如下:

```java
public class Person {

    /**
     * 需要乘坐一辆公交车上班
     */
```

```
    private Bus bus;

    /**
     * 人拥有一辆自行车
     */
    private Bike bike;

    /**
     * 创建对象,同时创建一辆公交车
     */
    public Person () {
        //bike = new Bike();
        bus = new Bus();
    }

    /**
     * 上班去
     */
    public void go () {
        //System.out.println(bike.go("骑车去上班"));
        System.out.println(bus.go("乘坐公交去上班"));
    }
}
```

如果员工买车,就不需要坐公交上班,这时对于 Person 要加入新的对象如 Car 来满足新的需求。

如图 2-4 所示,按照以上的案例设计,对于每一次的需求更新,对 Person 类的修改太多,几乎每一次需求变更,都要对 Person 进行一次重构,显然这种做法不合适。通过分析可以得知,Person 是依赖于 Bike、Bus 的。Person 类的 go()方法功能完全依赖于 bike 和 bus 属性的 go()方法。直接依赖于具体实现,带来的后果就是每次需求变更,必然要对代码进行重构。

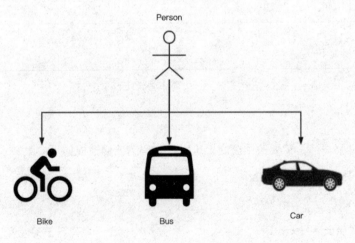

图 2-4 传统 Java 开发方式

接下来对上述的设计方法进行优化，对 Person 引进抽象。将 Person 需要使用的交通工具抽象成一个接口 Movable，接口中有个 go()方法。具体代码如下：

```
public interface Movable {
    /**
     * 出发
     * @param content
     * @return
     */
    String go(String content);
}
```

让 Person 依赖 Movable 接口的实例对象，当需求变更后，对 Person 的改动很小。Person 依赖 Movable，对 Person 的改动如下：

```
public class Person {
    /**
     * 某一种交通工具
     */
    private Movable movable;

    public Person () {
        //修改交通工具，只需要修改这里
        movable = new Bike();
    }

    /**
     * 上班去
     */
    public void go () {
        System.out.println(movable.go("乘交通工具去上班"));
    }
}
```

Person 中的 go()方法依赖 Movable 抽象，在 Person 内部不确定指明其依赖的具体某种交通工具（如 Bike、Bus、Car 等），至此可以说，Person 这个上层类，已经不直接依赖于底层类了，实现了依赖抽象的设计风格。

Movable 是抽象，其代表了一类行为。Bike、Bus、Car 是具体的实现细节，它们有自己的个性化部分——Bike 需要用脚踩，Bus 需要投币，Car 需要加油等，但是它们都有一个共性——都可以用作上班时的交通工具。抽象不应该依赖于具体实现，具体实现应该依赖于抽象，即定义交通工具 Movable 时，不应该依赖于每种具体交通工具的具体功能，反之，每种交通工具（Bike、Car、Bus 等具体的实现），应该依赖抽象 Movable。如图 2-5 所示，当用户想要乘坐地铁上班时，按照依赖倒置的软件设计风格，Person 类可以轻松地实现扩展。其实依赖倒置是软件工程中面向接口编程的一种具体实现。

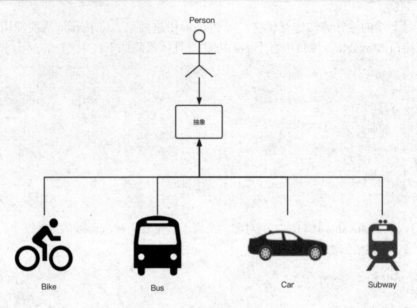

图 2-5 依赖倒置的软件设置风格

2.1.3 依赖注入

通过以上小节的分析，软件设计虽然采用依赖倒置的原则，但是，对象所依赖的其他对象，还是需要研发人员手动管理，即 Person 依赖的 Bike、Bus、Car，还是需要开发人员主观创建这些依赖对象。有没有办法实现无须开发人员手动管理依赖关系呢？这就是依赖注入。

依赖注入（DI，Dependency Injection）是 Spring 实现 IoC 容器的一种重要手段。依赖注入将对象间的依赖的控制权从开发人员转移到了容器，降低了开发成本。此时，Person 不再需要开发人员手动维护其依赖项，而是通过 Spring 将依赖关系注入 Person 中。一个简单的依赖注入代码如下：

```java
@Service("bike")
public class Bike implements Movable {
    /**
     * 出发
     *
     * @return
     */
    @Override
    public String go(String content) {
        return content;
    }
}
```

"@Service("bike")" 表示将 Bike 交给 IoC 容器管理，容器中会存放一个符合单例模式的 Bike 对象。当 Person 依赖 Bike 时，无须开发人员手工维护 Person 和 Bike 的依赖关系，IoC 容器很好地解决了这个问题，代码如下：

```
@Component
public class Person {
    /**
     * 某一种交通工具
     */
    @Resource(name = "bike")
    private Movable movable;

    public Person () {
        //修改交通工具，只需要修改这里
        movable = new Bike();
    }

    /**
     * 上班去
     */
    public void go () {
        System.out.println(movable.go("乘交通工具去上班"));
    }
}
```

图 2-6 可以更好地帮助理解依赖注入。开发人员手工管理依赖，类似于顾客主动向服务员要取菜单，"服务员，把菜单给我，我要点餐"。通过依赖注入这种设计方案，顾客无需向服务员索要菜单，服务员将会走到顾客身边，将菜单递给顾客。回到软件开发场景中，通过依赖注入，Person 类不再需要创建其依赖的交通工具，当 Person 需要借助交通工具去上班时，直接使用 Movable 对象就可以实现，具体是通过何种交通工具上班，开发人员无须再关心。

图 2-6　依赖倒置示意图

依赖注入减少了开发人员维护大量依赖关系的工作量，提高了开发人员的工作效率。

控制反转、依赖倒置和依赖注入三者之间的关系是：控制反转是一种软件设计模式，其遵循了软件工程中的依赖倒置原则；依赖注入是 Spring 框架实现控制反转的一种方式。

2.2　Spring IoC 的实现方式

2.2.1　XML 方式实现

用构造器方式实现 IoC 分为无参构造器和有参构造器两种。下面以 User 和 Order 为例说明，

User 使用无参构造器的方式，Order 使用有参构造器的方式，分别实现无参构造器和有参构造器的 IoC。

User 类的实现如下：

```java
/**
 * 无参构造器实现 IoC
 */
public class User implements Speakable {
    /**
     * 无参构造器
     */
    public User () {

    }

    /**
     * 说话的方法
     */
    @Override
    public void say() {
        System.out.println("大家好");
    }
}
```

在 spring-chapter2.xml 文件中，通过 bean 标签将 User 类交给 IoC 容器管理，代码如下：

```xml
<!-- User 无参构造器 -->
<bean id="user" class="com.test.ioc.constructor.User"/>
```

与 User 类不同的是，Order 类是没有无参构造器的，Order 类含有一个带有两个参数——订单号和订单金额的有参构造器。Order 类的定义如下：

```java
/**
 * 有参构造器实现 IoC
 */
public class Order implements Deliverable {
    /**
     * 订单号
     */
    private long orderId;
    /**
     * 订单金额
     */
    private double amount;

    /**
     * 有参构造器
```

```java
 * @param orderId
 * @param amount
 */
public Order (long orderId, double amount) {
    this.orderId = orderId;
    this.amount = amount;
}

/**
 * 订单发货方法
 */
@Override
public void delivery() {
    System.out.printf("订单号%s, 金额%s, 已发货!", orderId, amount);
}
}
```

在 spring-chapter2.xml 文件中通过 bean 标签将 User 类交给 IoC 容器管理。具体配置如下：

```xml
<!--Order 有参构造器-->
<bean id="order" class="com.test.ioc.xml.Order">
    <constructor-arg index="0" value="201808121706"/>
    <constructor-arg index="1" value="1000"/>
</bean>
```

在单元测试类 XmlTest 中，通过依赖注入得到 Speakable 的对象 User 和 Deliverable 的对象 Order，单元测试代码如下：

```java
/**
 * 测试 XML 方式的 IoC
 */
@RunWith(SpringJUnit4ClassRunner.class)
@ContextConfiguration("classpath:spring-chapter2.xml")
public class XmlTest {
    //Spring 容器注入依赖的 Speakable 对象
    @Autowired
    private Speakable speakable;
    //Spring 容器注入依赖的 Deliverable 对象
    @Autowired
    private Deliverable deliverable;
    @Test
    public void test() {
        speakable.say();
        deliverable.delivery();
    }
}
```

其中@RunWith 这个注解指定了让单元测试运行于 Spring 的环境中，@ContextConfiguration 这个注解指定 Spring 加载的配置文件。执行单元测试，测试结果如下。

```
大家好
订单号201808121706，金额1000.0，已发货！
```

2.2.2 通过注解方式实现

除了通过构造器实现 IoC，还可以通过 Spring 提供的注解方法实现 IoC，这也是企业开发过程中最常用的一种 IoC 实现方式。下面通过学生类 Student 阐述注解的方式实现 IoC。

Student 类的定义如下：

```java
@Service
public class Student implements HomeWork {
    /**
     * 写家庭作业
     */
    @Override
    public void doHomeWork() {
        System.out.println("我是学生，我要写家庭作业");
    }
}
```

注意此时的 Student 类上加了一个@Service 注解，这告诉 Spring，让其管理这个类的对象，因此开发人员就不再需要管理 Student 对象了。

与 XML 方式实现的 IoC 不同的是，注解方式除了配置@Service 注解外，还需要指定 Spring 对需要管理的 bean 目录，否则 Spring 不能定位其需要管理的 bean。具体配置如下：

```xml
<!--spring 管理的bean的路径-->
<context:component-scan base-package="com.test.ioc"></context:component-scan>
```

接下来在测试类 AnnotationTest 中通过依赖注入，将 HomeWork 对象注入到 AnnotationTest 测试类中，测试代码如下：

```java
/**
 * 测试注解方式的 IoC
 */
@RunWith(SpringJUnit4ClassRunner.class)
@ContextConfiguration("classpath:spring-chapter2.xml")
public class AnnotationTest {
    @Autowired
    private HomeWork homeWork;
    //Spring 容器注入依赖的 Deliverable 对象
    @Test
    public void test() {
```

```
        homeWork.doHomeWork();
    }
}
```

运行单元测试，测试结果如下：

> 我是学生，我要写家庭作业

除了例中的注解@Service 可以实现 Bean 的 IoC 以外，Spring 还提供了很多其他的注解来实现 IoC。

（1）@Component 将 Java 类标记成一个 Spring Bean 组件。
（2）@Service 将业务层实现类标记成一个 Spring Bean 组件。
（3）@Controller 将控制层类标记成一个 Spring Bean 组件。
（4）@Repository 将一个持久层实现类标记成一个 Spring Bean 组件。

2.3 Spring IoC 实现原理解析

本节将介绍 Spring IoC 容器是如何实现的。这是一个非常庞大的问题，会涉及很多 Spring 底层代码的解析。但是任何复杂的框架，其底层原理都是很简单的。为了能使读者明白代码的实现，下面将一些 Spring IoC 中重要的概念提炼出来。

2.3.1 BeanFactory 代码解析

这是整个 IoC 容器最顶层接口，其定义了一个 IoC 容器的基本规范，以下是 Spring BeanFactory 的接口定义。限于篇幅原因，删除了大量的代码注释。BeanFactory 是一个低配版的 IoC 容器，其定义了 IoC 容器基本的功能。BeanFactory 具体代码如下：

```java
public interface BeanFactory {
    String FACTORY_BEAN_PREFIX = "&";

    Object getBean(String name) throws BeansException;

    <T> T getBean(String name, Class<T> requiredType) throws BeansException;

    Object getBean(String name, Object... args) throws BeansException;

    <T> T getBean(Class<T> requiredType) throws BeansException;

    <T> T getBean(Class<T> requiredType, Object... args) throws BeansException;

    boolean containsBean(String name);

    boolean isSingleton(String name) throws NoSuchBeanDefinitionException;
```

```
  boolean isPrototype(String name) throws NoSuchBeanDefinitionException;

  boolean isTypeMatch(String name, ResolvableType typeToMatch)
throws NoSuchBeanDefinitionException;

  boolean isTypeMatch(String name, @Nullable Class<?> typeToMatch)
throws NoSuchBeanDefinitionException;

  Class<?> getType(String name) throws NoSuchBeanDefinitionException;

  String[] getAliases(String name);
}
```

2.3.2 ApplicationContext 代码解析

相比于低配版的 IoC 容器 BeanFactory，ApplicationContext 是高配版的 IoC 容器了。从 ApplicationContext 接口的定义可以看出，其在原有的 BeanFactory 接口上实现了更加复杂的扩展功能。ApplicationContext 代码如下：

```
public interface ApplicationContext extends EnvironmentCapable,
    ListableBeanFactory, HierarchicalBeanFactory,
    MessageSource, ApplicationEventPublisher, ResourcePatternResolver {

  String getId();

  String getApplicationName();

  String getDisplayName();

  long getStartupDate();

  ApplicationContext getParent();

  AutowireCapableBeanFactory getAutowireCapableBeanFactory()
                    throws IllegalStateException;
}
```

上面 ApplicationContext 接口的定义并不能很直观地反应出 ApplicationContext 接口对 BeanFactory 接口的扩展。如图 2-7 给出了更加直观的说明。

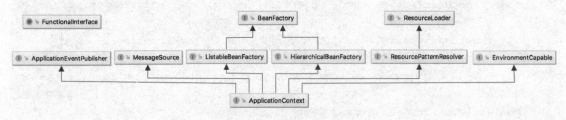

图 2-7 BeanFactory 和 ApplicationContext 继承关系

从图 2-7 可以看到，ListableBeanFactory 和 HierarchicalBeanFactory 这两个接口都继承了

BeanFactory，ApplicationContext 则继承了这两个接口，可以证明 ApplicationContext 接口是扩展了 BeanFactory 接口的。

2.3.3 BeanDefinition 代码解析

由于 BeanDefinition 代码比较长，限于篇幅，此处只截取部分代码如下：

```java
public interface BeanDefinition extends AttributeAccessor, BeanMetadataElement {
    ......
    int ROLE_APPLICATION = 0;
    int ROLE_SUPPORT = 1;
    int ROLE_INFRASTRUCTURE = 2;
    void setParentName(@Nullable String parentName);
    String getParentName();
    void setBeanClassName(@Nullable String beanClassName);
    String getBeanClassName();
    void setScope(@Nullable String scope);
    String getScope();
    void setLazyInit(boolean lazyInit);
    boolean isLazyInit();
    void setDependsOn(@Nullable String... dependsOn);
    String[] getDependsOn();
    ......
}
```

熟悉 Spring 配置的读者都知道，对于每个 JavaBean，都有各自的类名、属性、类型和是否单例等信息。Spring 是如何管理这些 JavaBean 信息的呢？其实 Spring 就是通过 BeanDefinition 来管理各种 JavaBean 及 JavaBean 相互之间的依赖关系的。BeanDefinition 抽象了开发人员对 JavaBean 的定义，Spring 将开发人员定义的 JavaBean 的数据结构转化为内存中的 BeanDefinition 数据结构进行维护。

2.3.4 Spring IoC 代码分析

由于 Spring IoC 的容器实现很多，这里通过图 2-8 所示的其中一个标准的 IoC 容器 FileSystemXmlApplicationContext 为例，分析整个 IoC 容器的启动过程。

整个 IoC 容器的启动可以概括为以下两步。

（1）创建 BeanFactory。
（2）实例化 Bean 对象。

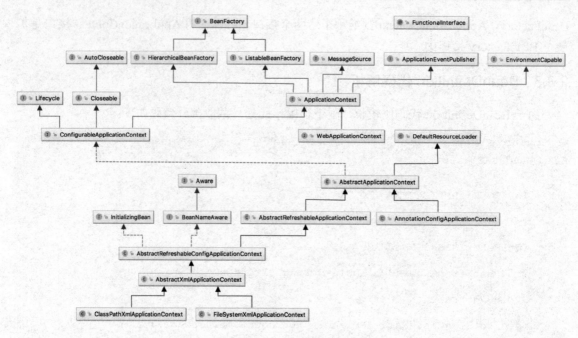

图 2-8 各容器类图

下面将对以上两步做具体的分析。为了能更加直观地查看 IoC 启动过程，本书通过一个简单的示例代码，一步步走进 Spring 底层代码。创建一个出版社类 PressServiceImpl 和图书类 BookServiceImpl，PressServiceImpl 依赖 BookServiceImpl，PressServiceImpl 类源代码如下：

```
/**
 * @Author zhouguanya
 * @Date 2018/8/17
 * @Description 出版社类
 */
public class PressServiceImpl implements PressService {
    /**
     * 依赖 BookService
     */
    private BookService bookService;

    /**
     * 依赖注入的地方
     * @param bookService
     */
    public void setBookService(BookService bookService) {
        this.bookService = bookService;
    }

    @Override
    public String say() {
```

```
        return "本书的价格是:" + bookService.getBookPrice();
    }
}
```

BookServiceImpl 类如下所示:

```
/**
 * @Author zhouguanya
 * @Date 2018/8/17
 * @Description 图书
 */
public class BookServiceImpl implements BookService {
    @Override
    public double getBookPrice() {
        return 58.8;
    }
}
```

本例用一个最简单最基础的通过 XML 配置的方式,阐述 IoC 容器的启动过程,PressServiceImpl 和 BookServiceImpl 两者依赖关系配置如下:

```xml
<!--BookService-->
<bean id="bookService" class="com.test.sourcecodelearning.BookServiceImpl">
</bean>

<!--PressService-->
<bean id="pressService" class="com.test.sourcecodelearning.PressServiceImpl">
    <!--设置依赖关系:pressService 依赖 pressService-->
    <property name="bookService" ref="bookService"></property>
</bean>
```

通过简单的测试代码,从 Spring IoC 容器中获取出版社对象,然后打印图书的价格。测试代码如下:

```java
/**
 * @Author zhouguanya
 * @Date 2018/8/14
 * @Description IoC 代码解析
 */
public class SourceCodeLearning {
    public static void main(String[] args) {
        //ApplicationContext:Spring 的上下文。通过对代码的类的集成关系可以看出,
        // FileSystemXmlApplicationContext 是 ApplicationContext 的一个标准实现
        ApplicationContext applicationContext = new FileSystemXmlApplicationContext("classpath:spring-chapter2.xml");
        //从容器中获取名字为 user 的 bean
```

```
        PressService pressService = (PressService) applicationContext.getBean
("pressService");
        //调用 bean 的方法
        String price = pressService.say();
        System.out.println(price);
    }
}
```

在 SourceCodeLearning 类中设置好断点后,下面将一步步进入 Spring 底层代码。首先从如下代码行开始进入 Spring 代码:

```
ApplicationContext applicationContext =
    new FileSystemXmlApplicationContext("classpath:spring-chapter2.xml");
```

断点跟踪进入到 FileSystemXmlApplicationContext 的构造器中后,会发现其最终调用的是如下构造器:

```
public FileSystemXmlApplicationContext(
    String[] configLocations, boolean refresh, @Nullable ApplicationContext parent)
    throws BeansException {
    super(parent);
    setConfigLocations(configLocations);
    if (refresh) {
      refresh();
    }
}
```

通过 FileSystemXmlApplicationContext 跟踪上述构造器可以发现,其主要完成了以下三个步骤:

(1) 初始化父容器 AbstractApplicationContext。
(2) 设置资源文件的位置 setConfigLocations。
(3) 使用核心方法 refresh(),其实是在超类 AbstractApplicationContext 中定义的一个模板方法(模板方法设计模式参见附录)。

下面将重点介绍其核心方法 refresh()。

首先找到 refresh()方法的定义——ConfigurableApplicationContext 接口中定义了该方法。方法定义如下:

```
void refresh() throws BeansException, IllegalStateException;
```

由图 2-8 可知,ConfigurableApplicationContext 的基类是 BeanFactory。AbstractApplicationContext 类实现了 ConfigurableApplicationContext 接口,重写了 refresh()方法。下面将分析 FileSystemXmlApplicationContext 类的超类 AbstractApplicationContext 中的 refresh()方法,由于方法很长,只截取了部分重要代码:

```java
@Override
public void refresh() throws BeansException, IllegalStateException {
    synchronized (this.startupShutdownMonitor) {
        // Prepare this context for refreshing.
        prepareRefresh();
        // Tell the subclass to refresh the internal bean factory.
        ConfigurableListableBeanFactory beanFactory = obtainFreshBeanFactory();
        // Prepare the bean factory for use in this context.
        prepareBeanFactory(beanFactory);
        try {
            // Allows post-processing of the bean factory in context subclasses.
            postProcessBeanFactory(beanFactory);
            // Invoke factory processors registered as beans in the context.
            invokeBeanFactoryPostProcessors(beanFactory);
            // Register bean processors that intercept bean creation.
            registerBeanPostProcessors(beanFactory);
            // Initialize message source for this context.
            initMessageSource();
            // Initialize event multicaster for this context.
            initApplicationEventMulticaster();
            // Initialize other special beans in specific context subclasses.
            onRefresh();
            // Check for listener beans and register them.
            registerListeners();
            // Instantiate all remaining (non-lazy-init) singletons.
            finishBeanFactoryInitialization(beanFactory);
            // Last step: publish corresponding event.
            finishRefresh();
        }
```

AbstractApplicationContext.refresh()方法是个模板方法，定义了需要执行的一些步骤。并不是实现了所有的逻辑，只是充当了一个模板，由其子类去实现更多个性化的逻辑。

模板方法 refresh()中最核心的两步如下。

（1）创建 BeanFactory：

```java
ConfigurableListableBeanFactory beanFactory = obtainFreshBeanFactory();
```

（2）实例化 Bean：

```java
finishBeanFactoryInitialization(beanFactory);
```

1. 创建 BeanFactory

创建 BeanFactory 重点分析 AbstractApplicationContext.obtainFreshBeanFactory()方法。以下是 AbstractApplicationContext.obtainFreshBeanFactory()方法的实现：

```
protected ConfigurableListableBeanFactory obtainFreshBeanFactory() {
    refreshBeanFactory();
    ConfigurableListableBeanFactory beanFactory = getBeanFactory();
    if (logger.isDebugEnabled()) {
        logger.debug("Bean factory for " + getDisplayName() + ": " + beanFactory);
    }
    return beanFactory;
}
```

从以上代码可以发现，AbstractApplicationContext.obtainFreshBeanFactory()方法分为以下两步：

（1）刷新 BeanFactory，即 refreshBeanFactory()。

（2）获取 BeanFactory，即 getBeanFactory()。

这两步骤中刷新 BeanFactory 的方法 refreshBeanFactory()是核心，接下来进一步分析 refreshBeanFactory()方法。其方法定义如下：

```
protected abstract void refreshBeanFactory()
                       throws BeansException, IllegalStateException;
```

这个方法的定义是在 AbstractApplicationContext 中，是一个抽象方法，也是一个模版方法，需要 AbstractApplicationContext 的子类来实现逻辑。其具体实现是在其子类 AbstractRefreshableApplicationContext 中完成的。refreshBeanFactory()方法实现的部分代码如下：

```
@Override
protected final void refreshBeanFactory() throws BeansException {
    if (hasBeanFactory()) {
        destroyBeans();
        closeBeanFactory();
    }
    try {
        DefaultListableBeanFactory beanFactory = createBeanFactory();
        beanFactory.setSerializationId(getId());
        customizeBeanFactory(beanFactory);
        loadBeanDefinitions(beanFactory);
    }
}
```

可以发现，在 refreshBeanFactory()方法的实现中，首先检查当前上下文是否已经存在 BeanFactory。如果已经存在 BeanFactory，先销毁 Bean 和 BeanFactory，然后创建新的 BeanFactory。

DefaultListableBeanFactory beanFactory = createBeanFactory();这行代码只是创建了一个空的 BeanFactory，其中没有任何 Bean。因此 refreshBeanFactory() 方法的核心功能是在 loadBeanDefinitions(beanFactory);这行代码中实现的。进入 loadBeanDefinitions(beanFactory)方法进行分析。

首先来看 loadBeanDefinitions 方法的定义：

```java
    protected abstract void loadBeanDefinitions(DefaultListableBeanFactory beanFactory)
        throws BeansException, IOException;
```

此方法是抽象方法，需要其子类实现。其具体实现是在 AbstractXmlApplicationContext 类中。其方法实现如下所示：

```java
    @Override
    protected void loadBeanDefinitions(DefaultListableBeanFactory beanFactory) throws BeansException, IOException {
        // Create a new XmlBeanDefinitionReader for the given BeanFactory.
        XmlBeanDefinitionReader beanDefinitionReader = new XmlBeanDefinitionReader(beanFactory);

        // Configure the bean definition reader with this context's
        // resource loading environment.
        beanDefinitionReader.setEnvironment(this.getEnvironment());
        beanDefinitionReader.setResourceLoader(this);
        beanDefinitionReader.setEntityResolver(new ResourceEntityResolver(this));

        // Allow a subclass to provide custom initialization of the reader,
        // then proceed with actually loading the bean definitions.
        initBeanDefinitionReader(beanDefinitionReader);
        loadBeanDefinitions(beanDefinitionReader);
    }
```

loadBeanDefinitions(DefaultListableBeanFactory beanFactory)方法中，通过上一步创建的空的 BeanFactory 来创建一个 XmlBeanDefinitionReader 对象。XmlBeanDefinitionReader 是用来解析 XML 中定义的 bean 的。下面重点讲解 loadBeanDefinitions(beanDefinitionReader)方法，这是一个重载的方法，这个方法的入参是刚刚生成的 XmlBeanDefinitionReader 对象。下面将进入重载的 loadBeanDefinitions 方法进行分析，代码如下：

```java
    protected void loadBeanDefinitions(XmlBeanDefinitionReader reader) throws BeansException, IOException {
        Resource[] configResources = getConfigResources();
        if (configResources != null) {
            reader.loadBeanDefinitions(configResources);
        }
        String[] configLocations = getConfigLocations();
        if (configLocations != null) {
            reader.loadBeanDefinitions(configLocations);
        }
    }
```

这个方法主要功能是解析资源文件的位置，然后调用 XmlBeanDefinitionReader 对象的 loadBeanDefinitions 方法解析 Bean 的定义。

下面将对 reader.loadBeanDefinitions(configLocations);这段代码进行分析。如图 2-9 所示，XmlBeanDefinitionReader 是 AbstractBeanDefinitionReader 的子类，所以 reader.loadBeanDefinitions(configLocations)会调用其父类的方法 loadBeanDefinitions。

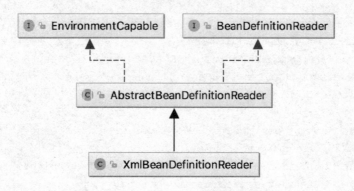

图 2-9　XmlBeanDefinitionReader 相关的类图

下面将分析 AbstractBeanDefinitionReader 的方法 loadBeanDefinitions，其方法实现如下：

```java
@Override
public int loadBeanDefinitions(String... locations)
throws BeanDefinitionStoreException {
  Assert.notNull(locations, "Location array must not be null");
  int counter = 0;
  for (String location : locations) {
    counter += loadBeanDefinitions(location);
  }
  return counter;
}
```

可以发现 loadBeanDefinitions(String... locations)方法会遍历资源数组，最终会调用重载方法 loadBeanDefinitions(String location, @Nullable Set<Resource> actualResources)，重载方法的部分实现代码如下：

```java
public int loadBeanDefinitions(String location, @Nullable Set<Resource> actualResources) throws BeanDefinitionStoreException {
    ResourceLoader resourceLoader = getResourceLoader();
    ......
    if (resourceLoader instanceof ResourcePatternResolver) {
      // Resource pattern matching available.
      try {
        Resource[] resources = ((ResourcePatternResolver) resourceLoader).getResources(location);
        int loadCount = loadBeanDefinitions(resources);
```

```
      ......
    }
  }
  else {
    // Can only load single resources by absolute URL.
    Resource resource = resourceLoader.getResource(location);
    int loadCount = loadBeanDefinitions(resource);
    ......
    return loadCount;
  }
}
```

这个方法会解析资源文件的路径，得到 Resource[] 资源数组，核心逻辑是调用 loadBeanDefinitions(resource)方法，进入这个方法查看其代码如下：

```
@Override
public int loadBeanDefinitions(Resource... resources)
throws BeanDefinitionStoreException {
  Assert.notNull(resources, "Resource array must not be null");
  int counter = 0;
  for (Resource resource : resources) {
    counter += loadBeanDefinitions(resource);
  }
  return counter;
}
```

loadBeanDefinitions 内部工作原理是遍历每个资源，依次调用 loadBeanDefinitions(Resource resource)重载的方法。该重载的方法定义在顶层接口 BeanDefinitionReader 中，其方法定义如下：

```
int loadBeanDefinitions(Resource resource) throws
BeanDefinitionStoreException;
```

如图 2-9 所示，loadBeanDefinitions 方法的实现是在 XmlBeanDefinitionReader 中。具体实现如下：

```
@Override
public int loadBeanDefinitions(Resource resource) throws
BeanDefinitionStoreException {
    return loadBeanDefinitions(new EncodedResource(resource));
}
```

该方法实现会调用重载方法 loadBeanDefinitions(EncodedResource encodedResource)。因该重载方法很长，现截取部分代码如下：

```
public int loadBeanDefinitions(EncodedResource encodedResource)
throws BeanDefinitionStoreException {
    ......
    try {
```

```
          InputStream inputStream = encodedResource.getResource().
getInputStream();
      try {
        InputSource inputSource = new InputSource(inputStream);
        if (encodedResource.getEncoding() != null) {
          inputSource.setEncoding(encodedResource.getEncoding());
        }
        return doLoadBeanDefinitions(inputSource, encodedResource.
getResource());
      }
    }
    ......
  }
```

loadBeanDefinitions(EncodedResource encodedResource)方法以流的方式读取资源文件，调用doLoadBeanDefinitions()方法。doLoadBeanDefinitions()是载入定义 Bean 的核心方法。进入doLoadBeanDefinitions()方法，查看其部分代码如下：

```
    protected int doLoadBeanDefinitions(InputSource inputSource, Resource resource)
        throws BeanDefinitionStoreException {
      try {
        Document doc = doLoadDocument(inputSource, resource);
        return registerBeanDefinitions(doc, resource);
      }
    ......
    }
```

从 doLoadBeanDefinitions(InputSource inputSource, Resource resource)方法的定义可以看出，最终注册 Bean 的地方是在 registerBeanDefinitions(doc, resource);这一行代码。进入 registerBeanDefinitions(doc, resource)方法，查看其代码如下：

```
    public int registerBeanDefinitions(Document doc, Resource resource)
      throws BeanDefinitionStoreException {
      BeanDefinitionDocumentReader documentReader = createBeanDefinitionDocumentReader();
      int countBefore = getRegistry().getBeanDefinitionCount();
      documentReader.registerBeanDefinitions(doc, createReaderContext(resource));
      return getRegistry().getBeanDefinitionCount() - countBefore;
    }
```

registerBeanDefinitions(Document doc, Resource resource) 方法的核心逻辑是在 documentReader.registerBeanDefinitions(doc, createReaderContext(resource));这一行，这里发生了对 Bean 的注册。进入 registerBeanDefinitions(Document doc, XmlReaderContext readerContext)方法查看其方法实现如下：

```
@Override
public void registerBeanDefinitions(Document doc, XmlReaderContext 
readerContext) {
    this.readerContext = readerContext;
    logger.debug("Loading bean definitions");
    Element root = doc.getDocumentElement();
    doRegisterBeanDefinitions(root);
}
```

registerBeanDefinitions(Document doc, XmlReaderContext readerContext)方法是在 DefaultBeanDefinitionDocumentReader 中实现的。核心是通过 doRegisterBeanDefinitions()方法实现的。进入 doRegisterBeanDefinitions(Element root)方法的代码，查看其部分实现代码如下：

```
protected void doRegisterBeanDefinitions(Element root) {
    .......
    preProcessXml(root);
    parseBeanDefinitions(root, this.delegate);
    postProcessXml(root);
    this.delegate = parent;
}
```

doRegisterBeanDefinitions(Element root)方法的核心逻辑在 parseBeanDefinitions(root, this.delegate);这个方法中处理。进入 parseBeanDefinitions(Element root, BeanDefinitionParserDelegate delegate)方法的代码，查看其方法部分实现如下：

```
protected void parseBeanDefinitions(Element root, 
BeanDefinitionParserDelegate delegate) {
    if (delegate.isDefaultNamespace(root)) {
        NodeList nl = root.getChildNodes();
        for (int i = 0; i < nl.getLength(); i++) {
            Node node = nl.item(i);
            if (node instanceof Element) {
                Element ele = (Element) node;
                if (delegate.isDefaultNamespace(ele)) {
                    parseDefaultElement(ele, delegate);
                }
                .......
            }
        }
    }
}
```

parseBeanDefinitions(Element root, BeanDefinitionParserDelegate delegate)方法的核心逻辑是依赖 parseDefaultElement(ele, delegate);方法实现的，进入 parseDefaultElement(Element ele, BeanDefinitionParserDelegate delegate)方法的代码，查看其代码如下：

```
private void parseDefaultElement(Element ele, BeanDefinitionParserDelegate 
delegate) {

    if (delegate.nodeNameEquals(ele, IMPORT_ELEMENT)) {
```

```
      importBeanDefinitionResource(ele);
    }
    else if (delegate.nodeNameEquals(ele, ALIAS_ELEMENT)) {
      processAliasRegistration(ele);
    }
    else if (delegate.nodeNameEquals(ele, BEAN_ELEMENT)) {
      processBeanDefinition(ele, delegate);
    }
    else if (delegate.nodeNameEquals(ele, NESTED_BEANS_ELEMENT)) {
      // recurse
      doRegisterBeanDefinitions(ele);
    }
  }
}
```

根据不同 bean 的配置不同，进入不同分支执行。本书的示例是进入 processBeanDefinition(ele, delegate)方法，下面进入 processBeanDefinition(Element ele, BeanDefinitionParserDelegate delegate) 方法，查看其实现如下：

```
protected void processBeanDefinition(Element ele,
BeanDefinitionParserDelegate delegate) {
  BeanDefinitionHolder bdHolder = delegate.parseBeanDefinitionElement(ele);
  if (bdHolder != null) {
    bdHolder = delegate.decorateBeanDefinitionIfRequired(ele, bdHolder);
    try {
      // Register the final decorated instance.
      BeanDefinitionReaderUtils.registerBeanDefinition(bdHolder,
getReaderContext().getRegistry());
    }
    catch (BeanDefinitionStoreException ex) {
      getReaderContext().error("Failed to register bean definition with name
'" + bdHolder.getBeanName() + "'", ele, ex);
    }
    // Send registration event.
    getReaderContext().fireComponentRegistered(new
BeanComponentDefinition(bdHolder));
  }
}
```

从 processBeanDefinition(Element ele, BeanDefinitionParserDelegate delegate)方法的代码可知，最关键的是 BeanDefinitionReaderUtils.registerBeanDefinition(bdHolder,getReaderContext().getRegistry()); 的调用。这是注册 Bean 的关键代码，查看其代码如下：

```
public static void registerBeanDefinition(

    BeanDefinitionHolder definitionHolder, BeanDefinitionRegistry registry)
    throws BeanDefinitionStoreException {
```

```java
    // Register bean definition under primary name.
    String beanName = definitionHolder.getBeanName();
    registry.registerBeanDefinition(beanName, definitionHolder.getBeanDefinition());

    // Register aliases for bean name, if any.
    String[] aliases = definitionHolder.getAliases();
    if (aliases != null) {
      for (String alias : aliases) {
        registry.registerAlias(beanName, alias);
      }
    }
  }
```

registry.registerBeanDefinition(beanName, definitionHolder.getBeanDefinition());这一行是将 Bean 的名字和 BeanDefinition 对象进行注册的地方。该方法的定义是在 BeanDefinitionRegistry 中。其定义如下：

```java
void registerBeanDefinition(String beanName, BeanDefinition beanDefinition)
throws BeanDefinitionStoreException;
```

本例将进入 BeanDefinitionRegistry 接口的实现类 DefaultListableBeanFactory 中，查看该方法，部分实现代码如下：

```java
@Override
  public void registerBeanDefinition(String beanName, BeanDefinition beanDefinition)
      throws BeanDefinitionStoreException {
    ......
    BeanDefinition existingDefinition = this.beanDefinitionMap.get(beanName);
    if (existingDefinition != null) {
      if (!isAllowBeanDefinitionOverriding()) {
        throw new BeanDefinitionStoreException(beanDefinition.getResourceDescription(), beanName, "Cannot register bean definition [" + beanDefinition + "] for bean '" + beanName +"': There is already [" + existingDefinition + "] bound.");
      }
      ......
      this.beanDefinitionMap.put(beanName, beanDefinition);
      this.beanDefinitionNames.add(beanName);
      this.manualSingletonNames.remove(beanName);
    }
    ......
  }
```

从 registerBeanDefinition(String beanName, BeanDefinition beanDefinition)方法代码可以看出，

先从 beanDefinitionMap 这个 ConcurrentHashMap 对象根据 beanName 查找是否已经有同名的 bean，如果不存在，则会调用 beanDefinitionMap.put(beanName, beanDefinition)方法，以 beanName 为 key，beanDefinition 为 value 注册，将这个 Bean 注册到 BeanFactory 中，并将所有的 BeanName 保存到 beanDefinitionNames 这个 ArrayList 中。

到此，完成了 IoC 第一部分——创建 BeanFactory 的代码解析。但是，此时 Bean 只是完成了 Bean 名称和 BeanDefinition 对象的注册，并没有实现 Bean 的实例化和依赖注入。下面将要分析 IoC 的第二个关键部分 Bean 的初始化。

2. 实例化 Bean

在创建 BeanFactory 的过程中，BeanDefinition 注册到了 BeanFactory 中的一个 ConcurrentHashMap 对象中了，并且以 BeanName 为 key，BeanDefinition 为 value 注册。下面将要分析实例化 Bean 的过程，即从上文提到的 AbstractApplicationContext 类的 refresh()方法中的 finishBeanFactoryInitialization(ConfigurableListableBeanFactory beanFactory)方法开始向底层分析。

首先进入 finishBeanFactoryInitialization(ConfigurableListableBeanFactory beanFactory)方法，查看其部分代码如下：

```
protected void finishBeanFactoryInitialization
(ConfigurableListableBeanFactory beanFactory) {
    ......
    // Stop using the temporary ClassLoader for type matching.
    beanFactory.setTempClassLoader(null);

    // Allow for caching all bean definition metadata, not expecting further changes.
    beanFactory.freezeConfiguration();

    // Instantiate all remaining (non-lazy-init) singletons.
    beanFactory.preInstantiateSingletons();
}
```

从上述 finishBeanFactoryInitialization(ConfigurableListableBeanFactory beanFactory)方法的部分代码看到，beanFactory.preInstantiateSingletons();这行代码是实例化 Bean 的。打开 preInstantiateSingletons()方法的代码如下：

```
public void preInstantiateSingletons() throws BeansException {
    ......
    List<String> beanNames = new ArrayList<>(this.beanDefinitionNames);
    // Trigger initialization of all non-lazy singleton beans...
    for (String beanName : beanNames) {
        RootBeanDefinition bd = getMergedLocalBeanDefinition(beanName);
        if (!bd.isAbstract() && bd.isSingleton() && !bd.isLazyInit()) {
            ......
            if (isEagerInit) {
                getBean(beanName);
            }
```

```
            }
          }
          else {
            getBean(beanName);
          }
        }
      }
      ......
      }
    }
```

该方法遍历 beanDefinitionNames 这个 ArrayList 对象中的 BeanName，循环调用 getBean(beanName)方法。该方法实际上就是创建 Bean 并递归构建 Bean 间的依赖关系。getBean(beanName)方法最终会调用 doGetBean(name, null, null, false)，进入该方法查看 doGetBean 方法的代码，由于这个方法特别长，故下面只挑选最关键的代码作解析：

```
String[] dependsOn = mbd.getDependsOn();
    if (dependsOn != null) {
      ......
        registerDependentBean(dep, beanName);
        try {
          getBean(dep);
        }
        ......
      }
    }
    // Create bean instance.
    if (mbd.isSingleton()) {
      sharedInstance = getSingleton(beanName, () -> {
        try {
          return createBean(beanName, mbd, args);
        }
        ......
      }
```

可以看到，该方法首先会获取当前 Bean 依赖关系 mbd.getDependsOn();接着根据依赖的 BeanName 递归调用 getBean()方法，直到调用 getSingleton()方法返回依赖 Bean，即当前正在创建的 Bean，不断探寻依赖其的 Bean，直到依赖关系最底层的 Bean 没有依赖的对象了，至此整个递归过程结束。getSingleton() 方法的参数是 createBean() 方法返回值。createBean() 是在 AbstractAutowireCapableBeanFactory 中实现的。createBean(String beanName, RootBeanDefinition mbd, @Nullable Object[] args)方法部分代码如下：

```
@Override
protected Object createBean(String beanName, RootBeanDefinition mbd,
    @Nullable Object[] args)
```

```
      throws BeanCreationException {
    ......
    try {
      Object beanInstance = doCreateBean(beanName, mbdToUse, args);
      if (logger.isDebugEnabled()) {
        logger.debug("Finished creating instance of bean '" + beanName + "'");
      }
      return beanInstance;
    }
    ......
  }
```

createBean(String beanName, RootBeanDefinition mbd, @Nullable Object[] args)方法的核心是 doCreateBean(beanName, mbdToUse, args)这个方法，doCreateBean 将会返回 Bean 对象的实例。查看 doCreateBean 的部分代码如下：

```
protected Object doCreateBean(final String beanName, final RootBeanDefinition mbd, final @Nullable Object[] args) throws BeanCreationException {
    // Instantiate the bean.
    BeanWrapper instanceWrapper = null;
    if (mbd.isSingleton()) {
      instanceWrapper = this.factoryBeanInstanceCache.remove(beanName);
    }
    if (instanceWrapper == null) {
      instanceWrapper = createBeanInstance(beanName, mbd, args);
    }
    ......
    // Initialize the bean instance.
    Object exposedObject = bean;
    try {
      populateBean(beanName, mbd, instanceWrapper);
      exposedObject = initializeBean(beanName, exposedObject, mbd);
    }
    }
    ......
    return exposedObject;
}
```

这个方法很长，这里挑选重要的两行代码进行讲解：

（1）instanceWrapper = createBeanInstance(beanName, mbd, args)用来创建实例。

（2）方法 populateBean(beanName, mbd, instanceWrapper)用于填充 Bean，该方法可以说就是发生依赖注入的地方。

先看方法 createBeanInstance(String beanName, RootBeanDefinition mbd, @Nullable Object[] args)，其核心实现如下：

```
    if (resolved) {
        if (autowireNecessary) {
            return autowireConstructor(beanName, mbd, null, null);
        }
        else {
            return instantiateBean(beanName, mbd);
        }
    }
}
```

createBeanInstance(String beanName, RootBeanDefinition mbd, @Nullable Object[] args)方法会调用 instantiateBean(beanName, mbd) 方法，进入该方法，其部分实现如下：

```
protected BeanWrapper instantiateBean(final String beanName, final RootBeanDefinition mbd) {
    try {
        Object beanInstance;
        ......
        else {
            beanInstance = getInstantiationStrategy().instantiate(mbd, beanName, parent);
        }
        BeanWrapper bw = new BeanWrapperImpl(beanInstance);
        initBeanWrapper(bw);
        return bw;
    }
    ......
}
```

instantiateBean(final String beanName, final RootBeanDefinition mbd)方法核心逻辑是 beanInstance = getInstantiationStrategy().instantiate(mbd, beanName, parent)，发挥作用的策略对象是 SimpleInstantiationStrategy，在该方法内部调用了静态方法 BeanUtils.instantiateClass(constructorToUse)，这个方法的部分实现如下：

```
    ......
    Assert.notNull(ctor, "Constructor must not be null");
    try {
        ReflectionUtils.makeAccessible(ctor);
        return (KotlinDetector.isKotlinType(ctor.getDeclaringClass()) ?
            KotlinDelegate.instantiateClass(ctor, args) :
            ctor.newInstance(args));
    }
    ......
```

该方法会判断是否是 Kotlin 类型。如果不是 Kotlin 类型，则调用 Constructor 的 newInstance 方法，也就是最终使用反射创建了该实例。

到这里，Bean 的实例已经创建完成。但是 Bean 实例的依赖关系还没有设置，下面回到

doCreateBean()方法中的populateBean(beanName, mbd, instanceWrapper)方法,该方法用于填充Bean,该方法可以说就是发生依赖注入的地方。回到 AbstractAutowireCapableBeanFactory 类中看一下 populateBean()方法的实现。populateBean()部分代码如下：

```
......
PropertyValues pvs = (mbd.hasPropertyValues() ? mbd.getPropertyValues() : null);
......
if (pvs != null) {
    applyPropertyValues(beanName, mbd, bw, pvs);
}
```

整个方法的核心逻辑是 PropertyValues pvs = (mbd.hasPropertyValues() ? mbd.getPropertyValues() : null);这一行代码,即获取该 bean 的所有属性,就是配置 property 元素,即依赖关系。最后执行 applyPropertyValues(beanName, mbd, bw, pvs)方法,其实现如下：

```
protected void applyPropertyValues(String beanName, BeanDefinition mbd,
BeanWrapper bw, PropertyValues pvs) {
    ......
    for (PropertyValue pv : original) {
        if (pv.isConverted()) {
            deepCopy.add(pv);
        }
        else {
            String propertyName = pv.getName();
            Object originalValue = pv.getValue();
            Object resolvedValue = valueResolver.resolveValueIfNecessary(pv, originalValue);
            ......
            bw.setPropertyValues(new MutablePropertyValues(deepCopy));
            ......
        }
    }
}
```

关键代码 Object resolvedValue = valueResolver.resolveValueIfNecessary(pv, originalValue);该方法是获取 property 对应的值。resolveValueIfNecessary(pv, originalValue)方法部分代码如下：

```
public Object resolveValueIfNecessary(Object argName, @Nullable Object value){
    // We must check each value to see whether it requires a runtime reference
    // to another bean to be resolved.
    if (value instanceof RuntimeBeanReference) {
        RuntimeBeanReference ref = (RuntimeBeanReference) value;
        return resolveReference(argName, ref);
    }
}
```

resolveValueIfNecessary(Object argName, @Nullable Object value)方法的核心是 resolveReference(argName, ref),该方法是解决 Bean 依赖关系的。进入该方法的代码,查看其部分实现如下：

```
private Object resolveReference(Object argName, RuntimeBeanReference ref) {
    try {
        Object bean;
        String refName = ref.getBeanName();
        refName = String.valueOf(doEvaluate(refName));
        if (ref.isToParent()) {
            if (this.beanFactory.getParentBeanFactory() == null) {
                throw new BeanCreationException(
                    this.beanDefinition.getResourceDescription(), this.beanName,
                    "Can't resolve reference to bean '" + refName +
                    "' in parent factory: no parent factory available");
            }
            bean = this.beanFactory.getParentBeanFactory().getBean(refName);
            ......
}
```

这段代码的核心是以下这一行：

```
bean = this.beanFactory.getParentBeanFactory().getBean(refName)
```

这里将会发生递归调用，根据依赖的名称，从 BeanFactory 中递归得到依赖。到这段结束，就可以获取到依赖的 Bean。回到 applyPropertyValues 入口处，获取到依赖的对象值后，将会调用 bw.setPropertyValues(new MutablePropertyValues(deepCopy))方法，这是将依赖值注入的地方。此方法会调用 AbstractPropertyAccessor 类的 setPropertyValues 方法，查看 AbstractPropertyAccessor.setPropertyValues 方法的实现，其部分代码如下：

```
@Override
public void setPropertyValues(PropertyValues pvs, boolean ignoreUnknown,
    boolean ignoreInvalid) throws BeansException {
    List<PropertyAccessException> propertyAccessExceptions = null;
    List<PropertyValue> propertyValues = (pvs instanceof MutablePropertyValues ?
        ((MutablePropertyValues) pvs).getPropertyValueList() :
Arrays.asList(pvs.getPropertyValues()));
    for (PropertyValue pv : propertyValues) {
        try {
            // This method may throw any BeansException, which won't be caught
            // here, if there is a critical failure such as no matching field.
            // We can attempt to deal only with less serious exceptions.
            setPropertyValue(pv);
        }
        ......
}
```

该方法会循环 Bean 的属性列表，循环中调用 setPropertyValue(PropertyValue pv)方法，该方法是通过调用 AbstractNestablePropertyAccessor.setPropertyValue (String propertyName, @Nullable

Object value)方法来实现的,进入该方法的代码,其部分实现如下:

```
@Override
  public void setPropertyValue(String propertyName, @Nullable Object value)
throws BeansException {
    ......
    PropertyTokenHolder tokens = getPropertyNameTokens(getFinalPath(nestedPa,
propertyName));
    nestedPa.setPropertyValue(tokens, new PropertyValue(propertyName,
value));
  }
```

其核心是最后一行 nestedPa.setPropertyValue(tokens, new PropertyValue(propertyName, value)) 代码,进入 setPropertyValue 方法查看其部分实现如下:

```
protected void setPropertyValue(PropertyTokenHolder tokens, PropertyValue pv)
throws BeansException {
    ......
    else {
      processLocalProperty(tokens, pv);
    }
}
```

进入 processLocalProperty(tokens, pv) 方法的代码,该方法非常复杂,其核心实现如下:

```
    private void processLocalProperty(PropertyTokenHolder tokens,
PropertyValue pv) {
      ......
      Object oldValue = null;
      try {
        Object originalValue = pv.getValue();
        Object valueToApply = originalValue;
        ........
        ph.setValue(valueToApply);
      }
      ........
    }
```

上述代码调用的 ph.setValue(valueToApply)方法是 BeanWrapperImpl.setValue(final @Nullable Object value)方法,进入这个方法的代码,查看其部分实现如下:

```
    public void setValue(final @Nullable Object value) throws Exception {
      final Method writeMethod = (this.pd instanceof
GenericTypeAwarePropertyDescriptor ?
        ((GenericTypeAwarePropertyDescriptor)this.pd).
getWriteMethodForActualAccess() :
          this.pd.getWriteMethod());
        if (System.getSecurityManager() != null) {
```

```
    .....
  }
  else {
    ReflectionUtils.makeAccessible(writeMethod);
    writeMethod.invoke(getWrappedInstance(), value);
  }
}
```

该方法是最后一步，这里可以看到该方法会找到属性的 set 方法，然后调用 Method 的 invoke 方法，完成属性注入。至此 IoC 容器的启动过程完毕。

下面总结一下 IoC 的底层原理实现：Spring IoC 容器启动分为两步——创建 BeanFactory 和实例化 Bean。Spring 的 Bean 在内存中的状态就是 BeanDefinition，在 Bean 的创建和依赖注入的过程中，需要根据 BeanDefinition 的信息来递归地完成依赖注入，从代码中可以看到，这些递归都是以 getBean()为入口的，一个递归是在上下文体系中查找需要的 Bean 和创建 Bean 的依赖关系，另一个递归是在依赖注入时，通过递归调用容器的 getBean 方法得到当前的依赖 Bean，同时也触发对依赖 Bean 的创建和注入。在对 Bean 的属性进行依赖注入时，解析过程也是一个递归过程，这样根据依赖关系，一层一层地完成 Bean 的创建和注入，直到与当前 Bean 相关的整个依赖链的注入完成。由于整个 IoC 容器启动过程比较复杂，本书限于篇幅，无法显示流程图。简化后的流程图如图 2-10 所示。更多更详细的 IoC 流程图请参考本书 GitHub 代码（https://github.com/ online-demo/spring5projectdemo），即第 2 章相关内容。

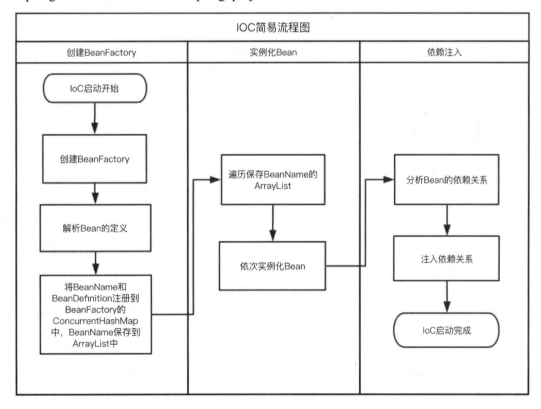

图 2-10　IoC 容器启动简易流程图

2.4　Spring IoC 容器中 Bean 的生命周期

Spring IoC 容器管理的 Bean 默认都是单利设计模式的（参见本书附录），即每个 Bean 只有一个实例化的 Bean 对象存在于 Spring IoC 容器中，因此 Spring IoC 容器需要负责管理 Bean 的产生、使用和销毁等生命周期。

Spring IoC 容器中的 Bean 的生命周期可以分为以下 4 类：

- Bean 自身方法。
- Bean 生命周期接口方法。
- 容器级生命周期接口方法。
- 工厂后处理器接口方法。

各个阶段涉及的具体接口和方法如表 2-1 所示。

表 2-1　各阶段具体接口和方法

Spring Bean 生命周期各阶段	相关接口及方法
Bean 自身方法	Bean 本身业务的方法； 配置文件中 init-method 和 destroy-method 指定的方法
Bean 生命周期接口方法	InitializingBean 接口 DiposableBean 接口 BeanNameAware 接口 ApplicationContextAware 接口 BeanFactoryAware 接口 其他
容器级生命周期接口方法 （一般称为"后处理器"）	InstantiationAwareBeanPostProcessor 接口实现 BeanPostProcessor 接口实现
工厂级生命周期接口方法 （也可以归为容器级的）	AspectJWeavingEnabler ConfigurationClassPostProcessor CustomAutowireConfigurer 等

下面先以 Bean 自身方法和 Bean 生命周期接口方法为例，演示其各个生命周期的执行时序。

- init-method：指定某个方法在 Bean 实例化完成，依赖关系设置结束后执行。
- destroy-method：指定某个方法在 Bean 销毁之前被执行。
- InitializingBean 接口：指定在 Bean 实例化完成，依赖关系设置结束后执行（在 init-method 之前执行）。
- DiposableBean 接口：指定某个方法在 Bean 销毁之前被执行（在 destory-method 之前执行）。
- ApplicationContextAware 接口：在实例化 Bean 时，为 Bean 注入 ApplicationContext。
- BeanNameAware 接口：在实例化 Bean 时，为 Bean 注入 beanName。

以下代码 BeanLifecycle 将实现上述 4 个接口 InitializingBean、DiposableBean、ApplicationContextAware 和 BeanNameAware，并通过在 XML 文件中配置该 Bean 的 init-method 和 destroy-method。通过 BeanLifecycle 例子将可以更加清晰地阐述 Bean 自身方法和 Bean 生命周期接口方法的生命周期。

```java
/**
 * @Author zhouguanya
 * @Date 2018/8/18
 * @Description Bean 生命周期
 */
public class BeanLifecycle implements BeanNameAware, ApplicationContextAware,
InitializingBean, DisposableBean {
    /**
     * 1. 构造器
     */
    public BeanLifecycle() {
        System.out.println("1.【Bean 级别】构造器执行了");
    }

    /**
     * 2. BeanNameAware 接口方法实现
     */
    @Override
    public void setBeanName(String name) {
        System.out.println("2.【Bean 级别】setBeanName 方法执行了");
    }

    /**
     * 3. ApplicationContextAware 接口方法实现
     */
    @Override
    public void setApplicationContext(ApplicationContext applicationContext)
throws BeansException {
        System.out.println("3.【Bean 级别】setApplicationContext 方法执行了");
    }

    /**
     * 4. InitializingBean 接口方法实现
     */
    @Override
    public void afterPropertiesSet() throws Exception {
        System.out.println("4.【Bean 级别】afterPropertiesSet 方法执行了");
    }

    /**
     * 5. init-method 属性指定的方法
     */
```

```java
    public void lifecycleInit() {
        System.out.println("5.【Bean级别】init-method指定的方法执行了");
    }

    /**
     * 6. Bean中的业务方法
     */
    public void sayHello() {
        System.out.println("6.【Bean级别】sayHello方法执行了");
    }

    /**
     * 7. DisposableBean接口方法实现
     */
    @Override
    public void destroy() throws Exception {
        System.out.println("7.【Bean级别】destroy方法执行了");
    }

    /**
     * 8. destroy-method属性指定的方法
     */
    public void lifecycleInitDestroy() {
        System.out.println("8.【Bean级别】destroy-method属性指定的方法执行了");
    }
}
```

如上述代码中的注释所示，BeanLifecycle这个Bean的生命周期将按照序号1~8顺序执行。下面是对BeanLifecycle准备的测试代码：

```java
/**
 * @Author zhouguanya
 * @Date 2018/8/18
 * @Description Bean生命周期测试
 */
@RunWith(SpringRunner.class)
@ContextConfiguration("classpath:spring-chapter2-beanlifecycle.xml")
public class BeanLifecycleTest {
    @Autowired
    private BeanLifecycle beanLifecycle;

    @Test
    public void test() {
        beanLifecycle.sayHello();
    }
}
```

这段测试代码很简单，就是通过从 Spring IoC 容器中注入的 BeanLifecycle 对象，调用其 sayHello()方法。测试代码运行后的结果如图 2-11 所示。

```
1. 【Bean级别】构造器执行了
2. 【Bean级别】setBeanName方法执行了
3. 【Bean级别】setApplicationContext方法执行了
4. 【Bean级别】afterPropertiesSet方法执行了
5. 【Bean级别】init-method指定的方法执行了
6. 【Bean级别】sayHello方法执行了
7. 【Bean级别】destroy方法执行了
8. 【Bean级别】destroy-method属性指定的方法执行了
```

图 2-11　Bean 自身方法和 Bean 生命周期接口方法生命周期测试图

通过执行结果可以得到 Bean 自身方法和 Bean 生命周期接口方法的执行时序：

（1）执行构造器。

（2）执行 BeanNameAware 接口的 setBeanName(String name)方法。

（3）执行 ApplicationContextAware 接口的 setApplicationContext(ApplicationContextapplicationContext)方法。

（4）执行 InitializingBean 接口的 afterPropertiesSet()方法。

（5）执行 init-method 指定的方法。

（6）执行运行时 Bean 中的业务方法。

（7）执行 DisposableBean 接口的 destroy()方法。

（8）执行 destroy-method 指定的方法。

Bean 自身方法和 Bean 生命周期接口方法执行的生命周期时序图如图 2-12 所示。

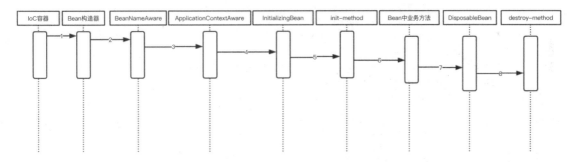

图 2-12　Bean 自身方法和 Bean 生命周期接口方法生命周期时序图

下面将介绍容器级生命周期接口方法的执行时序。容器级生命周期接口方法有 InstantiationAwareBeanPostProcessor 和 BeanPostProcessor 这两个接口，一般也将其实现类称为后处理器。容器级生命周期接口的实现独立于 Spring IoC 容器中的 Bean，其是以容器扩展的形式注册到 Spring 中的。无论 Spring IoC 管理任何的 Bean，这些后处理器都会发生作用。因此后处理器影响范围是全局的 Spring IoC 容器中的 Bean。用户可以通过编写合理的后处理器来实现感兴趣的 Bean 加工处理逻辑。

- BeanPostProcessor 接口：此接口的方法可以对 Bean 的属性进行更改。
- InstantiationAwareBeanPostProcessor 接口：此接口可以在 Bean 实例化前、Bean 实例化后分别进行操作，也可以对 Bean 实例化之后进行属性操作（为 BeanPostProcessor 的子接口）。
- InstantiationAwareBeanPostProcessorAdapter：适配器类。

BeanPostProcessor、InstantiationAwareBeanPostProcessor 和 InstantiationAwareBeanPostProcessorAdapter 三者的关系如图 2-13 所示。

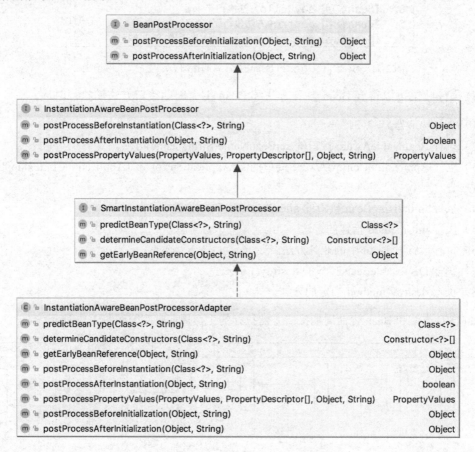

图 2-13　BeanPostProcessor 相关类图

如图 2-13 所示，InstantiationAwareBeanPostProcessorAdapter 最终实现了 BeanPostProcessor 这个顶级接口。下面以 InstantiationAwareBeanPostProcessorAdapter 为例，讲解容器级生命周期接口方法的执行时序。下面代码将通过 ContainerLifecycle 类继承 InstantiationAwareBeanPostProcessorAdapter 来阐述容器级生命周期接口方法的执行时序，具体代码如下：

```
/**
 * @Author zhouguanya
 * @Date 2018/8/19
 * @Description Bean 级生命周期+容器级生命周期
 */
public class ContainerLifecycle extends
```

```java
InstantiationAwareBeanPostProcessorAdapter {
    /**
     * 构造器
     */
    public ContainerLifecycle() {
        System.out.println("① 【容器级别】ContainerLifecycle 构造器执行了");
    }

    /**
     * 接口方法和实例化 Bean 之前调用
     * @param beanClass
     * @param beanName
     * @return
     */
    @Override
    public Object postProcessBeforeInstantiation(Class beanClass, String beanName) {
        System.out.println("② 【容器级别】postProcessBeforeInstantiation 方法执行了，class=" + beanClass);
        return null;
    }

    /**
     * 设置某个属性时调用
     * @param pvs
     * @param pds
     * @param bean
     * @param beanName
     * @return
     */
    @Override
    public PropertyValues postProcessPropertyValues(PropertyValues pvs, PropertyDescriptor[] pds, Object bean, String beanName) {
        System.out.println("③ 【容器级别】postProcessPropertyValues 方法执行了，beanName=" + bean.getClass());
        return pvs;
    }

    /**
     * 接口方法和实例化 Bean 之后调用
     * @param bean
     * @param beanName
     * @return
     */
    @Override
```

```
        public Object postProcessAfterInitialization(Object bean, String 
beanName){
        System.out.println("④【容器级别】postProcessAfterInitialization方法执
行了,beanName=" + bean.getClass());
        return null;
    }
}
```

如上代码段所示,ContainerLifecycle 类继承了 InstantiationAwareBeanPostProcessorAdapter,重写了其中 postProcessBeforeInstantiation、postProcessPropertyValues 和 postProcessAfterInitialization 方法。其执行顺序如下。

- ContainerLifecycle:构造器最先执行。
- postProcessBeforeInstantiation:接口方法和实例化 Bean 之前调用。
- postProcessPropertyValues:设置某个属性时调用。
- postProcessAfterInitialization:接口方法和实例化 Bean 之后调用。

测试代码中,还是以上例的 BeanLifecycle 类为例,调用 BeanLifecycle 类的 sayHello()方法,测试代码如下:

```
/**
 * @Author zhouguanya
 * @Date 2018/8/19
 * @Description 容器级生命周期测试
 */
public class ContainerLifecycleTest {
    public static void main(String[] args) {
        ClassPathXmlApplicationContext context = new 
ClassPathXmlApplicationContext("classpath:spring-chapter2-beanlifecycle.xml","
classpath:spring-chapter2-containerlifecycle.xml");
        BeanLifecycle beanLifecycle = 
context.getBean("beanLifecycle",BeanLifecycle.class);
        beanLifecycle.sayHello();
        context.close();
    }
}
```

运行单元测试,测试结果如图 2-14 所示。

下面将介绍的是工厂级生命周期接口方法,工厂级生命周期接口方法涉及到的有 BeanFactoryPostProcessor 接口。下面将通过实现 BeanFactoryPostProcessor 接口来分析。

```
① 【容器级别】ContainerLifecycle构造器执行了
② 【容器级别】postProcessBeforeInstantiation方法执行了，class=class com.test.lifecycle.beanlifcycle.BeanLifecycle
1. 【Bean级别】构造器执行了
③ 【容器级别】postProcessPropertyValues方法执行了，beanName=class com.test.lifecycle.beanlifcycle.BeanLifecycle
2. 【Bean级别】setBeanName方法执行了
3. 【Bean级别】setApplicationContext方法执行了
4. 【Bean级别】afterPropertiesSet方法执行了
5. 【Bean级别】init-method指定的方法执行了
④ 【容器级别】postProcessAfterInitialization方法执行了，beanName=class com.test.lifecycle.beanlifcycle.BeanLifecycle
6. 【Bean级别】sayHello方法执行了
7. 【Bean级别】destroy方法执行了
8. 【Bean级别】destroy-method属性指定的方法执行了
```

图 2-14 容器级生命周期接口方法结果

工厂级生命周期接口的生命周期，实现代码如下：

```
/**
 * @Author zhouguanya
 * @Date 2018/8/19
 * @Description 工厂级生命周期
 */
public class FactoryLifecycle implements BeanFactoryPostProcessor {

    /**
     * 构造器
     */
    public FactoryLifecycle () {
        System.out.println("一 【工厂级别】FactoryLifecycle 构造器执行了");
    }

    /**
     * Bean 实例化之前
     */
    @Override
    public void postProcessBeanFactory(ConfigurableListableBeanFactory beanFactory) throws BeansException {
        System.out.println("二 【工厂级别】postProcessBeanFactory 方法执行了");
    }
}
```

测试代码将所有级别生命周期接口进行统一测试，以方便观察完整的 Bean 生命周期的执行时序：

```
/**
 * @Author zhouguanya
 * @Date 2018/8/19
 * @Description Bean 级生命周期 + 容器级生命周期 + 工厂级生命周期测试
 */
public class FactoryLifecycleTest {
```

```java
    public static void main(String[] args) {
        ClassPathXmlApplicationContext context = new ClassPathXmlApplicationContext("classpath:spring-chapter2-beanlifecycle.xml","classpath:spring-chapter2-containerlifecycle.xml","classpath:spring-chapter2-factorybeanlifecycle.xml");
        BeanLifecycle beanLifecycle = context.getBean("beanLifecycle",BeanLifecycle.class);
        beanLifecycle.sayHello();
        context.close();
    }
}
```

测试效果图如图 2-15 所示。

```
一【工厂级别】FactoryLifecycle构造器执行了
二【工厂级别】postProcessBeanFactory方法执行了
①【容器级别】ContainerLifecycle构造器执行了
②【容器级别】postProcessBeforeInstantiation方法执行了, class=class com.test.lifecycle.beanlifcycle.BeanLifecycle
1.【Bean级别】构造器执行了
③【容器级别】postProcessPropertyValues方法执行了, beanName=class com.test.lifecycle.beanlifcycle.BeanLifecycle
2.【Bean级别】setBeanName方法执行了
3.【Bean级别】setApplicationContext方法执行了
4.【Bean级别】afterPropertiesSet方法执行了
5.【Bean级别】init-method指定的方法执行了
④【容器级别】postProcessAfterInitialization方法执行了, beanName=class com.test.lifecycle.beanlifcycle.BeanLifecycle
6.【Bean级别】sayHello方法执行了
7.【Bean级别】destroy方法执行了
8.【Bean级别】destroy-method属性指定的方法执行了
```

图 2-15 完整的生命周期执行顺序

2.5 小　　结

本章主要介绍了 Spring 框架最核心的概念之一——IoC，并通过案例讲解了 IoC 的实现方式，从 Spring 代码入手，分析了 Spring IoC 容器的启动过程，并通过案例讲解了 Spring IoC 容器中 Bean 的生命周期，至此 Spring 核心 IoC 分析完毕。下一章将讲解 Spring 框架的另一个核心概念——AOP。

第 3 章

Spring AOP 揭秘

本章将介绍 Spring 框架另一个核心概念——AOP（Aspect Oriented Programming，面向切面编程）。本章将从 AOP 的理论基础开始介绍，通过案例走进 Spring AOP 的实现，并将 Spring AOP 与 AspectJ AOP 进行对比，本章最后将通过代码剖析 Spring AOP 的实现原理。

3.1 AOP 前置知识

3.1.1 JDK 动态代理

动态代理是相对于静态代理（参见本书附录）而提出的设计模式。在 Spring 中，有两种方式可以实现动态代理——JDK 动态代理和 CGLIB 动态代理。本节将介绍 JDK 动态代理。

对于静态代理，一个代理类只能代理一个对象，如果有多个对象需要被代理，就需要很多代理类，造成代码的冗余。JDK 动态代理，从字面意思就可以看出，JDK 动态代理的对象是动态生成的。

JDK 动态代理的条件是被代理对象必须实现接口。

下面以一个简单的案例说明 JDK 动态代理的实现方式。如一个 Animal 接口，接口中定义一个方法 eat，表示动物需要吃饭。Animal 接口定义如下：

```
/**
 * @Author zhouguanya
 * @Date 2018/8/20
 * @Description 接口
 */
public interface Animal {
    /**
```

```
    * 接口方法
    */
   void eat();
}
```

然后需要一个 Dog 类实现 Animal 接口，需要重写 eat()方法：

```
/**
 * @Author zhouguanya
 * @Date 2018/8/20
 * @Description 接口实现类
 */
public class Dog implements Animal {
    /**
     * 接口方法
     */
    @Override
    public void eat() {
        System.out.println("Dog 要吃骨头");
    }
}
```

需要创建动态代理类，动态代理类需要实现 InvocationHandler 接口。具体动态代理类如下：

```
/**
 * @Author zhouguanya
 * @Date 2018/8/20
 * @Description 动态代理类
 */
public class AnimalInvocationHandler implements InvocationHandler {
    /**
     * 被代理对象
     */
    private Object target;

    /**
     * 绑定业务对象并返回一个代理类
     * @param target
     * @return
     */
    public Object bind(Object target) {
        this.target = target;
        //通过反射机制，创建一个代理类对象实例并返回。用户进行方法调用时使用
        return Proxy.newProxyInstance(target.getClass().getClassLoader(),
target.getClass().getInterfaces(), this);
    }
```

```java
/**
 * 接口方法
 */
@Override
public Object invoke(Object proxy, Method method, Object[] args) throws Throwable {
    Object result=null;
    //方法执行前加一段逻辑
    System.out.println("————调用前处理————");
    //调用真正的业务方法
    result=method.invoke(target, args);
    //方法执行前加一段逻辑
    System.out.println("————调用后处理————");
    return result;
}
```

在动态代理类中,在被代理的方法前后各加了一段输出逻辑,而不必破坏原方法。下面将用一个测试类,证明动态代理生效。测试类如下:

```java
/**
 * @Author zhouguanya
 * @Date 2018/8/20
 * @Description 测试
 */
public class JDKDynamicProxyDemo {
    public static void main(String[] args) {
        //被代理对象
        Dog dog = new Dog();
        //动态代理类对象
        AnimalInvocationHandler animalInvocationHandler = new AnimalInvocationHandler();
        //代理对象
        Animal proxy = (Animal) animalInvocationHandler.bind(dog);
        proxy.eat();
    }
}
```

在这段测试代码中,首先创建原对象 Dog 和动态代理类 AnimalInvocationHandler,然后用原对象生成代理对象 animalInvocationHandler.bind(dog),最后通过调用代理对象的 invoke 方法实现业务逻辑。测试结果如图 3-1 所示。

————调用前处理————
Dog需要吃骨头
————调用后处理————

图 3-1 JDK 动态代理测试效果图

这证明动态代理生效了,想要在 dog 对象的 eat()方法前后加上额外的逻辑,可以不直接修改 eat()方法,通过以上编程方式就可以实现如图 3-1 所示的逻辑。

以上就是 Spring AOP 的基本原理,只是 Spring 不需要开发人员自己维护代理类,其已帮开发人员生成了代理类。Spring AOP 的实现是通过在程序运行时,根据具体的类对象和方法等信息动态地生成了一个代理类的 class 文件的字节码,再通过 ClassLoader 将代理类加载到内存中,最后通过生成的代理对象进行程序的方法调用。

3.1.2 CGLIB 动态代理

从上一节对 JDK 动态代理的实现可以发现,JDK 动态代理有一个缺点,即被代理类必须实现接口。这显然不能满足开发过程中的需要。有没有可能不实现接口,直接就对 Java 类进行代理呢?这就需要 CGLIB 发挥作用了。

下面将以一个简单案例说明 CGLIB 是如何实现动态代理的。在本例中,实现一个 Cat 类,其有一个 cry()方法。Cat 实现代码如下:

```java
/**
 * @Author zhouguanya
 * @Date 2018/8/21
 * @Description 被代理类
 */
public class Cat {
    /**
     * 方法
     */
    public void cry() {
        System.out.println("喵喵喵");
    }
}
```

CGLIB 动态代理的实现需要实现 MethodInterceptor 接口,重写 intercept()方法。本例中接口的实现类代码如下:

```java
/**
 * @Author zhouguanya
 * @Date 2018/8/21
 * @Description 实现 MethodInterceptor 接口
 */
public class CatMethodInterceptor implements MethodInterceptor {

    /**
     * 生成方法拦截器
     * @param o 要进行增强的对象
     * @param method 拦截的方法
     * @param objects 参数列表
     * @param methodProxy 方法的代理
```

```
     * @return
     * @throws Throwable
     */
    @Override
    public Object intercept(Object o, Method method, Object[] objects,
MethodProxy methodProxy) throws Throwable {
        System.out.println("————调用前处理————");
        //对被代理对象方法的调用
        Object object = methodProxy.invokeSuper(o, objects);
        System.out.println("————调用后处理————");
        return object;
    }
}
```

如上代码注释所示，在调用被代理对象的方法前后各加入一段输出打印逻辑以观察拦截的效果。测试代码如下：

```
/**
 * @Author zhouguanya
 * @Date 2018/8/21
 * @Description 测试Cglib
 */
public class CglibDynamicProxyDemo {

    public static void main(String[] args) {
        Enhancer enhancer = new Enhancer();
        //被代理类：Cat
        enhancer.setSuperclass(Cat.class);
        //设置回调
        enhancer.setCallback(new CatMethodInterceptor());
        //生成代理对象
        Cat cat = (Cat) enhancer.create();
        //调用代理类的方法
        cat.cry();
    }
}
```

运行测试代码后的执行结果如图 3-2 所示。

————调用前处理————
喵喵喵
————调用后处理————

图 3-2 CGLIB 动态代理测试效果图

3.1.3 AOP 联盟

面向切面编程（AOP）是一种编程技术，可以增强几个现有的中间件环境（例如 J2EE）或开发环境（例如 JBuilder，Eclipse）。

AOP 联盟定义了一套用于规范 AOP 实现的底层 API，通过这些统一的底层 API，可以使得各个 AOP 实现工具之间实现相互兼容。现在 AOP 联盟已有几个项目提供了与 AOP 相关的技术，如通用代理，拦截器或字节码转换器。

- ASM：轻量级字节码转换器。
- AspectJ：一个面向切面的框架，扩展了 Java 语言。
- AspectWerkz：一个面向切面的框架，基于字节码级别的动态织入和配置。
- BCEL：字节码转换器。
- CGLIB：用于类工件操作和方法拦截的高级 API。
- Javassist：具有高级 API 的字节码转换器。
- JBoss-AOP：拦截和基于元数据的 AO 框架。

除了以上列举的 AOP 联盟的相关项目之外，还有很多其他项目，此处不再一一列举。所有这些项目都有其各自的目标和特点。但是，一些基本组件对于构建完整的面向切面的系统是必需的。例如，一个能够在基础组件上添加元数据的组件，一个拦截框架，一个能够执行代码转换以便为类提供 advice 的组件，一个 weaver 组件，一个配置组件等。

3.2 AOP 概述

3.2.1 AOP 基本概念

AOP（Aspect Oriented Programming）是 OOP（Object Oriented Programming）即面向对象编程的一种补充和完善。

以 Java 语言为例，其提供了封装、继承和多态等概念，实现了面向对象编程。开发人员可以使用 Java 语言将现实世界中各种事物抽象成 Java 语言中的对象，一类对象有共同的行为和特性，这就是 Java 语言相对于 C 语言或汇编语言而言，被称为高级语言的本质。

虽然面向对象编程语言实现了纵向的对每个对象的行为进行归类和划分，实现了高度的抽象，但是，不同对象间的共性却不适合用面向对象编程的方式实现。如学生对象和汽车对象，都要实现与其自身业务逻辑无关的监控，在这种场景下，使用面向对象编程的方式可能最好的解决方案就是让学生对象和汽车对象都集成监控接口，然后学生对象和汽车对象分别实现监控方法。这种编程方式的缺点是，由于监控逻辑并不是对象本身的核心功能，并且不同对象的监控逻辑实现基本上相同——都是监控某个时间发生了某件事，只是记录的对象不同而已，这会导致监控对象行为的代码逻辑散落在系统的各个地方，并且几乎都是重复的代码，与对象的核心功能并无很强的关联性。这样的设计会导致大量的代码重复，并且不利于模块的复用。

AOP 的出现，恰好解决了这个棘手的问题。其提供"横向"的切面逻辑，将与多个对象有关的公共模块分装成一个可重用模块，并将这个模块整合成为 Aspect，即切面。切面就是对与具体的业务逻辑无关的，却是许多业务模块共同的特性或职责的一种抽象，其减少了系统中的重复代码，因此降低了模块的耦合度，更加有利于扩展。

AOP 将软件系统分为两部分：核心逻辑和横切逻辑。核心逻辑主要处理系统正常的业务逻辑，横切逻辑不关注系统核心逻辑，其只关注与系统核心逻辑并非强相关的逻辑。核心逻辑和横切逻辑的关系如图 3-3 所示。

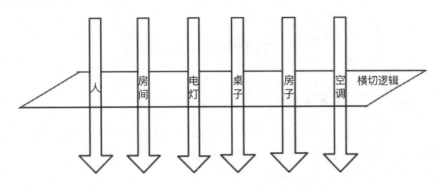

图 3-3 核心逻辑和横切逻辑关系图

3.2.2 Spring AOP 相关概念

下面介绍与 Spring AOP 相关的一些概念。

1. 横切关注点

一些具有横切多个不同软件模块的行为，通过传统的软件开发方法不能够有效地实现模块化的一类特殊关注点。横切关注点可以对某些方法进行拦截，拦截后对原方法进行增强处理。

2. 切面（Aspect）

切面就是对横切关注点的抽象，这个关注点可能会横切多个对象。

3. 连接点（JoinPoint）

连接点是在程序执行过程中某个特定的点，比如某方法调用的时候或者处理异常的时候。由于 Spring 只支持方法类型的连接点，所以在 Spring AOP 中一个连接点总是表示一个方法的执行。

4. 切入点（Pointcut）

切入点是匹配连接点的拦截规则，在满足这个切入点的连接点上运行通知。切入点表达式如何和连接点匹配是 AOP 的核心，Spring 默认使用 AspectJ 切入点语法。

5. 通知（Advice）

在切面上拦截到某个特定的连接点之后执行的动作。

6. 目标对象（Target Object）

目标对象，被一个或者多个切面所通知的对象，即业务中需要进行增强的业务对象。

7. 织入（Weaving）

织入是把切面作用到目标对象，然后产生一个代理对象的过程。

8. 引入（Introduction）

引入是用来在运行时给一个类声明额外的方法或属性，即不需为类实现一个接口，就能使用接口中的方法。

3.3 Spring AOP 实现

3.3.1 基于 JDK 动态代理实现

Spring AOP 的实现方式有两种，分别是基于 JDK 动态代理的实现和基于 CGLIB 的动态代理实现。本节将讲解基于 JDK 动态代理的方式实现 Spring AOP。

基于 JDK 动态代理的方式实现 Spring AOP 有两种方式，分别是基于 XML 配置的方式和注解的方式，下面将先以 XML 配置的方式讲解。

Spring AOP 的一个特点是被代理的对象需要实现一个接口。下面以一个 Fruit 接口为例，验证基于 XML 配置的 Spring AOP 实现，Fruit 接口的定义如下：

```
/**
 * @Author: zhouguanya
 * @Date: 2018/8/25 18:54
 * @Description: 水果接口
 */
public interface Fruit {
    /**
     * 吃水果
     */
    void eat();
}
```

接下来，实现被代理的对象，分别用 Apple 类和 Banana 类实现这个接口，这两个类的实现如下：

```
/**
 * @Author: zhouguanya
 * @Date: 2018/8/25 19:07
 * @Description: 苹果
 */
public class Apple implements Fruit {
    /**
     * 吃水果
     */
    @Override
```

```java
    public void eat() {
        try {
            //模拟吃苹果的过程
            Thread.sleep(1000);
        } catch (InterruptedException e) {
            e.printStackTrace();
        }
        System.out.println("吃苹果");
    }
}

/**
 * @Author: zhouguanya
 * @Date: 2018/8/25 19:08
 * @Description: 香蕉
 */
public class Banana implements Fruit {
    /**
     * 吃水果
     */
    @Override
    public void eat() {
        try {
            //模拟吃香蕉的过程
            Thread.sleep(1000);
        } catch (InterruptedException e) {
            e.printStackTrace();
        }
        System.out.println("吃香蕉");
    }
}
```

以下代码是对以上两个被代理对象加的横切关注点逻辑，横切逻辑打印吃水果的时间和水果吃完的时间：

```java
/**
 * @Author: zhouguanya
 * @Date: 2018/8/25 19:10
 * @Description: 横切关注点,打印吃水果的时间
 */
public class FruitHandler {

    /**
     * 打印开始吃水果的时间
     */
    public void startEatFruitDate() {
```

```java
        SimpleDateFormat simpleDateFormat = new SimpleDateFormat("yyyy-MM-dd hh:mm:ss");
        String startEatDate = simpleDateFormat.format(new Date());
        System.out.println("开始吃水果的时间是:" + startEatDate);
    }

    /**
     * 打印吃完吃水果的时间
     */
    public void endEatFruitDate() {
        SimpleDateFormat simpleDateFormat = new SimpleDateFormat("yyyy-MM-dd hh:mm:ss");
        String endEatDate = simpleDateFormat.format(new Date());
        System.out.println("结束吃水果的时间是:" + endEatDate);
    }
}
```

验证方式是通过从 Spring IoC 容器中获取 Apple 和 Banana 对象，分别调用对象的 eat()方法，测试代码如下：

```java
/**
 * @Author: zhouguanya
 * @Date: 2018/8/25 19:27
 * @Description: xml aop 测试
 */
public class SpringAopXmlDemo {

    public static void main(String[] args) throws InterruptedException {
        ApplicationContext applicationContext = new ClassPathXmlApplicationContext("classpath:spring-chapter3-xmlaop.xml");
        Fruit apple = (Fruit) applicationContext.getBean("apple");
        Fruit banana = (Fruit) applicationContext.getBean("banana");
        apple.eat();
        System.out.println("-----休息一会儿-----");
        Thread.sleep(1000);
        banana.eat();
    }
}
```

测试类中使用的配置文件是 spring-chapter3-xmlaop.xml，具体配置如下：

```xml
<context:component-scan base-package="com.test.aop.jdk.xml"></context:component-scan>
<!--apple-->
<bean id="apple" class="com.test.aop.jdk.xml.Apple"></bean>
<!--banana-->
<bean id="banana" class="com.test.aop.jdk.xml.Banana"></bean>
```

```xml
<!--fruitHandler-->
<bean id="fruitHandler" class="com.test.aop.jdk.xml.FruitHandler"></bean>
<aop:config>
    <aop:aspect id="datelog" ref="fruitHandler">
        <aop:pointcut id="eatFruit" expression="execution(* com.test.aop.jdk.xml.Fruit.*(..))" />
        <aop:before method="startEatFruitDate" pointcut-ref="eatFruit" />
        <aop:after method="endEatFruitDate" pointcut-ref="eatFruit" />
    </aop:aspect>
</aop:config>
```

执行测试代码，可以看到运行结果如图 3-4 所示。

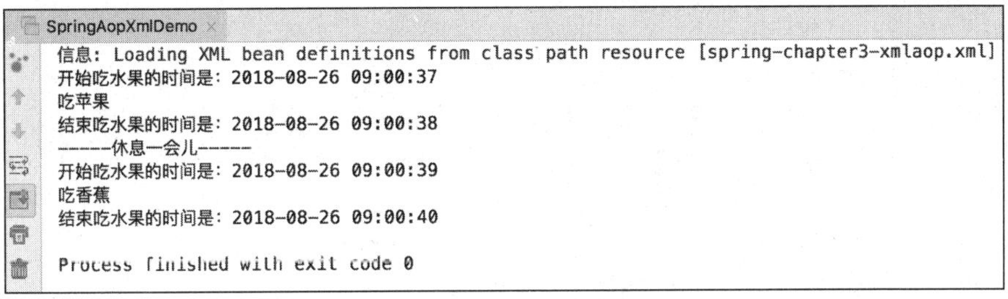

图 3-4 基于 JDK 动态代理的方式实现的 Spring AOP 测试结果

从图 3-4 执行结果可以发现，在没有修改 Apple 类和 Banana 类代码的情况下，每次执行 eat() 方法，都会输出"开始吃水果的时间"和"结束吃水果的时间"这样的日志记录逻辑，验证了 Spring AOP 是可以正常使用的。

下面将讲解基于注解方式实现 Spring AOP。

还是以上例中的 Fruit 为例，此时横切关注点修改如下：

```java
/**
 * @Author: zhouguanya
 * @Date: 2018/8/25 19:10
 * @Description: 横切关注点,打印吃水果的时间
 */
@Component
@Aspect
public class FruitAnnotationHandler {
    /**
     * 定义切点
     */
    @Pointcut("execution(* com.test.aop.jdk.xml.Fruit.*(..))")
    public void eatFruit() {

    }

    /**
     * 前置通知
```

```java
     * 打印开始水果的时间
     */
    @Before("eatFruit()")
    public void startEatFruitDate() {
        SimpleDateFormat simpleDateFormat = new SimpleDateFormat("yyyy-MM-dd hh:mm:ss");
        String startEatDate = simpleDateFormat.format(new Date());
        System.out.println("开始吃水果的时间是：" + startEatDate);
    }

    /**
     * 后置通知
     * 打印吃完吃水果的时间
     */
    @After("eatFruit()")
    public void endEatFruitDate() {
        SimpleDateFormat simpleDateFormat = new SimpleDateFormat("yyyy-MM-dd hh:mm:ss");
        String endEatDate = simpleDateFormat.format(new Date());
        System.out.println("结束吃水果的时间是：" + endEatDate);
    }
}
```

通过注解 "@Aspect" 用来声明其是切面，注解 "@Before" 用来表明前置通知，"@After" 用来表明后置通知。

```java
/**
 * @Author: zhouguanya
 * @Date: 2018/8/25 19:27
 * @Description: xml aop 测试
 */
public class SpringAopAnnotationDemo {

    public static void main(String[] args) throws InterruptedException {
        ApplicationContext applicationContext = new ClassPathXmlApplicationContext("classpath:spring-chapter3-annotationaop.xml");
        Fruit apple = (Fruit) applicationContext.getBean("apple");
        Fruit banana = (Fruit) applicationContext.getBean("banana");
        apple.eat();
        System.out.println("-----休息一会儿-----");
        Thread.sleep(1000);
        banana.eat();
    }
}
```

测试代码配置文件更改为 spring-chapter3-annotationaop.xml。具体配置如下：

```xml
<!-- 开启注解扫描 -->
<context:component-scan base-package="com.test.aop.jdk.annotation"/>
<!--apple-->
<bean id="apple" class="com.test.aop.jdk.xml.Apple"></bean>
<!--banana-->
<bean id="banana" class="com.test.aop.jdk.xml.Banana"></bean>
<!--fruitHandler-->
<bean id="fruitHandler" class="com.test.aop.jdk.xml.FruitHandler"></bean>
<!-- 开启 aop 注解方式，此步骤不能少 -->
<aop:aspectj-autoproxy/>
```

运行测试代码，其效果如图 3-4 所示。

3.3.2 基于 CGLIB 动态代理实现

3.3.1 小节已经阐述了基于 JDK 动态代理实现的 Spring AOP，可以发现，JDK 动态代理的一个缺点是被代理对象必须实现一个接口。这种严苛的条件并不能满足日常开发的全部需求，毕竟 Java 中并不是所有的类都必须继承接口。那么有没有方法实现对没有实现接口的类进行代理呢？这就是本节将要介绍的基于 CGLIB 方式实现的 Spring AOP 编程。

下面将以 XML 的形式介绍基于 CGLIB 的方式实现的 Spring AOP，这个例子中有 Desk 和 Table 两个类，分别打印各自的位置，代码如下：

```java
/**
 * @Author zhouguanya
 * @Date 2018/8/27
 * @Description 课桌
 */
public class Desk {
    /**
     * 打印位置信息
     */
    public void location() throws InterruptedException {
        //模拟耗时，方便观察输出结果
        Thread.sleep(1000);
        System.out.println("我是课桌，我被放在教室中");
    }
}

/**
 * @Author zhouguanya
 * @Date 2018/8/27
 * @Description 桌子
 */
public class Table {
    /**
```

```
 * 打印位置信息
 */
public void location() throws InterruptedException {
    //模拟耗时,方便观察输出结果
    Thread.sleep(1000);
    System.out.println("我是餐桌,我被放在厨房中");
}
}
```

下面定义一个横切关注点 CglibXmlAspect,在 Desk 和 Table 两个类前后分别打印时间,以观察两个类在执行 location()方法的时间,具体代码如下:

```
/**
 * @Author: zhouguanya
 * @Date: 2018/8/27 19:50
 * @Description: 横切关注点,打印开始和结束的时间
 */
public class CglibXmlAspect {

    /**
     * 打印事件开始的时间
     */
    public void startDate() {
        SimpleDateFormat simpleDateFormat = new SimpleDateFormat("yyyy-MM-dd hh:mm:ss");
        String startEatDate = simpleDateFormat.format(new Date());
        System.out.println("开始的时间是: " + startEatDate);
    }

    /**
     * 打印事件结束的时间
     */
    public void endDate() {
        SimpleDateFormat simpleDateFormat = new SimpleDateFormat("yyyy-MM-dd hh:mm:ss");
        String endEatDate = simpleDateFormat.format(new Date());
        System.out.println("结束的时间是: " + endEatDate);
    }
}
```

相关的 Bean 全部通过 XML 的方式进行配置,配置如下:

```
<!--desk-->
<bean id="desk" class="com.test.aop.cglib.xml.Desk"></bean>
<!--table-->
<bean id="table" class="com.test.aop.cglib.xml.Table"></bean>
<!--切面-->
```

```xml
<bean id="cglibXmlAspect"
class="com.test.aop.cglib.xml.CglibXmlAspect"></bean>

<aop:config>
    <aop:aspect id="datelog" ref="cglibXmlAspect">
        <aop:pointcut id="location" expression="execution(* com.test.aop.cglib.xml.*.*(..))" />
        <aop:before method="startDate" pointcut-ref="location" />
        <aop:after method="endDate" pointcut-ref="location" />
    </aop:aspect>
</aop:config>
<!-- 强制使用 CGLIB，此步骤不能少 -->
<aop:aspectj-autoproxy proxy-target-class="true"/>
```

接下来的测试代码中，从 Spring IoC 容器中获取 Desk 和 Table 类的对象，分别调用对象的 location()方法，测试代码如下：

```java
/**
 * @Author zhouguanya
 * @Date 2018/8/27
 * @Description cglib aop 测试
 */
public class CgLibAnnotationDemo {
    public static void main(String[] args) throws InterruptedException {
        ApplicationContext applicationContext = new ClassPathXmlApplicationContext("classpath:spring-chapter3-xmlcglib.xml");
        Desk desk = (Desk) applicationContext.getBean("desk");
        Table table = (Table) applicationContext.getBean("table");
        desk.location();
        System.out.println("-----分割线-----");
        Thread.sleep(1000);
        table.location();
    }
}
```

运行测试代码，测试效果如图 3-5 所示。

```
CgLibXmlDemo
八月 28, 2018 7:31:02 下午 org.springframework.context.support.AbstractApplicationContext prepareRe
信息: Refreshing org.springframework.context.support.ClassPathXmlApplicationContext@78e03bb5: star
八月 28, 2018 7:31:02 下午 org.springframework.beans.factory.xml.XmlBeanDefinitionReader loadBeanDe
信息: Loading XML bean definitions from class path resource [spring-chapter3-xmlcglib.xml]
开始的时间是: 2018-08-28 07:31:03
我是课桌，我被放在教室中
结束的时间是: 2018-08-28 07:31:04
-----分割线-----
开始的时间是: 2018-08-28 07:31:05
我是餐桌，我被放在厨房中
结束的时间是: 2018-08-28 07:31:06

Process finished with exit code 0
```

图 3-5 基于 CGLIB 方式实现的 Spring AOP 测试结果

下面是用注解方式演示基于 CGLIB 方式实现的 Spring AOP。Desk 和 Table 分别用 @Component 注解修饰，代码如下：

```java
/**
 * @Author zhouguanya
 * @Date 2018/8/27
 * @Description 课桌
 */
@Component
public class Desk {
    /**
     * 打印位置信息
     */
    public void location () throws InterruptedException {
        //模拟耗时，方便观察输出结果
        Thread.sleep(1000);
        System.out.println("我是课桌，我被放在教室中");
    }
}
/**
 * @Author zhouguanya
 * @Date 2018/8/27
 * @Description 桌子
 */
@Component
public class Table {
    /**
     * 打印位置信息
     */
    public void location() throws InterruptedException {
        //模拟耗时，方便观察输出结果
        Thread.sleep(1000);
        System.out.println("我是餐桌，我被放在厨房中");
    }
}
```

用注解配置横切关注点，"@Aspect"用来声明切面，"@Pointcut"用来定义切点，"@Before"用来设置前置通知，"@After"用来设置后置通知，具体代码如下：

```java
/**
 * @Author: zhouguanya
 * @Date: 2018/8/25 19:10
 * @Description: 横切关注点,打印开始和结束的时间
 */
@Component
@Aspect
```

```java
public class CglibAnnotationHandler {
    /**
     * 定义切点
     */
    @Pointcut("execution(* com.test.aop.cglib.annotation.*.*(..))")
    public void location() {

    }

    /**
     * 前置通知
     * 打印开始时间
     */
    @Before("location()")
    public void startEatFruitDate() {
        SimpleDateFormat simpleDateFormat = new SimpleDateFormat("yyyy-MM-dd hh:mm:ss");
        String startEatDate = simpleDateFormat.format(new Date());
        System.out.println("开始的时间是:" + startEatDate);
    }

    /**
     * 后置通知
     * 打印结束的时间
     */
    @After("location()")
    public void endEatFruitDate() {
        SimpleDateFormat simpleDateFormat = new SimpleDateFormat("yyyy-MM-dd hh:mm:ss");
        String endEatDate = simpleDateFormat.format(new Date());
        System.out.println("结束的时间是:" + endEatDate);
    }
}
```

配置文件修改为 spring-chapter3-annotationcglibaop.xml 中不再需要单独定义的 Bean 和切面，只需要很少的配置：

```xml
<!-- 开启注解扫描 -->
<context:component-scan base-package="com.test.aop.cglib.annotation"/>
<!-- 强制使用CGLIB,此步骤不能少 -->
<aop:aspectj-autoproxy proxy-target-class="true"/>
```

测试代码如下：

```
/**
 * @Author zhouguanya
 * @Date 2018/8/27
```

```java
 * @Description cglib aop测试
 */
public class CgLibAnnotationDemo {
    public static void main(String[] args) throws InterruptedException {
        ApplicationContext applicationContext= new ClassPathXmlApplicationContext("classpath:spring-chapter3-annotationcglibaop.xml");
        Desk desk = (Desk) applicationContext.getBean("desk");
        Table table = (Table) applicationContext.getBean("table");
        desk.location();
        System.out.println("-----分割线-----");
        Thread.sleep(1000);
        table.location();
    }
}
```

测试结果如图 3-5 所示。

注：以上测试代码中 execution 表达式各个部分含义说明：

execution(<修饰符模式>?<返回类型模式><方法名模式>(<参数模式>)<异常模式>?)

其中，除了返回类型模式、方法名模式和参数模式外，其他项都是可选的。以表达式 execution(* com.test.aop.cglib.annotation.*.*(..))为例，其含义是匹配 com.test.aop.cglib.annotation 这个 package 下任意类的任意方法名、任意方法入参和任意方法返回值的这部分方法。

3.4 基于 Spring AOP 的实战

3.4.1 增强类型

AOP 联盟为增强定义了 org.aopalliance.aop.Advice 接口，Spring 支持 5 种类型的增强。本章 3.3 节中使用到的@Before、@After 等注解是基于 AspectJ 实现的增强类型。其实 Spring 也支持很多增强类型，Spring AOP 按照增强在目标类方法中的连接点位置可以分为 5 种。

- 前置增强：表示在目标方法执行前实施增强。
- 后置增强：表示在目标方法执行后实施增强。
- 环绕增强：表示在目标方法执行前后实施增强。
- 异常抛出增强：表示在目标方法抛出异常后实施增强。
- 引介增强：表示在目标类中添加一些新的方法和属性。

以下将依次介绍每种增量类型，由于基于 XML 和基于注解的配置其本质都是相同的，因此下面将只通过注解的方式演示 Spring 各种增强类型，并观察各种增强类型的执行时序。

3.4.2 前置增强

Spring 的前置增强主要接口是 MethodBeforeAdvice，其顶级接口是 AOP 联盟中的 Advice 接口。从如图 3-6 所示的类图可以发现，Spring 的前置增强扩展了 Advice 接口。

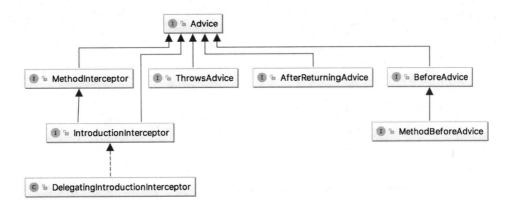

图 3-6　MethodBeforeAdvice 相关类图

下面将通过实现 MethodBeforeAdvice 接口来新增一个前置增强实现类，然后通过案例阐述使用 Spring 的前置增强类型的编程方式。前置增强实现类的代码如下：

```
/**
 * @Author zhouguanya
 * @Date 2018/9/2
 * @Description 前置增强实现类
 */
@Component
public class SpringBeforeAdvice implements MethodBeforeAdvice {
    @Override
    public void before(Method method, Object[] args, Object target) throws Throwable {
        String methodName = method.getName();
        System.out.printf("MethodBeforeAdvice 增强的方法是%s%n", methodName);
        System.out.printf("MethodBeforeAdvice 增强的方法的参数是%s%n", args[0]);
        System.out.printf("MethodBeforeAdvice 增强的对象是%s%n", target);
    }
}
```

下面将创建一个被增强类 Waiter，用于被 SpringBeforeAdvice 类增强，Waiter 类的代码如下：

```
@Component
public class Waiter {
    /**
     * 服务
     * @param name
```

```java
    */
    public String serve(String name) {
        System.out.println(name + ",您好,很高兴为您服务。");
        SimpleDateFormat format = new SimpleDateFormat("yyyy-MM-dd hh:mm:ss");
        return name + ",您好,现在是北京时间" + format.format(new Date());
    }

    /**
     * 开车
     * @param name
     */
    public void driving(String name) {
        throw new RuntimeException(name + ",您好,禁止酒后驾车!");
    }
}
```

测试代码中需要创建被代理对象和前置增强的对象,并通过 Spring 生成代理对象,测试代码如下:

```java
/**
 * @Author zhouguanya
 * @Date 2018/9/2
 * @Description Spring前置增强测试
 */
public class SpringBeforeAdviceDemo {

    public static void main(String[] args) {
        ApplicationContext context = new ClassPathXmlApplicationContext("classpath:spring-chapter3-springaoptype.xml");
        Waiter waiter = (Waiter) context.getBean("waiter");
        SpringBeforeAdvice advice = (SpringBeforeAdvice) context.getBean("springBeforeAdvice");
        //Spring 提供的代理工厂
        ProxyFactory pf = new ProxyFactory();
        //设置代理目标
        pf.setTarget(waiter);
        pf.addAdvice(advice);
        //生成代理实例
        Waiter proxy = (Waiter)pf.getProxy();
        proxy.serve("Michael");
        proxy.serve("Tommy");
    }
}
```

ProxyFactory 生成的代理对象 proxy 就是被增强后的对象,运行测试代码,得到的运行结果如图 3-7 所示。

```
MethodBeforeAdvice增强的方法是serve
MethodBeforeAdvice增强的方法的参数是Michael
MethodBeforeAdvice增强的对象是com.test.springaoptype.Waiter@54c562f7
Michael,您好,很高兴为您服务。
MethodBeforeAdvice增强的方法是serve
MethodBeforeAdvice增强的方法的参数是Tommy
MethodBeforeAdvice增强的对象是com.test.springaoptype.Waiter@54c562f7
Tommy,您好,很高兴为您服务。
```

图 3-7 前置增强测试效果图

由测试结果可以看出,在 Waiter 类的 serve 方法执行之前,前置增强的逻辑执行了。前置增强即在目标方法执行前实施增强逻辑。

3.4.3 后置增强

Spring 的后置增强主要接口是 AfterReturningAdvice,其类图如图 3-6 所示。

下面将通过实现 AfterReturningAdvice 接口来新增一个后置增强实现类,然后通过 3.4.2 节中的 Waiter 案例阐述使用 Spring 的后置增强类型的编程方式。后置增强实现类的代码如下:

```java
/**
 * @Author zhouguanya
 * @Date 2018/9/2
 * @Description 后置增强
 */
@Component
public class SpringAfterReturningAdvice implements AfterReturningAdvice {
    @Override
    public void afterReturning(Object returnValue, Method method, Object[] args, Object target) throws Throwable {
        String methodName = method.getName();
        System.out.printf("AfterReturningAdvice 增强的方法返回值是:%s%n", returnValue);
        System.out.printf("AfterReturningAdvice 增强的方法是:%s%n", methodName);
        System.out.printf("AfterReturningAdvice 增强的方法的参数是:%s%n", args[0]);
        System.out.printf("AfterReturningAdvice 增强的对象是:%s%n", target);
    }
}
```

测试代码中只需修改一行代码:

```java
SpringAfterReturningAdvice advice = (SpringAfterReturningAdvice) context.getBean("springAfterReturningAdvice");
```

测试结果如图 3-8 所示。

```
Michael,您好，很高兴为您服务。
AfterReturningAdvice增强的方法返回值是:Michael,您好，现在是北京时间2018-09-02 04:31:03
AfterReturningAdvice增强的方法是:serve
AfterReturningAdvice增强的方法的参数是:Michael
AfterReturningAdvice增强的对象是:com.test.springaoptype.Waiter@433d61fb
Tommy,您好，很高兴为您服务。
AfterReturningAdvice增强的方法返回值是:Tommy,您好，现在是北京时间2018-09-02 04:31:03
AfterReturningAdvice增强的方法是:serve
AfterReturningAdvice增强的方法的参数是:Tommy
AfterReturningAdvice增强的对象是:com.test.springaoptype.Waiter@433d61fb
```

图 3-8 后置增强测试效果图

3.4.4 环绕增强

Spring 的环绕增强主要接口是 MethodInterceptor，其类图如图 3-6 所示。

下面将通过实现 MethodInterceptor 接口来新增一个环绕增强实现类，然后通过 3.4.2 节中的 Waiter 案例阐述使用 Spring 环绕增强类型的编程方式。环绕增强实现类的代码如下：

```java
/**
 * @Author zhouguanya
 * @Date 2018/9/2
 * @Description 环绕增强
 */
@Service
public class SpringMethodInterceptor implements MethodInterceptor {
    @Override
    public Object invoke(MethodInvocation invocation) throws Throwable {
        // 前置增强
        System.out.println("前置增强执行了");
        // 通过反射机制调用目标方法
        Object obj = invocation.proceed();
        // 后置增强
        System.out.println("后置增强执行了");
        return obj;
    }
}
```

测试代码只需要修改使用环绕增强实现类 SpringMethodInterceptor，测试结果如图 3-9 所示。

```
前置增强执行了
Michael,您好，很高兴为您服务。
后置增强执行了
前置增强执行了
Tommy,您好，很高兴为您服务。
后置增强执行了
```

图 3-9 环绕增强测试效果图

3.4.5 异常抛出增强

Spring 的异常抛出增强主要接口是 ThrowsAdvice，其类图如图 3-6 所示。

下面将通过实现 ThrowsAdvice 接口来新增一个异常抛出增强实现类，然后通过 3.4.2 节中的 Waiter 案例阐述使用 Spring 的异常抛出增强类型的编程方式。异常抛出增强实现类的代码如下：

```java
/**
 * @Author zhouguanya
 * @Date 2018/9/2
 * @Description 异常抛出增强
 */
@Component
public class SpringThrowsAdvice implements ThrowsAdvice {

    public void afterThrowing(Exception e) throws Throwable{
        System.out.printf("异常抛出增强执行：%s%n", e);
    }
}
```

异常抛出增强的测试代码执行效果如图 3-10 所示。

```
异常抛出增强执行: java.lang.RuntimeException: Michael, 您好, 禁止酒后驾车!
Exception in thread "main" java.lang.RuntimeException: Michael, 您好, 禁止酒后驾车!
    at com.test.springaoptype.Waiter.driving(Waiter.java:30)
    at com.test.springaoptype.Waiter$$FastClassBySpringCGLIB$$fbd843f0.invoke(<generated>)
    at org.springframework.cglib.proxy.MethodProxy.invoke(MethodProxy.java:204)
    at org.springframework.aop.framework.CglibAopProxy$CglibMethodInvocation.invokeJoinpoint(CglibAopProxy.java:746)
    at org.springframework.aop.framework.ReflectiveMethodInvocation.proceed(ReflectiveMethodInvocation.java:163)
    at org.springframework.aop.framework.adapter.ThrowsAdviceInterceptor.invoke(ThrowsAdviceInterceptor.java:112)
    at org.springframework.aop.framework.ReflectiveMethodInvocation.proceed(ReflectiveMethodInvocation.java:185)
    at org.springframework.aop.framework.CglibAopProxy$DynamicAdvisedInterceptor.intercept(CglibAopProxy.java:688)
    at com.test.springaoptype.Waiter$$EnhancerBySpringCGLIB$$551f1efd.driving(<generated>)
    at com.test.springaoptype.throwsadvice.SpringAfterThrowsAdviceDemo.main(SpringAfterThrowsAdviceDemo.java:26)
```

图 3-10　异常抛出增强测试效果图

3.4.6 引介增强

引介增强的目标是在目标类中添加一些新的方法和属性。以 Waiter 类为例，现在想给其添加一个 Management 接口中的 manage()方法而不修改 Waiter 类的代码。Management 代码如下所示：

```java
/**
 * @Author zhouguanya
 * @Date 2018/9/2
 * @Description
 */
public interface Management {
    /**
     * 管理
     * @param name
```

```
     */
    void manage(String name);
}
```

Spring 的引介增强主要接口是 IntroductionInterceptor，通过图 3-11 可以看出，Spring 已经提供了 IntroductionInterceptor 接口的实现类 DelegatingIntroductionInterceptor。

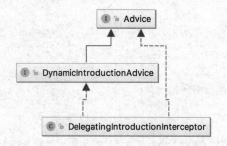

图 3-11　IntroductionInterceptor 相关类图

下面将通过扩展 DelegatingIntroductionInterceptor 来实现引介增强。通过 Manager 类继承 DelegatingIntroductionInterceptor 并实现 Management 接口，Manager 代码如下：

```
/**
 * @Author zhouguanya
 * @Date 2018/9/2
 * @Description 经理类
 */
public class Manager extends DelegatingIntroductionInterceptor implements Management {

    @Override
    public void manage(String name) {
        System.out.println(name + ",您好,我是经理,负责管理服务员");
    }
}
```

此时需要修改配置文件，需要指定引介增强所在的实现接口并需要将 proxyTargetClass 属性设置为 true。具体配置文件如下：

```
<context:component-scan base-package="com.test.springaoptype"/>
<bean id="waiter" class="com.test.springaoptype.Waiter"/>

<bean id="manager" class="com.test.springaoptype.introductioninterceptor.Manager"/>

<bean id="waiterProxy" class="org.springframework.aop.framework.ProxyFactoryBean"
    p:interfaces="com.test.springaoptype.introductioninterceptor.Management"
```

```
    p:interceptorNames="manager"
    p:target-ref = "waiter"
    p:proxyTargetClass="true"/>
```

测试代码中，从 Spring 上下文中获取代理对象 waiterProxy，将其强制转化为一个 Management 对象。修改后测试代码如下：

```
/**
 * @Author zhouguanya
 * @Date 2018/9/2
 * @Description Spring 引介增强测试
 */
public class SpringIntroductionInterceptorDemo {

    public static void main(String[] args) {
        ApplicationContext context = new ClassPathXmlApplicationContext
("classpath:spring-chapter3-springintroductioninterceptor.xml");
        Waiter waiterProxy = (Waiter) context.getBean("waiterProxy");
        Management manager = (Management)waiterProxy;
        manager.manage("Michael");
    }
}
```

运行测试结果如图 3-12 所示，发现 Waiter 类的代理对象多了一个新的功能，可以调用 Management 接口的 manage 方法。

```
九月 02, 2018 4:58:28 下午 org.springframework.context
信息: Refreshing org.springframework.context.support.
九月 02, 2018 4:58:28 下午 org.springframework.beans.f
信息: Loading XML bean definitions from class path re
九月 02, 2018 4:58:28 下午 org.springframework.beans.f
信息: Overriding bean definition for bean 'waiter' wi
Michael,您好，很高兴为您服务。
Michael，您好，我是经理，负责管理服务员
```

图 3-12　引介增强测试效果图

3.4.7　切入点类型

如 3.2.2 节所述，切入点是匹配连接点的拦截规则。之前的案例中使用的是注解@Pointcut，该注解是 AspectJ 中的。除了这个注解之外，Spring 也提供了其他一些切入点类型：

- 静态方法切入点 StaticMethodMatcherPointcut
- 动态方法切入点 DynamicMethodMatcherPointcut
- 注解切入点 AnnotationMatchingPointcut
- 表达式切入点 ExpressionPointcut
- 流程切入点 ControlFlowPointcut
- 复合切入点 ComposablePointcut
- 标准切入点 TruePointcut

各种切入点的类图如图 3-13 所示。

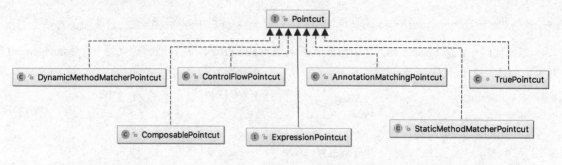

图 3-13　切入点各类的类图

3.5　Spring 集成 AspectJ 实战

本章 3.4 节中阐述了使用 Spring 的方式实现 AOP 编程，本节将以 AspectJ 相关注解的方式来实现 AOP 编程。

AspectJ 是一个面向切面的框架，其可以生成遵循 Java 字节码规范的 Class 文件。

Spring AOP 和 AscpectJ 之间的关系：Spring 使用了和 AspectJ 一样的注解，并使用 AspectJ 来做切入点解析和匹配。但是 Spring AOP 运行时并不依赖于 AspectJ 的编译器或者织入器等特性。

3.5.1　使用 AspectJ 方式配置 Spring AOP

本节将通过 AspectJ 的方式实现 AOP 编程，并通过案例阐述使用不同的 AspectJ 注解实现各种类型的通知。

在 AspectJ 中使用 "@Aspect" 注解来标示一个切面；使用 "@Pointcut" 注解标示切入点；各种通知类型通过 "@Before（前置通知）" "@Around（环绕通知）" "@AfterReturning（后置通知）" 和 "@AfterThrowing（异常通知）" 等注解来实现。

下面将通过案例阐述 AspectJ 的各种注解的使用。本例中有一个 Person 类，其包含一个说话方法 say()，Person 类的代码如下：

```
/**
 * @Author: zhouguanya
 * @Date: 2018/9/1
 * @Description: 一个Spring Bean
 */
@Component
public class Person {

    /**
     * 说话的方法
     */
    public void say() {
```

```java
        System.out.println("Hello Spring 5");
    }
}
```

定义一个切面 AllAspect，切面中实现各种通知，切面 AllAspect 的代码如下：

```java
/**
 * @Author: zhouguanya
 * @Date: 2018/9/1
 * @Description: 包含各种增强类型的切面
 */
@Component
@Aspect
public class AllAspect {

    /**
     * 切入点
     */
    @Pointcut("execution(* com.test.aspectj.advicetype.*.*(..))")
    public void allAointCut() {

    }

    /**
     * 前置增强
     */
    @Before("allAointCut()")
    public void before() {
        System.out.println("before advice");
    }

    /**
     * 环绕增强
     * @param proceedingJoinPoint
     */
    @Around("allAointCut()")
    public void around(ProceedingJoinPoint proceedingJoinPoint) throws Throwable {
        System.out.println("around advice 1");
        proceedingJoinPoint.proceed();
        System.out.println("around advice 2");
    }

    /**
     * 后置增强
     */
    @AfterReturning("allAointCut()")
    public void afterReturning() {
```

```java
        System.out.println("afterReturning advise");
    }

    /**
     * 异常抛出增强
     */
    @AfterThrowing("allAointCut()")
    public void afterThrowing() {
        System.out.println("afterThrowing advise");
    }

    /**
     * 后置增强
     */
    @After("allAointCut()")
    public void after() {
        System.out.println("after advise");
    }
}
```

下面创建一个测试类,从 Spring 上下文中获取 Person 对象,并调用 Person 的 say()方法。测试代码如下:

```java
/**
 * @Author: zhouguanya
 * @Date: 2018/9/1
 * @Description: 测试各种类型的增强
 */
public class AllAspectDemo {

    public static void main(String[] args) {
        ApplicationContext context = new ClassPathXmlApplicationContext("classpath:spring-chapter3-aoptype.xml");
        Person person = (Person) context.getBean("person");
        person.say();
    }
}
```

测试结果如下:

```
around advice 1
before advice
Hello Spring 5
around advice 2
after advise
afterReturning advise
```

测试结果与使用 3.4 节中 Spring AOP 的结果类似,此处不再阐述。

3.5.2 AspectJ 各种切点指示器

Spring 中支持若干个 AspectJ 切点指示器，它们用不同的方式描述目标类的连接点，表 3-1 所示是 Spring 中常见的几种 AspectJ 切点指示器。

表 3-1 Spring 中常见的 AspectJ 切点指示器

AspectJ 指示器	功能描述
args()	通过判断目标类方法运行时入参对象的类型定义指定连接点
@args()	通过判断目标方法运行时入参对象的类是否标注特定注解来指定连接点
execution()	匹配满足某一匹配条件的目标方法的连接点
this()	代理类按类型匹配于指定类，则被代理的目标类所有连接点匹配切点
target()	限制连接点匹配目标对象为指定类型的类
@target()	限制连接点匹配目标对象为被特定注解标注的类
within()	匹配特定域下的所有连接点
@within()	限制匹配特定注解标注的类的连接点
@annotation()	限制匹配带有指定注解的连接点

下面将通过案例讲解每一种指示器的使用和其对应的效果。

3.5.3 args() 与 "@args()"

args() 匹配的是方法的入参类型，该函数接收一个类名，表示目标类方法入参对象是指定类（包含子类）时切点匹配。

比如 args(com.test.Waiter) 表示运行时入参是 Waiter 类型的方法，args 与 execution 的区别在于 execution 是针对类方法的签名而言的，而 args 是针对运行时的入参类型而言。

"@args()" 函数接收一个注解类的类名，当方法的运行时入参对象标注了指定的注解时，匹配切点。

下面通过案例阐述两者的使用。创建一个 Factory 接口，用 FoodFactory 和 PhoneFactory 两个类分别实现 Factory。三者的代码如下。

Factory 接口中定义了两个方法，做产品的 make 方法和运输产品的 delivery 方法。

```
/**
 * @Author zhouguanya
 * @Date 2018/9/10
 * @Description 工厂接口
 */
public interface Factory {
    /**
     * 制作产品
     */
    void make();
```

```
    /**
     * 运输
     */
    void delivery(String address);
}
```

FoodFactory 实现 Factory 接口,并添加额外的 testArgsAnnotation()方法,代码如下:

```
/**
 * @Author zhouguanya
 * @Date 2018/9/10
 * @Description 食品工厂
 */
@Component
public class FoodFactory implements Factory {
    /**
     * 制作产品的方法
     */
    @Override
    public void make() {
        System.out.println("生产食品");
    }

    /**
     * 运输
     *
     * @param address
     */
    @Override
    public void delivery(String address) {
        System.out.println("销售食品至" + address);
    }

    /**
     * 测试@args 注解
     */
    public void testArgsAnnotation(FreshFoodFactory freshFoodFactory) {

    }
}
```

PhoneFactory 实现 Factory 接口,重写 make()和 delivery()方法,代码如下:

```
/**
 * @Author zhouguanya
 * @Date 2018/9/10
 * @Description 手机工厂
```

```java
 */
@Component
public class PhoneFactory implements Factory {
    /**
     * 制作产品的方法
     */
    @Override
    public void make() {
        System.out.println("生产手机");
    }

    /**
     * 运输手机的方法
     */
    @Override
    public void delivery(String address) {
        System.out.println("运输手机至" + address);
    }
}
```

创建一个监听功能的自定义注解"@Listen",其目的是为了被"@args"匹配。自定义注解的实现如下:

```java
/**
 * @Author zhouguanya
 * @Date 2018/9/10
 * @Description 监听注解
 */
@Retention(RetentionPolicy.RUNTIME)
@Target({ ElementType.TYPE, ElementType.METHOD })
@Documented
public @interface Listen {
    String value() default "";
}
```

另有如下两个类FreshFoodFactory和FrozenFoodFactory。FreshFoodFactory继承自FoodFactory,FrozenFoodFactory继承自FreshFoodFactory。其中需要注意的是,在FreshFoodFactory类上加上注解@Listen。

FreshFoodFactory继承自FoodFactory。

```java
/**
 * @Author zhouguanya
 * @Date 2018/9/10
 * @Description 新鲜食品工厂
 */
```

```
@Listen
@Component
public class FreshFoodFactory extends FoodFactory {

}
```

FrozenFoodFactory 继承自 FreshFoodFactory。

```
/**
 * @Author zhouguanya
 * @Date 2018/9/10
 * @Description 冷冻食品工厂
 */
@Component
public class FrozenFoodFactory extends FreshFoodFactory {

}
```

下面自定义一个 ArgsAspect 切面，其中前置增强匹配字符串类型的方法入参，后置增强匹配被"@Listen"标注的类。

```
/**
 * @Author zhouguanya
 * @Date 2018/9/10
 * @Description args 和@args 切面逻辑
 */
@Aspect
public class ArgsAspect {
    @Before("args(java.lang.String)")
    public void before() {
        System.out.println("args 匹配方法入参是 String 的方法");
    }

    @After("@args(com.test.aspectj.expression.args.Listen)")
    public void after() {
        System.out.println("@args 匹配到方法实行了");
    }
}
```

用一个测试类 AspectJExpressionDemo 来验证 args()和@args()函数。测试代码如下：

```
/**
 * @Author zhouguanya
 * @Date 2018/9/10
 * @Description 测试 args()
 */
public class AspectJExpressionDemo {
    public static void main(String[] args) {
```

```
        ApplicationContext context = new ClassPathXmlApplicationContext
("classpath:spring-chapter3-aspectjargsexpression.xml");
        FoodFactory foodFactory = (FoodFactory) context.getBean
("foodFactory");
        foodFactory.delivery("上海");
        System.out.println("-----分割线-----");
        Factory phoneFactory = (Factory) context.getBean("phoneFactory");
        phoneFactory.delivery("北京");
        System.out.println("-----分割线-----");
        FreshFoodFactory freshFoodFactory = (FreshFoodFactory)
context.getBean("freshFoodFactory");
        freshFoodFactory.testArgsAnnotation(freshFoodFactory);
        System.out.println("-----分割线-----");
        FrozenFoodFactory frozenFoodFactory = (FrozenFoodFactory)
context.getBean("frozenFoodFactory");
        frozenFoodFactory.testArgsAnnotation(frozenFoodFactory);
    }
}
```

其中的配置文件 spring-chapter3-aspectjargsexpression.xml 的配置如下：

```
<context:component-scan base-package="com.test.aspectj.expression"/>
<bean id="annotationAspect"
class="com.test.aspectj.expression.args.ArgsAspect"/>
```

运行测试代码，得到的测试结果如下：

```
args 匹配方法入参是 String 的方法
销售食品至上海
-----分割线-----
args 匹配方法入参是 String 的方法
运输手机至北京
-----分割线-----
@args 匹配到方法执行了
-----分割线-----
@args 匹配到方法执行了
```

从测试结果可以看出，args() 匹配了 FoodFactory 和 PhoneFactory 类中的入参是 String 类型的 delivery 方法，"@args()" 匹配到了 FreshFoodFactory 和 FrozenFoodFactory 类中的方法 testArgsAnnotation(FreshFoodFactory freshFoodFactory)。

值得一提的是，本例中 testArgsAnnotation(FreshFoodFactory freshFoodFactory) 的方法签名为入参类型点，被 "@Listen" 注解标记的 FreshFoodFactory 称为注解点。按图 3-14 所示的从上到下的继承关系，当注解点"低于"入参类型点时，那么入参类型点的所有子孙类都可以被 "@args()" 匹配，否则将不会被 "@args()" 匹配。

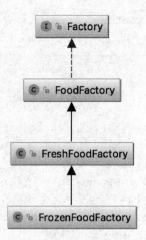

图 3-14 测试案例相关类继承结构图

如果修改此例中的"@Listen"注解点的位置到 FoodFactory 类上,那么"@args()"是没办法匹配到 testArgsAnnotation() 方法执行的。此时的 FoodFactory 代码如下:

```
@Listen
@Component
public class FoodFactory implements Factory {
    /**
     * 制作产品的方法
     */
    @Override
    public void make() {
        System.out.println("生产食品");
    }

    /**
     * 运输
     *
     * @param address
     */
    @Override
    public void delivery(String address) {
        System.out.println("销售食品至" + address);
    }

    /**
     * 测试@args注解
     */
    public void testArgsAnnotation(FreshFoodFactory freshFoodFactory) {

    }
}
```

删除 FreshFoodFactory 上的"@Listen"注解,此时的 FreshFoodFactory 代码如下:

```
@Component
public class FreshFoodFactory extends FoodFactory {

}
```

再次执行测试代码,将会发现此时"@args()"注解匹配不到 testArgsAnnotation()方法的执行,运行测试代码将得到如下结果:

```
args 匹配方法入参是 String 的方法
销售食品至上海
-----分割线-----
args 匹配方法入参是 String 的方法
运输手机至北京
-----分割线-----
-----分割线-----
```

3.5.4 @annotation()

"@annotation"匹配被指定注解标记的所有方法。

新建一个自定义注解"@Log"表示用于记录日志,将"@Log"加在 3.5.3 节的案例中的 PhoneFactory 类的 make()方法上。自定义注解"@Log"的代码如下:

```
/**
 * @Author zhouguanya
 * @Date 2018/9/10
 * @Description 自定义日志注解
 */
@Retention(RetentionPolicy.RUNTIME)
@Target(ElementType.METHOD)
public @interface Log {
    boolean value() default true;
}
```

被"@Log"注解后 PhoneFactory 的 make()方法如下:

```
/**
 * @Author zhouguanya
 * @Date 2018/9/10
 * @Description 手机工厂
 */
@Component
public class PhoneFactory implements Factory {
    /**
     * 制作产品的方法
     */
    @Override
```

```java
    @Log
    public void make() {
        System.out.println("生产手机");
    }

    /**
     * 运输手机的方法
     */
    @Override
    public void delivery(String address) {
        System.out.println("运输手机至" + address);
    }
}
```

定义切面逻辑,使用"@annotation()"来为所有加了"@Log"注解的方法织入增强,定义的切面 AnnotationAspect 代码如下:

```java
/**
 * @Author zhouguanya
 * @Date 2018/9/10
 * @Description 使用 @annotation() 来为所有加了 @Log 注解的方法织入增强
 */
@Aspect
public class AnnotationAspect {
@AfterReturning("@annotation(com.test.aspectj.expression.annotation.Log)")
    public void log() {
        System.out.println("打印日志");
    }
}
```

编写测试代码,并在测试代码中调用 PhoneFactory 的 make()方法,观察 make()方法是否被增强。测试代码如下:

```java
/**
 * @Author zhouguanya
 * @Date 2018/9/10
 * @Description 测试 ExecutionAspect 切面
 */
public class AspectJExpressionDemo {
    public static void main(String[] args) {
        ApplicationContext context = new ClassPathXmlApplicationContext("classpath:spring-chapter3-aspectjannotationexpression.xml");
        Factory foodFactory = (Factory) context.getBean("foodFactory");
        foodFactory.make();
        System.out.println("-----分割线-----");
        Factory phoneFactory = (Factory) context.getBean("phoneFactory");
```

```
        phoneFactory.make();
    }
}
```

配置文件 spring-chapter3-aspectjannotationexpression.xml 中的相关配置如下：

```xml
<context:component-scan base-package="com.test.aspectj.expression"/>
<bean id="annotationAspect"
class="com.test.aspectj.expression.annotation.AnnotationAspect"/>
<aop:aspectj-autoproxy/>
```

运行测试代码，测试结果如下：

```
生产食品
-----分割线-----
生产手机
打印日志
```

从测试结果可以证明，foodFactory 对象中的 make()方法因没有被 "@Log" 注解，因此没有被增强；phoneFactory 中的 make()方法加了 "@Log" 注解，所以在 phoneFactory 对象的 make()方法执行后得到了增强，执行了切面 AnnotationAspect 中的 log()方法。

3.5.5　execution

execution 是最常用的切点函数，其具体语法如下：

```
execution(<修饰符模式>?<返回类型模式><方法名模式>(<参数模式>)<异常模式>?)
```

execution 是匹配某些类的某些方法执行的。下面定义一个切面 ExecutionAspect 使用 execution 函数，并匹配 3.5.3 节 Factory 中所有方法的执行，切面 ExecutionAspect 代码如下：

```java
/**
 * @Author zhouguanya
 * @Date 2018/9/10
 * @Description 使用 execution 来为所有 Factory 接口的实现类植入增强
 */
@Aspect
public class ExecutionAspect {
    @AfterReturning("execution(* com.test.aspectj.expression.Factory.*(..))")
    public void make() {
        System.out.println("make 方法执行了");
    }
}
```

重点分析 execution 表达式的含义：

在表达式 "* com.test.aspectj.expression.Factory.*(..)" 中，第 1 个 "*" 表示任意的方法返回值类型，"com.test.aspectj.expression.Factory.*" 表示 Factory 中的所有的方法，(..)表示任意类型参数且参数个数不限。因此整个表达式的含义就是匹配 Factory 中的任意返回值、任意入参的所有方法。

测试代码如下：

```java
/**
 * @Author zhouguanya
 * @Date 2018/9/10
 * @Description 测试 execution 增强
 */
public class AspectJExpressionDemo {
    public static void main(String[] args) {
        ApplicationContext context = new ClassPathXmlApplicationContext("classpath:spring-chapter3-aspectjexecutionexpression.xml");
        Factory foodFactory = (Factory) context.getBean("foodFactory");
        foodFactory.make();
        System.out.println("-----分割线-----");
        Factory phoneFactory = (Factory) context.getBean("phoneFactory");
        phoneFactory.make();
    }
}
```

执行测试代码，得到测试结果如下：

```
生产食品
make 方法执行了
-----分割线-----
生产手机
make 方法执行了
```

从测试结果可以看出，Factory 中的 make() 方法执行后，都被增强了。

表达式的写法除了本例中的以外，还有很多不同的写法。下面详细介绍每种表达式写法的含义，如表 3-2 所示。

表 3-2 execution 表达式

表 达 式	功能描述
execution(public * *(..))	匹配所有目标类的所有 public 方法
execution(* pre*(..))	匹配所有目标类所有以 pre 为前缀的方法
execution(* com.test. Factory.*(..))	匹配 Factory 中的所有方法
类模式表达式中的.*	匹配包中的所有类，不包括子孙包中的类
类模式表达式中的..*	匹配包中以及子孙包的所有类
方法入参表达式中的*	匹配任意类型参数
方法入参表达式中的**	匹配任意类型参数且参数不限个数
execution(* make(int,String))	匹配 make(int,String)方法

3.5.6 target()与"@target()"

target()表示目标类型是指定的类型时，目标类型的所有方法都匹配到。target()可以匹配所有实现类及其子孙类中的所有方法。

"@target()"匹配标注了指定注解的类。

下面通过代码演示 target()和"@target()"的使用。先创建一个注解"@Run"，代码如下：

```java
/**
 * @Author zhouguanya
 * @Date 2018/9/13
 * @Description 执行的注解
 */
@Retention(RetentionPolicy.RUNTIME)
@Target({ ElementType.TYPE, ElementType.METHOD })
@Documented
public @interface Run {
    String value() default "";
}
```

再创建一个 HuaweiPhoneFactory 类，该类继承 PhoneFactory，并用注解"@Run"标注 HuaweiPhoneFactory 类。

```java
/**
 * @Author zhouguanya
 * @Date 2018/9/13
 * @Description 华为手机工厂
 */
@Run
@Component
public class HuaweiPhoneFactory extends PhoneFactory {

}
```

接着定义切面 TargetAspect，该注解中分别使用 target()和"@target()"函数，切面代码如下：

```java
/**
 * @Author zhouguanya
 * @Date 2018/9/10
 * @Description
 */
@Aspect
public class TargetAspect {

    @Before("target(com.test.aspectj.expression.PhoneFactory)")
    public void before() {
        System.out.println("target 匹配到，方法执行前增强");
    }

    @After("@target(com.test.aspectj.expression.target.Run)")
    public void after() {
        System.out.println("@target 匹配到，方法执行后增强");
    }

}
```

在测试代码中，调用 HuaweiPhoneFactory 类的 make()方法，并观察输出结果：

```java
/**
 * @Author zhouguanya
 * @Date 2018/9/10
 * @Description 测试
 */
public class TargetExpressionDemo {
    public static void main(String[] args) {
        ApplicationContext context = new ClassPathXmlApplicationContext("classpath:spring-chapter3-aspectjtargetexpression.xml");
        HuaweiPhoneFactory huaweiPhoneFactory = (HuaweiPhoneFactory) context.getBean("huaweiPhoneFactory");
        huaweiPhoneFactory.make();
    }
}
```

配置文件 spring-chapter3-aspectjtargetexpression.xml 中的主要配置如下：

```xml
<context:component-scan base-package="com.test.aspectj.expression"/>
<bean id="annotationAspect" class="com.test.aspectj.expression.target.TargetAspect"/>
<aop:aspectj-autoproxy proxy-target-class="true"/>
```

执行测试代码，得到如下的测试结果：

```
target 匹配到，方法执行前增强
生产手机
@target 匹配到，方法执行后增强
```

从测试结果可以证明，target(com.test.aspectj.expression.PhoneFactory)匹配到了其子类 HuaweiPhoneFactory 的 make()方法的执行，@target(com.test.aspectj.expression.target.Run)匹配到了已加注解"@Run"的类 HuaweiPhoneFactory。

3.5.7　this()

this()与 target()几乎是等效的，两者在引介切面的场景下略有差别。下面通过案例分析两者的区别。

创建一个接口 Listener，在接口中定义一个监听方法 listen()，Listener 接口如下：

```java
/**
 * @Author zhouguanya
 * @Date 2018/9/13
 * @Description 引介切面要实现的接口
 */
public interface Listener {
    /**
     * 监听
```

```
     */
    void listen();
}
```

创建一个 Listener 接口的实现类 DefaultListener，其中重写了 Listener 接口的 listen()方法，DefaultListener 类的实现如下：

```
/**
 * @Author zhouguanya
 * @Date 2018/9/13
 * @Description Listener 接口实现类
 */
public class DefaultListener implements Listener {
    /**
     * 监听
     */
    @Override
    public void listen() {
        System.out.println("开始监听");
    }
}
```

定义引介切面 ListenerAspect，其为 FoodFactory 植入 Listener 接口，ListenerAspect 的实现如下：

```
/**
 * @Author zhouguanya
 * @Date 2018/9/13
 * @Description 为FoodFactory添加 Listener 接口的切面
 */
@Aspect
public class ListenerAspect implements Ordered {

    /**
     * 为FoodFactory添加接口实现,要实现的接口是Listener,接口的默认实现是
DefaultListener
     */
    @DeclareParents(value = "com.test.aspectj.expression.FoodFactory",
defaultImpl = DefaultListener.class)
    public static Listener listener;

    /**
     * 如果有多个切面,注意多切面织入的顺序
     */
    @Override
    public int getOrder() {
```

```
        return 2;
    }
}
```

除了上面的引介切面外,还需要一个切面 ThisAspect,这个切面中分别使用 this()和 target()函数,ThisAspect 的实现如下:

```
/**
 * @Author zhouguanya
 * @Date 2018/9/13
 * @Description 测试 this 和 target 的切面
 */
@Aspect
public class ThisAspect implements Ordered {
    /**
     * 织入运行期对象为 Listener 类型的 Bean 中
     */
    @AfterReturning("this(com.test.aspectj.expression.thisexpression.Listener)")
    public void after() {
        System.out.println("ThisAspect after 方法执行了");
    }

    @Before("target(com.test.aspectj.expression.thisexpression.Listener)")
    public void before() {
        System.out.println("ThisAspect before 方法执行了");
    }

    /**
     * 如果有多个切面,注意多切面织入的顺序
     */
    @Override
    public int getOrder() {
        return 1;
    }
}
```

创建测试类,观察 this()和 target()函数的区别,测试代码如下:

```
/**
 * @Author zhouguanya
 * @Date 2018/9/10
 * @Description 测试
 */
public class ThisExpressionDemo {
    public static void main(String[] args) {
        ApplicationContext context = new ClassPathXmlApplicationContext
```

```
("classpath:spring-chapter3-aspectjthisexpression.xml");
        FoodFactory foodFactory = (FoodFactory) context.getBean("foodFactory");
        foodFactory.make();
        System.out.println("-----分割线-----");
        Listener listener = (Listener) foodFactory;
        listener.listen();
    }
}
```

配置文件 spring-chapter3-aspectjthisexpression.xml 主要配置如下：

```
<context:component-scan base-package="com.test.aspectj.expression"/>
<bean id="listenerAspect"
class="com.test.aspectj.expression.thisexpression.ListenerAspect"/>
<bean id="thisAspect"
class="com.test.aspectj.expression.thisexpression.ThisAspect"/>
<aop:aspectj-autoproxy />
```

运行测试代码，测试结果如下：

```
生产食品
ThisAspect after 方法执行了
-----分割线-----
ThisAspect before 方法执行了
开始监听
ThisAspect after 方法执行了
```

从测试结果可以看到，在调用 make()方法时，ThisAspect 类中的 before()方法并未执行，即 target 没有匹配到 make()方法的执行；当调用 listen()方法时，ThisAspect 类中的 before()方法执行了。

可以得出以下结论。

this(com.test.aspectj.expression.thisexpression.Listener)不仅可以匹配 Listener 接口中定义的方法，而且还可以匹配 FoodFactory 中的方法；target(com.test.aspectj.expression.thisexpression.Listener)仅仅匹配 Listener 中定义的方法。

3.5.8　within()与"@within()"

within()与 execution()的功能类似，两者的区别是，within()定义的连接点的最小范围是类级别的，而 execution()定义的连接点的最小范围可以精确到方法的入参，因此可以认为 execution()涵盖了 within()的功能。

"@within()"匹配标注了指定注解的类及其子孙类。

下面通过案例阐述 within()和"@within()"的使用。

首先创建一个表示监控的注解"@Monitor"，代码如下：

```
/**
 * @Author zhouguanya
 * @Date 2018/9/12
```

```
 * @Description 监控
 */
@Retention(RetentionPolicy.RUNTIME)
@Target({ ElementType.TYPE, ElementType.METHOD })
@Documented
public @interface Monitor {
    String value() default "";
}
```

下面修改 PhoneFactory，在其中加入 testWithin()方法，修改后的 PhoneFactory 代码如下：

```
/**
 * @Author zhouguanya
 * @Date 2018/9/10
 * @Description 手机工厂
 */

@Component
public class PhoneFactory implements Factory {
    /**
     * 制作产品的方法
     */
    @Override
    @Log
    public void make() {
        System.out.println("生产手机");
    }

    /**
     * 运输手机的方法
     */
    @Override
    public void delivery(String address) {
        System.out.println("运输手机至" + address);
    }
    /**
     * 测试@Within注解
     */
    public void testWithin() {

    }
}
```

接着创建 MobilePhoneFactory 类，使其继承 PhoneFactory，并重写父类 PhoneFactory 中的 testWithin()方法，并使用"@Monitor"注解标注：

```java
/**
 * @Author zhouguanya
 * @Date 2018/9/13
 * @Description 手机工厂
 */
@Monitor
@Component
public class MobilePhoneFactory extends PhoneFactory {
    @Override
    public void testWithin() {

    }
}
```

下面创建 IPhoneFactory 类继承 PhoneFactory，代码如下：

```java
/**
 * @Author zhouguanya
 * @Date 2018/9/13
 * @Description iPhone 手机工厂
 */
@Component(value = "iPhoneFactory")
public class IPhoneFactory extends MobilePhoneFactory {

}
```

下面创建切面类 WithinAspect，要分别使用 within() 和 "@within()"，代码如下：

```java
/**
 * @Author zhouguanya
 * @Date 2018/9/10
 * @Description 测试 within() 和 @ within() 的切面
 */
@Aspect
public class WithinAspect {
    @Before("within(com.test.aspectj.expression.FoodFactory)")
    public void before() {
        System.out.println("方法执行前增强");
    }

    @After("@within(com.test.aspectj.expression.within.Monitor)")
    public void after() {
        System.out.println("@within 匹配到，执行增强");
    }
}
```

测试代码中分别调用了 FoodFactory 和 PhoneFactory 的 make() 方法用以验证 within()，分别调用 IPhoneFactory 和 IPhoneFactory 的 testWithin() 方法用以验证 "@within()"。测试代码如下：

```java
/**
 * @Author zhouguanya
 * @Date 2018/9/10
 * @Description 测试
 */
public class WithinExpressionDemo {
    public static void main(String[] args) {
        ApplicationContext context = new ClassPathXmlApplicationContext
("classpath:spring-chapter3-aspectjwithinexpression.xml");
        Factory foodFactory = (Factory) context.getBean("foodFactory");
        foodFactory.make();
        System.out.println("-----分割线-----");
        IPhoneFactory iPhoneFactory = (IPhoneFactory) context.getBean
("iPhoneFactory");
        iPhoneFactory.testWithin();
        System.out.println("-----分割线-----");
        MobilePhoneFactory mobilePhoneFactory = (MobilePhoneFactory)
context.getBean("mobilePhoneFactory");
        mobilePhoneFactory.testWithin();
    }
}
```

执行测试代码,测试结果如下:

```
方法执行前增强
生产食品
-----分割线-----
@within 匹配到,执行增强
-----分割线-----
@within 匹配到,执行增强
```

从测试结果可以证明,within(com.test.aspectj.expression.FoodFactory)匹配到了 FoodFactory 类的 make()方法的执行;@within(com.test.aspectj.expression.within.Monitor)不仅匹配到了被@Monitor 标注的类 MobilePhoneFactory,而且还匹配到了 MobilePhoneFactory 的子类 IPhoneFactory。

3.6　Spring AOP 的实现原理

Spring AOP 的实现是通过创建目标对象的代理类,并对目标对象进行拦截来实现的。分析 Spring AOP 的底层实现,需要重点分析几个常用类,相关类图如 3-15 所示。

ProxyConfig 类是一个基类——数据类,主要为各种 AOP 代理工厂提供属性配置。

AdvisedSupport 类是 ProxyConfig 类的子类,其封装了 AOP 中对通知(Advice)和通知器(Advisor)的相关操作,这些操作对于不同的创建代理对象的类都是相同的,但是对于具体的 AOP 代理对象的生成需要 AdvisedSupport 各个子类去实现。

第 3 章 Spring AOP 揭秘

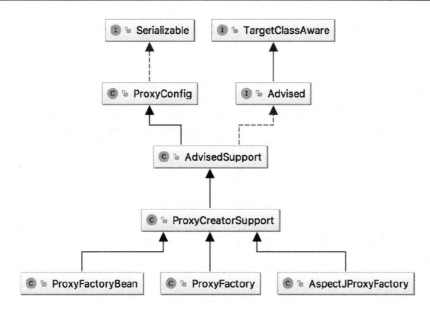

图 3-15 Spring AOP 核心类图

ProxyCreatorSupport 类是 AdvisedSupport 的子类——辅助类，不同子类的一些通用的操作都封装在 ProxyCreatorSupport 中。

ProxyFactoryBean，ProxyFactory 和 AspectJProxyFactory 是用于创建 AOP 代理对象的，这三个类的作用分别如下：

- ProxyFactoryBean 类：功能是创建声明式的代理对象。
- ProxyFactory 类：功能是创建编程式的代理对象。
- AspectJProxyFactory 类：功能是创建基于 AspectJ 的代理对象。

3.6.1 设计原理

下面以 ProxyFactoryBean 为例，分析 Spring AOP 的实现原理。

首先定义一个接口 Log，其中包含一个 printLog()方法，Log 接口的代码如下：

```
/**
 * @Author zhouguanya
 * @Date 2018/10/3
 * @Description 定义接口
 */
public interface Log {

    /**
     * 打印日志
     */
    void printLog();
}
```

再创建一个 Target 类，实现 Log 接口，重写 printLog 方法，Target 类的代码如下：

```java
/**
 * @Author zhouguanya
 * @Date 2018/10/3
 * @Description 目标对象
 */
public class Target implements Log {

    /**
     * 操作方法
     */
    @Override
    public void printLog() {
        try {
            //模拟一个耗时1秒的操作
            Thread.sleep(1000);
        } catch (InterruptedException e) {
            e.printStackTrace();
        }
        System.out.println("执行一些操作");
    }
}
```

然后创建一个通知类 LogAroundAdvice 并实现 MethodInterceptor 接口，重写 invoke()方法，在方法执行前后分别打印实现，LogAroundAdvice 的实现如下：

```java
/**
 * @Author zhouguanya
 * @Date 2018/10/3
 * @Description 通知
 */
public class LogAroundAdvice implements MethodInterceptor {

    @Override
    public Object invoke(MethodInvocation invocation) throws Throwable {
        SimpleDateFormat dateFormat = new SimpleDateFormat("yyyy-MM-dd hh:mm:ss SSS");
        System.out.println("方法执行开始时间: " + dateFormat.format(new Date()));
        invocation.proceed();
        System.out.println("方法执行结束时间: " + dateFormat.format(new Date()));
        return null;
    }
}
```

创建一个测试类，用于观察测试结果，测试代码如下：

```java
/**
 * @Author zhouguanya
 * @Date 2018/10/3
 * @Description 测试
 */
public class AopSourceCodeLearningDemo {
    public static void main(String[] args) {
        //加载配置文件
        ApplicationContext context = new ClassPathXmlApplicationContext("classpath:spring-chapter3-sourcecodelearning.xml");
        //获取目标对象
        Log target = (Log) context.getBean("proxyFactoryBean");
        //执行目标对象的方法
        target.printLog();
    }
}
```

文件 spring-chapter3-sourcecodelearning.xml 的配置如下：

```xml
<!--配置通知器，通知器的实现定义了需要对目标对象进行的增强行为-->
<bean id="logAdvisor" class="com.test.sourcecodelearning.LogAroundAdvice"/>
<!--目标对象-->
<bean id="sourceTarget" class="com.test.sourcecodelearning.Target"/>
<!--配置AOP代理，封装AOP功能的主要类-->
<bean id="proxyFactoryBean" class="org.springframework.aop.framework.ProxyFactoryBean">
    <!--AOP代理接口-->
    <property name="proxyInterfaces">
        <value>
            com.test.sourcecodelearning.Log
        </value>
    </property>
    <!--需要增强的对象-->
    <property name="target">
        <ref bean="sourceTarget"/>
    </property>
    <!--拦截器的名字，即通知器在AOP代理的配置下通过使用代理对象的拦截机制发挥作用-->
    <property name="interceptorNames">
        <list>
            <value>logAdvisor</value>
        </list>
    </property>
</bean>
```

运行测试代码，测试结果如下：

```
方法执行开始时间:2018-10-03 08:18:51 167
执行一些操作
方法执行结束时间:2018-10-03 08:18:52 172
```

从测试结果可以看出,在正常的调用 printLog()方法前后分别打印了日志,说明 AOP 已经实现了。下面将通过这个案例分析 ProxyFactoryBean 的实现逻辑。

打开 ProxyFactoryBean 的代码,其生成代理对象的核心方法是 getObject()方法,部分代码如下:

```java
public class ProxyFactoryBean extends ProxyCreatorSupport implements
FactoryBean<Object>, BeanClassLoaderAware, BeanFactoryAware {
    //标志通知器链是否已经完成初始化
    private boolean advisorChainInitialized = false;

    /** 单例模式的代理对象 */
    private Object singletonInstance;

    /**
     * 创建代理对象的入口
     */
    @Override
    @Nullable
    public Object getObject() throws BeansException {
        //初始化通知器链
        initializeAdvisorChain();
        //如果是单例模式
        if (isSingleton()) {
            //返回单利模式的代理对象
            return getSingletonInstance();
        }
        //如果不是单例模式
        else {
        ......
        }
        //每次创建一个新的代理对象
        return newPrototypeInstance();
        }
    }
```

下面分析 getObject()方法中的 initializeAdvisorChain()方法,initializeAdvisorChain()方法是初始化通知器链(或者叫拦截器链)的,其代码如下:

```java
private synchronized void initializeAdvisorChain() throws AopConfigException,
BeansException {
        //如果通知器链已经被初始化,则直接返回,即通知器链只在第一次获取代理对象时生成
        if (this.advisorChainInitialized){
```

```java
            return;
        }
        //如果ProxyFactoryBean中配置的通知器/拦截器名称不为空
        if (!ObjectUtils.isEmpty(this.interceptorNames)) {
            ......
            //遍历通知器链，向容器添加通知器
            for (String name : this.interceptorNames) {
                ......
                //如果通知器是全局的
                if (name.endsWith(GLOBAL_SUFFIX)) {
                    ......
                    //向容器中添加全局通知器
                    addGlobalAdvisor((ListableBeanFactory) this.beanFactory,
                    name.substring(0, name.length() - GLOBAL_SUFFIX.length()));
                }
                //如果通知器不是全局的
                else {
                    Object advice;
                    //如果通知器是单态模式
                    if (this.singleton || this.beanFactory.isSingleton(name)){
                        //从容器获取单态模式的通知器
                        advice = this.beanFactory.getBean(name);
                    }
                    //如果通知器是原型模式
                    else {
                        //创建一个新的通知器对象
                        advice = new PrototypePlaceholderAdvisor(name);
                    }
                    //添加通知器到通知器链上
                    addAdvisorOnChainCreation(advice, name);
                }
            }
        }
        //设置通知器链已初始化标识
        this.advisorChainInitialized = true;
    }
```

执行 initializeAdvisorChain()方法后，如果是单例模式，将会调用 getSingletonInstance()方法获取一个单例模式的代理对象，getSingletonInstance()方法代码如下：

```java
private synchronized Object getSingletonInstance() {
    //如果单例模式的代理对象还未被创建
    if (this.singletonInstance == null) {
        //获取代理的目标源
        this.targetSource = freshTargetSource();
```

```java
            if (this.autodetectInterfaces &&
                getProxiedInterfaces().length == 0 &&
                    !isProxyTargetClass()
                Class<?> targetClass = getTargetClass();
                if (targetClass == null) {
                    throw new FactoryBeanNotInitializedException("Cannot determine target class for proxy");
                }
                //设置被代理的接口
                setInterfaces(ClassUtils.getAllInterfacesForClass(targetClass, this.proxyClassLoader));
            }
            // 初始化共享的单例模式对象
            super.setFrozen(this.freezeProxy);
            //调用 ProxyFactory 生成代理对象调用
ProxyCreatorSupport.createAopProxy()
            this.singletonInstance = getProxy(createAopProxy());
        }
        //返回已创建的单例对象
        return this.singletonInstance;
    }
```

执行 initializeAdvisorChain() 方法后，如果是非单例模式即原型模式，将会调用 newPrototypeInstance() 方法获取一个新的原型模式的代理对象，newPrototypeInstance() 方法代码如下：

```java
    private synchronized Object newPrototypeInstance() {
        ......
        //创建一个 ProxyCreatorSupport 对象
        ProxyCreatorSupport copy = new ProxyCreatorSupport(getAopProxyFactory());
        // The copy needs a fresh advisor chain, and a fresh TargetSource.
        TargetSource targetSource = freshTargetSource();
        //从当前对象中复制 AOP 的配置，为了保持原型模式对象的独立性
        copy.copyConfigurationFrom(this, targetSource, freshAdvisorChain());
        if (this.autodetectInterfaces && getProxiedInterfaces().length == 0 &&
            !isProxyTargetClass()) {
            //设置代理接口
            Class<?> targetClass = targetSource.getTargetClass();
            if (targetClass != null) {
                copy.setInterfaces(ClassUtils.getAllInterfacesForClass(targetClass,this.proxyClassLoader));
            }
        }
        copy.setFrozen(this.freezeProxy);
```

```
            if (logger.isTraceEnabled()) {
                logger.trace("Using ProxyCreatorSupport copy: " + copy);
            }
            //生成代理对象调用 ProxyCreatorSupport.createAopProxy()
            return getProxy(copy.createAopProxy());
        }
```

可以发现，无论是单例模式还是原型模式，最终都是通过调用 getProxy()方法获取代理对象的，getProxy()的实现如下：

```
protected Object getProxy(AopProxy aopProxy) {
    //通过 AOPProxy 产生代理对象
    return aopProxy.getProxy(this.proxyClassLoader);
}
```

通过以上对 getSingletonInstance()方法和 newPrototypeInstance()方法的代码注释可以发现，这两个方法都会调用 ProxyCreatorSupport.createAopProxy()方法，ProxyCreatorSupport 类的核心代码如下：

```
public class ProxyCreatorSupport extends AdvisedSupport {
    //AOPProxy 工厂
    private AopProxyFactory aopProxyFactory;
    ......
    //当第一个 AOPProxy 代理对象被创建时，设置为 true
    private boolean active = false;
    /**
     * 默认使用 DefaultAopProxyFactory 的作用 AopProxyFactory
     */
    public ProxyCreatorSupport() {
        this.aopProxyFactory = new DefaultAopProxyFactory();
    }
    ......
    /**
     * 创建 AOPProxy 代理的入口
     */
    protected final synchronized AopProxy createAopProxy() {
        if (!this.active) {
            activate();
        }
        //调用 DefaultAopProxyFactory 的创建 AOPProxy 代理的方法
        return getAopProxyFactory().createAopProxy(this);
    }
    ......
```

从 createAopProxy()方法的代码可以看出，AopProxy 对象是在 DefaultAopProxyFactory 类的 createAopProxy()方法中生成的，DefaultAopProxyFactory.createAopProxy()方法的代码如下：

```
    public AopProxy createAopProxy(AdvisedSupport config) throws
AopConfigException {
        //如果AOP使用显式优化，或者配置了目标类，或者只使用Spring支持的代理接口
        if (config.isOptimize() || config.isProxyTargetClass() ||
hasNoUserSuppliedProxyInterfaces(config)) {
            //获取AOP配置的目标类
            Class<?> targetClass = config.getTargetClass();
            ......
            //如果配置的AOP目标类是接口，则使用JdkDynamicAopProxy生成代理对象
            if (targetClass.isInterface() ||
Proxy.isProxyClass(targetClass)){
                return new JdkDynamicAopProxy(config);
            }
            //否则使用ObjenesisCglibAopProxy生成代理对象
            return new ObjenesisCglibAopProxy(config);
        }
        //如果不满足if条件则使用JdkDynamicAopProxy生成代理对象
        else {
            return new JdkDynamicAopProxy(config);
        }
    }
```

从 DefaultAopProxyFactory.createAopProxy()使用的类的名称可以发现，如果是继承了接口的类，会使用 JDK 动态代理，即用 JdkDynamicAopProxy 类创建代理对象，否则将会使用 CGLIB 动态代理即用 ObjenesisCglibAopProxy 类创建代理对象，关于这两种动态代理的具体使用，请参考本章 3.1 节。

3.6.2　JdkDynamicAopProxy

JDK 动态代理只能针对接口起作用，Spring 中通过 JdkDynamicAopProxy 类使用 JDK 动态代理创建 AOPProxy 对象，JdkDynamicAopProxy 类的定义如下：

```
final class JdkDynamicAopProxy implements AopProxy, InvocationHandler,
Serializable
```

JdkDynamicAopProxy 类实现了 InvocationHandler 接口，因而可以使用 JDK 动态代理产生代理对象。

```
@Override
public Object getProxy() {
    return getProxy(ClassUtils.getDefaultClassLoader());
}
```

此处的 getProxy()方法是获取代理对象的入口，其是通过调用以下方法实现的：

```
@Override
public Object getProxy(@Nullable ClassLoader classLoader) {
```

```
        if (logger.isDebugEnabled()) {
            logger.debug("Creating JDK dynamic proxy: target source is " +
this.advised.getTargetSource());
        }
        //获取代理接口
        Class<?>[] proxiedInterfaces = AopProxyUtils.completeProxiedInterfaces
                                    (this.advised, true);
        //查找代理目标的接口中是否定义 equals()和 hashCode()方法
        findDefinedEqualsAndHashCodeMethods(proxiedInterfaces);
        //使用 JDK 的动态代理机制创建 AOP 代理对象
        return Proxy.newProxyInstance(classLoader, proxiedInterfaces, this);
    }
```

findDefinedEqualsAndHashCodeMethods()方法的功能是查找代理的接口是否有定义 equals()或 hashCode()方法。

```
    private void findDefinedEqualsAndHashCodeMethods(Class<?>[]
proxiedInterfaces) {
        //遍历代理接口
        for (Class<?> proxiedInterface : proxiedInterfaces) {
            //接口中所有声明的方法
            Method[] methods = proxiedInterface.getDeclaredMethods();
            for (Method method : methods) {
                //如果方法是 equals()方法，则设置当前对象 equalsDefined 属性
                if (AopUtils.isEqualsMethod(method)) {
                    this.equalsDefined = true;
                }
                //如果方法是 hashCode()方法，则设置当前对象 hashCodeDefined 属性
                if (AopUtils.isHashCodeMethod(method)) {
                    this.hashCodeDefined = true;
                }
                if (this.equalsDefined && this.hashCodeDefined) {
                    return;
                }
            }
        }
    }
```

通过在 3.1 节中的介绍可以得知，InvocationHandler 接口的 invoke()方法是代理对象执行方法调用和增强的地方，下面分析 JdkDynamicAopProxy 实现 InvocationHandler 接口重写 invoke()方法的代码：

```
    public Object invoke(Object proxy, Method method, Object[] args) throws
Throwable {
        MethodInvocation invocation;
        Object oldProxy = null;
```

```java
            boolean setProxyContext = false;
            //获取通知的相关信息
            TargetSource targetSource = this.advised.targetSource;
            Object target = null;
            try {
                //如果代理目标对象的接口中没有定义equals()方法，且当前调用的方法是equals()方法
                if (!this.equalsDefined && AopUtils.isEqualsMethod(method)) {
                    //调用JdkDynamicAopProxy中重写的equals()方法
                    return equals(args[0]);
                }
                //如果代理目标对象的接口中没有定义hashCode()方法，且当前调用的方法是hashCode()方法
                else if (!this.hashCodeDefined && AopUtils.isHashCodeMethod(method)){
                    //调用JdkDynamicAopProxy中重写的hashCode()方法
                    return hashCode();
                }
                ......

                Object retVal;
                ......
                //获取目标对象
                target = targetSource.getTarget();
                Class<?> targetClass = (target != null ? target.getClass() : null);
                //获取目标对象方法配置的拦截器(通知器)链
                List<Object> chain = this.advised.getInterceptorsAndDynamicInterceptionAdvice(method, targetClass);
                //如果没有配置任何通知
                if (chain.isEmpty()) {
                    //没有配置通知，使用反射直接调用目标对象的方法，并获取方法返回值
                    Object[] argsToUse = AopProxyUtils.adaptArgumentsIfNecessary(method, args);
                    retVal = AopUtils.invokeJoinpointUsingReflection(target, method, argsToUse);
                }
                //如果配置了通知
                else {
                    //创建MethodInvocation对象
                    invocation = new ReflectiveMethodInvocation(proxy, target, method, args, targetClass, chain);
                    //调用通知链，沿着通知器链调用所有配置的通知
                    retVal = invocation.proceed();
                }
                //如果方法有返回值，则将代理对象作为方法返回值
                Class<?> returnType = method.getReturnType();
                if (retVal != null && retVal == target &&returnType != Object.class &&
```

```
returnType.isInstance(proxy) &&!RawTargetAccess.class.isAssignableFrom
(method.getDeclaringClass())) {
            retVal = proxy;
        }
        ......
}
```

通过以上代码分析可知，最核心的功能都是在 invocation.proceed()方法中实现的，下面分析 ReflectiveMethodInvocation，代码如下：

```
public Object proceed() throws Throwable {
    //如果拦截器链中通知已经调用完毕
    if (this.currentInterceptorIndex ==
            this.interceptorsAndDynamicMethodMatchers.size() - 1) {
        //调用 invokeJoinpoint()方法
        return invokeJoinpoint();
    }
    //获取拦截器链中的通知器或通知
    Object interceptorOrInterceptionAdvice =
            this.interceptorsAndDynamicMethodMatchers.get(
                                ++this.currentInterceptorIndex);
    //如果获取的通知器或通知是动态匹配方法拦截器类型
    if (interceptorOrInterceptionAdvice instanceof
            InterceptorAndDynamicMethodMatcher) {
        //动态匹配方法拦截器
        InterceptorAndDynamicMethodMatcher dm =
(InterceptorAndDynamicMethodMatcher) interceptorOrInterceptionAdvice;
        //如果匹配，调用拦截器的方法
        if (dm.methodMatcher.matches(this.method, this.targetClass,
this.arguments)){
            return dm.interceptor.invoke(this);
        }
        //如果不匹配，递归调用 proceed()方法，直至拦截器链被全部调用为止
        else {
            return proceed();
        }
    }
    //如果不是动态匹配方法拦截器，则切入点在构造对象之前进行静态匹配，调用拦截器的方法
    else {
        return ((MethodInterceptor) interceptorOrInterceptionAdvice)
.invoke(this);
    }
}
```

invokeJoinpoint()方法是调用目标对象方法的地方，其实现如下：

```java
protected Object invokeJoinpoint() throws Throwable {
    return AopUtils.invokeJoinpointUsingReflection(this.target, this.method,
        this.arguments);
}
```

invokeJoinpoint()方法会调用 AopUtils.invokeJoinpointUsingReflection()方法，代码如下：

```java
public static Object invokeJoinpointUsingReflection(@Nullable Object target, Method me
    thod, Object[] args)
        throws Throwable {

    // Use reflection to invoke the method.
    try {
        ReflectionUtils.makeAccessible(method);
        return method.invoke(target, args);
    }
    catch (InvocationTargetException ex) {
        // Invoked method threw a checked exception.
        // We must rethrow it. The client won't see the interceptor.
        throw ex.getTargetException();
    }
    catch (IllegalArgumentException ex) {
        throw new AopInvocationException("AOP configuration seems to be invalid: tried calling method [" +method + "] on target [" + target + "]", ex);
    }
    catch (IllegalAccessException ex) {
        throw new AopInvocationException("Could not access method [" + method + "]", ex);
    }
}
```

可以看到 invokeJoinpointUsingReflection()方法最终是通过反射调用目标对象的方法。

通过对 JdkDynamicAopProxy 类的代码进行分析可以知道，JdkDynamicAopProxy 类实现了 InvocationHandler 接口，重写了 invoke()方法，当进行调用时，其实并不是调用目标对象，而是为目标对象创建一个代理对象，触发代理对象的 invoke()方法，在 invoke()方法中会通过反射调用目标对象的方法，Spring AOP 相关通知的调用也是在 invoke()方法中完成的。

3.6.3 CglibAopProxy

由于 JDK 动态代理只能针对接口生成代理对象，对于没有实现接口的目标对象，需要使用 CGLIB 产生代理对象，下面分析 CglibAopProxy 的代码。

回到 DefaultAopProxyFactory.createAopProxy()方法，如果目标对象没有实现接口，将会返回一个 ObjenesisCglibAopProxy 对象。ObjenesisCglibAopProxy 类的代码如下：

```java
class ObjenesisCglibAopProxy extends CglibAopProxy {
    private static final Log logger = LogFactory.getLog
(ObjenesisCglibAopProxy.class);
    private static final SpringObjenesis objenesis = new SpringObjenesis();
    /**
     * 调用父构造器
     */
        public ObjenesisCglibAopProxy(AdvisedSupport config) {
            super(config);
        }
    /**
     * 创建代理类并创建代理对象
     */
    @Override
    @SuppressWarnings("unchecked")
    protected Object createProxyClassAndInstance(Enhancer enhancer,
        Callback[] callbacks) {
            Class<?> proxyClass = enhancer.createClass();
            Object proxyInstance = null;
            if (objenesis.isWorthTrying()) {
                try {
                    proxyInstance = objenesis.newInstance(proxyClass,
enhancer.getUseCache());
                }
                ......
            }
            if (proxyInstance == null) {
                // Regular instantiation via default constructor...
                try {
                    Constructor<?> ctor = (this.constructorArgs != null ?
proxyClass.getDeclaredConstructor(this.constructorArgTypes):proxyClass.
getDeclaredConstructor());
                    ReflectionUtils.makeAccessible(ctor);
                    proxyInstance = (this.constructorArgs != null ?
ctor.newInstance(this.constructorArgs) : ctor.newInstance());
                }
                ......
            ((Factory) proxyInstance).setCallbacks(callbacks);
            return proxyInstance;
        }
    }
```

从代码可以看出，ObjenesisCglibAopProxy 继承了 CglibAopProxy，Objenesis 是一个轻量级的 Java 库，作用是绕过构造器创建一个实例。因此分析的重点还是 CglibAopProxy 类。

```java
    public Object getProxy(@Nullable ClassLoader classLoader) {
        ......
        try {
            //获取目标对象
            Class<?> rootClass = this.advised.getTargetClass();
            Assert.state(rootClass != null, "Target class must be available for creating a CGLIB proxy");
            //将目标对象本身作为基类
            Class<?> proxySuperClass = rootClass;
            //检查获取到的目标类是否为CGLIB产生的
            if (ClassUtils.isCglibProxyClass(rootClass)) {
                //如果目标类是有CGLIB产生的,获取目标类的基类
                proxySuperClass = rootClass.getSuperclass();
                //获取目标类的接口
                Class<?>[] additionalInterfaces = rootClass.getInterfaces();
                //将目标类的接口添加到容器AdvisedSupport中
                for (Class<?> additionalInterface : additionalInterfaces) {
                    this.advised.addInterface(additionalInterface);
                }
            }
            //校验代理基类
            validateClassIfNecessary(proxySuperClass, classLoader);
            //配置CGLIB的Enhancer类,Enhancer是CGLIB中的主要操作类
            Enhancer enhancer = createEnhancer();
            ......
            //设置enhancer的接口
            enhancer.setSuperclass(proxySuperClass);
            //设置接口
            enhancer.setInterfaces(AopProxyUtils.completeProxiedInterfaces(this.advised));
            enhancer.setNamingPolicy(SpringNamingPolicy.INSTANCE);
            enhancer.setStrategy(new ClassLoaderAwareUndeclaredThrowableStrategy(classLoader));
            //设置enhancer的回调方法
            Callback[] callbacks = getCallbacks(rootClass);
            Class<?>[] types = new Class<?>[callbacks.length];
            for (int x = 0; x < types.length; x++) {
                types[x] = callbacks[x].getClass();
            }
            enhancer.setCallbackFilter(new ProxyCallbackFilter(this.advised.getConfigurationOnlyCopy(),this.fixedInterceptorMap, this.fixedInterceptorOffset));
            //设置enhancer的回调类型
            enhancer.setCallbackTypes(types);
            //创建代理对象,由于该方法被子类重写了,因此会调用子类重写后的方法
```

```
            return createProxyClassAndInstance(enhancer, callbacks);
        }
        ......
}
```

由本章 3.1 节可知，CGLIB 的运行需要配合回调方法，实现 MethodInterceptor 接口，在 CglibAopProxy 中也是一样，下面分析获取回调方法 getCallbacks()的代码：

```
private Callback[] getCallbacks(Class<?> rootClass) throws Exception {
    // Parameters used for optimization choices...
    boolean exposeProxy = this.advised.isExposeProxy();
    boolean isFrozen = this.advised.isFrozen();
    boolean isStatic = this.advised.getTargetSource().isStatic();
    //根据AOP 配置创建一个动态通知拦截器，CGLIB 创建的动态代理会自动调用
    //DynamicAdvisedInterceptor 类的 intercept 方法对目标对象进行拦截处理
    Callback aopInterceptor = new DynamicAdvisedInterceptor(this.advised);
    // 创建目标分发器
    Callback targetInterceptor;
    if (exposeProxy) {
        targetInterceptor = (isStatic ?new StaticUnadvisedExposedInterceptor
(this.advised.getTargetSource().getTarget()) :new
DynamicUnadvisedExposedInterceptor(this.advised.getTargetSource()));
    }
    else {
        targetInterceptor = (isStatic ?new StaticUnadvisedInterceptor
(this.advised.getTargetSource().getTarget()) :new DynamicUnadvisedInterceptor
(this.advised.getTargetSource()));
    }
    // Choose a "direct to target" dispatcher (used for
    // unadvised calls to static targets that cannot return this).Callback
targetDispatcher = (isStatic ?new StaticDispatcher(this.advised.getTargetSource().
getTarget()) : new SerializableNoOp());
    Callback[] mainCallbacks = new Callback[] {
            aopInterceptor,  //普通通知
            targetInterceptor,
            new SerializableNoOp(),
            targetDispatcher, this.advisedDispatcher,
            new EqualsInterceptor(this.advised),
            new HashCodeInterceptor(this.advised)
    };
    Callback[] callbacks;
    //如果目标是静态的，并且通知链被冻结，则使用优化AOP 调用，直接对方法使用固定的通知链
    if (isStatic && isFrozen) {
        Method[] methods = rootClass.getMethods();
        Callback[] fixedCallbacks = new Callback[methods.length];
```

```
                this.fixedInterceptorMap = new HashMap<>(methods.length);
            for (int x = 0; x < methods.length; x++) {
                List<Object> chain = this.advised.
getInterceptorsAndDynamicInterceptionAdvice(methods[x], rootClass);
                fixedCallbacks[x] = new FixedChainStaticTargetInterceptor
(chain,this.advised.getTargetSource().getTarget(),this.advised.getTargetClass(
));
                this.fixedInterceptorMap.put(methods[x].toString(), x);
            }
            //将固定回调和主要回调复制到回调数组中
            callbacks = new Callback[mainCallbacks.length +
fixedCallbacks.length];
            System.arraycopy(mainCallbacks, 0, callbacks, 0,
mainCallbacks.length);
            System.arraycopy(fixedCallbacks, 0, callbacks, mainCallbacks.length,
fixedCallbacks.length);
            this.fixedInterceptorOffset = mainCallbacks.length;
        }
        //如果目标不是静态的，或者通知链不被冻结，则使用AOP主要的通知
        else {
            callbacks = mainCallbacks;
        }
        return callbacks;
    }
```

通过上面对 CGLIB 创建代理和获取回调通知的代码分析，可以了解到 CGLIB 在获取代理通知时，会创建 DynamicAdvisedInterceptor 类；当调用目标对象的方法时，不是直接调用目标对象，而是通过 CGLIB 创建的代理对象来调用目标对象；并且在调用目标对象的方法时，会触发 DynamicAdvisedInterceptor 的 intercept 回调方法对目标对象进行处理，CGLIB 回调拦截器链的代码如下：

```
    public Object intercept(Object proxy, Method method,
            Object[] args, MethodProxy methodProxy) throws Throwable {
        Object oldProxy = null;
        boolean setProxyContext = false;
        Object target = null;
        TargetSource targetSource = this.advised.getTargetSource();
        try {
            if (this.advised.exposeProxy) {
                // Make invocation available if necessary.oldProxy =
AopContext.setCurrentProxy(proxy);
                setProxyContext = true;
            }
            //获取目标对象
            target = targetSource.getTarget();
```

```java
            Class<?> targetClass = (target != null ? target.getClass() : null);
            //获取AOP配置的通知
            List<Object> chain = this.advised.
                getInterceptorsAndDynamicInterceptionAdvice(method, targetClass);
            Object retVal;
            //如果没有配置通知
            if (chain.isEmpty() && Modifier.isPublic(method.getModifiers())) {
                //直接调用目标对象的方法
                Object[] argsToUse = AopProxyUtils.adaptArgumentsIfNecessary(method, args);
                retVal = methodProxy.invoke(target, argsToUse);
            }
            //如果配置了通知
            else {
                //通过CglibMethodInvocation来启动配置的通知
                retVal = new CglibMethodInvocation(proxy, target, method, args, targetClass, chain, methodProxy).proceed();
            }
            //获取目标对象对象方法的回调结果,如果有必要则封装为代理
            retVal = processReturnType(proxy, target, method, retVal);
            return retVal;
        }
        ......
    }
```

这里的 CglibMethodInvocation 类继承了 ReflectiveMethodInvocation 类，CglibMethodInvocation.procceed()调用了父类的 ReflectiveMethodInvocation.proceed()方法，和 3.6.2 节中调用的方法是相同的，此处不再赘述。

3.7 小　　结

本章讲解了 Spring 核心功能 AOP 的使用，并通过对代码的分析，揭示了 JDK 动态代理和 CGLIB 动态代理的实现原理。下一章将介绍 Spring 5 的新特性。

第二篇

Spring 5新特性篇

本篇主要讲解 Spring 5 新的特性和功能。

第 4 章

Spring 5 新特性概述

本书完稿时 Spring 最新版本已经升级到了 Spring 5.1。本章主要概述 Spring 5.0 和 Spring 5.1 的新特性。

4.1 Spring 5.0 新特性

4.1.1 运行环境

Spring 5.0 正常运行时,需要以下环境:

- 整个 Spring 框架的代码基于 JDK 8 开发。当读者选择升级 Spring 框架时,需要先确认已经安装了 JDK 8 及以上的 JDK 版本,否则 Spring 5.0 将不能正常运行。
- Spring 5.0 通过使用泛型推断和 lambda 表达式等特性提高了代码的可阅读性。
- 支持使用 Java 8 编程。
- 支持 JDK 9 开发部署。
- 整个 Spring 5.0 框架在 JDK 9 环境下编译和测试通过(默认是运行在 JDK 8 上的)
- Spring 5.0 的相关特性需要 Java EE 7 API。
- 支持 Servlet 3.1、Bean Validation 1.1、JPA 2.1、JMS 2.0、Tomcat 8.5+、Jetty 9.4+、WildFly 10+。
- Spring 5.0 在运行时兼容 Java EE 8。
- 兼容 Servlet 4.0、Bean Validation 2.0、JPA 2.2、JSON Binding API 1.0、Tomcat 9.0、Hibernate Validator 6.0、Apache Johnzon 1.1。

4.1.2 删除的代码

涉及删除的地方有 beans.factory.access、jdbc.support.nativejdbc、mock.staticmock、web.view.tiles2、orm.hibernate3/hibernate4，另外 Spring 5.0 不再支持 Portlet、Velocity、JasperReports、XMLBeans、JDO、Guava，除此之外，Spring 5.0 将许多废弃的类和方法删除了。因此，读者在生产实践中升级 Spring 5 需要关注以上这些代码的删除是否对已有的业务有影响，做出适合自己的升级方案。

4.1.3 核心修改

Spring 5.0 核心修改如下。

- 基于 Java 8 反射增强的实现高效的方法参数访问。
- 选择性地对 Spring 核心接口使用 Java 8 默认方法的声明。
- 尽可能避免使用 JDK 9 废弃的 API。
- 通过构造函数实现一致的实例化（修改后的异常处理）。
- 对核心 JDK 类的反射防御性使用。
- 使用"@Nullable"明确注解可以为空的参数、字段和返回值。
- 访问资源 Resource 类提供 getFile 和 isFile 防御式抽象。
- Resource 接口中提供基于 NIO 的 readableChannel 的访问器。
- 通过 NIO 2.0 流进行文件系统访问（不再使用 BIO FileInput/OutputStream）。
- Spring 5 框架自带了通用的日志组件。
- spring-jcl 替代了通用的日志。
- 无需任何额外桥接即可自动检测 Log4j 2.x、SLF4J、JUL（java.util.logging）。
- spring-core 附带 ASM 6.0。

4.1.4 核心容器更新

Spring 5.0 的核心容器更新如下。

- 支持@Nullable 注解。
- GenericApplicationContext/AnnotationConfigApplicationContext 支持函数式风格编程。
- 基于 Supplier 的 bean 注册 API，可以为 bean 定制回调。
- 在接口层面使用 CGLIB 动态代理时，提供事物、缓存、异步注解检测。
- XML 配置命名空间简化为无版本化的模式，始终使用最新的 xsd 文件，不支持已弃用的功能，指定版本的声明仍然支持，但针对最新架构进行了验证。
- 支持候选组件索引（作为类路径扫描的替代方案）。

4.1.5 Spring Web MVC 更新

Spring Web MVC 更新如下。

- Spring 5.0 中的 Filter 实现了 Servlet 3.1 签名支持。
- 支持 Spring MVC 控制器方法中使用 Servlet 4.0 PushBuilder 参数。
- 通过委托 MediaTypeFactory 统一支持常见媒体类型，取代了 Java Activation Framework。
- 更新了不可变对象的数据绑定（Kotlin/Lombok/@ConstructorPorties）
- 支持 JSON 绑定 API（使用 Eclipse Yasson 或 Apache Johnzon 代替 Jackson 和 GSON）。
- 支持 Jackson 2.9。
- 支持 Protobuf 3。
- 支持 Reactor 3.1 Flux、Mono 和 RxJava 1.3/RxJava 2.1 作为 Spring MVC 控制器方法的返回值，目标是使用新的反应式 WebClient 或 Spring MVC 控制器中的 Spring Data Reactive 存储库。
- 用新的 ParsingPathMatcher 代替 AntPathMatcher，得到更高效的解析和扩展语法。
- @ExceptionHandler 方法允许使用 RedirectAttributes 参数（以及 flash 属性）。
- 支持 ResponseStatusException 作为 "@ResponseStatus" 的代替方案。
- 通过使用 ScriptEngine 中的 eval(String，Bindings)方法直接呈现脚本，以及通过新的 RenderingContext 参数在 ScriptTemplateView 中使用 i18n 和嵌套模板，支持不需要实现 Invocable 的脚本引擎。
- Spring 的 FreeMarker 宏（spring.ftl）现在使用 HTML 输出格式（需要 FreeMarker 2.3.24+）。

4.1.6　Spring WebFlux

Spring 5.0 新增加了 Spring WebFlux 模块，其特性如下。

- 新的 spring-webflux 模块是一个基于 reactive 的代替 spring-webmvc 的模块，完全的异步非阻塞，旨在使用 event-loop 执行模型替代传统的大线程池，每个线程处理一个请求的模型。
- spring-core 相关的基础组件，比如 Encode 和 Decoder 可以用来编码和解码数据流；DataBuffer 可以使用 Java 中的 ByteBuffer 或者 Netty 中的 ByteBuf 作为数据缓冲区；ReactiveAdapterRegistry 可以对相关的库提供传输层支持。
- 在 spring-web 包里包含 HttpMessageReade 和 HttpMessageWrite，其委托给 Encoder 和 Decoder "@Controller" 基于注解的编程模型，类似于 Spring MVC，在 WebFlux 中支持在反应堆栈上运行，例如能够支持反应类型作为控制器方法参数，非阻塞 IO，并可以在其他的非 Servlet 容器（如 Netty 和 Undertow）上运行。
- 新的函数式编程模型 WebFlux.fn 作为 "@Controller" 基于注解编程模型的替代方案，使用端点路由 API 进行最小化和透明化，在相同的反应堆栈和 WebFlux 基础架构上运行。
- 新的 WebClient 具有用于 HTTP 调用的功能和响应式 API，与 RestTemplate 相当，但通过流畅的 API，并且在基于 WebFlux 基础架构的非阻塞和流式方案中也表现出色，在 Spring 5 中，不推荐使用 AsyncRestTemplate，而是推荐使用 WebClient。

4.1.7　对 Kotlin 的支持

Spring 5.0 对 Kotlin 的支持如下。

- 使用 Kotlin1.1.50 或更高版本时，可以支持 Null 安全的 API。

- 支持带有可选参数和默认值的 Kotlin 不可变类。
- 支持使用 Kotlin DSL 定义函数式 Bean。
- 支持在 WebFlux 中使用有路由功能的 Kotlin DSL。
- 利用 Kotlin reified 的类型参数来避免在各种 API（如 RestTemplate 或 WebFlux API）中明确指定用于序列化/反序列化的 Class。
- 对@autowired、@Inject、@RequestParam 和@RequestHeader 等注解的 Kotlin null 安全支持，以确定注入点或处理程序方法参数是否合法。
- ScriptTemplateView 中的 Kotlin 脚本支持 Spring MVC 和 Spring WebFlux。
- 支持带有可选参数的 Kotlin 自动装配构造函数。
- Kotlin 反射用于确定接口方法参数。

4.1.8 测试改进

Spring 5.0 测试改进如下。

- 在 Spring TestContext Framework 中完全支持 JUnit 5 Jupiter 编程和扩展模型。
- SpringExtension：是 JUnit Jupiter 的多个扩展 API 的实现，它为 Spring TestContext Framework 的现有功能集提供完全支持。通过@ExtendWith(SpringExtension.class)启用此支持。
- @SpringJUnitConfig：一个复合注释，它将来自 JUnit Jupiter 的 "@ExtendWith(SpringExtension.class)" 与来自 Spring TestContext Framework 的 "@ContextConfiguration" 相结合。
- @SpringJUnitWebConfig：一个复合注释，它将来自 JUnit Jupiter 的 "@ExtendWith(SpringExtension.class)" 与来自 Spring TestContext Framework 的 "@ContextConfiguration" 和 "@WebAppConfiguration" 相结合。
- @EnabledIf：如果提供的 SpEL 表达式或属性占位符的计算结果为 true，则表示已启用带注释的测试类或测试方法。
- @DisabledIf：如果提供的 SpEL 表达式或属性占位符的计算结果为 true，则表示禁用带注释的测试类或测试方法。
- 支持 Spring TestContext Framework 执行并行测试。
- Spring TestContext Framework 新增测试之前和测试之后的执行回调功能。
- TestExecutionListener API 和 TestContextManager 新增 beforeTestExecution()和 afterTestExecution() 回调。
- MockHttpServletRequest 现在具有用于访问请求体的方法 getContentAsByteArray()和 getContentAsString()。
- 如果在模拟请求中设置了字符编码，则 Spring MVC Test 中的 print()和 log()方法现在会打印请求主体。
- Spring MVC Test 中的 redirectedUrl()和 forwardedUrl()方法现在支持具有可变参数扩展的 URI 模板。
- XMLUnit 支持升级到 XMLUnit 2.3。

4.2 Spring 5.1 新特性

4.2.1 核心修改

Spring 5.1 核心修改如下。

- 在类路径和模块路径上对 JDK 11 的无警告支持。
- 支持 Graal 原生图像约束。
- Reactor Core 升级到 Reactor Core 3.2 和 Reactor Netty 升级到 Reactor Netty 0.8（Reactor Californium）。
- ASM 升级到 ASM 7.0 和 CGLIB 升级到 CGLIB 3.2.8。
- 在 FileSystemResource 中提供 NIO 2.0 路径支持（取代 PathResource）。
- 核心类型和注释解析的性能改进。
- 可以通过标准的 Commons Logging 检测 Spring 的 JCL 桥。

4.2.2 核心容器更新

Spring 5.1 核心容器更新如下。

- 支持 "@Profile" 条件中的逻辑和/或表达式。
- 嵌套配置类的一致性检测。
- 优化 Kotlin bean 的 DSL，同一类型的多个 bean 的唯一隐式 bean 名称。
- 在 BeanFactory API 中统一地不暴露任何空的 bean。
- 通过 BeanFactory API 进行编程式的 ObjectProvider 检索。
- ObjectProvider 提供可迭代/流式访问。
- 支持在单个构造函数场景中的空集合/映射/数组注入。

4.2.3 Web 修改

Spring 5.1 中的 Web 修改如下。

- 在接口上也可以检测到控制器参数注释。
- 支持在 UriComponentsBuilder 中使用更严格的 URI 变量编码。
- spring-web 模块提供 FormContentFilter 拦截 HTTP 中的 PUT、PATCH 和 DELETE 请求。

4.2.4 Spring Web MVC 更新

Spring 5.1 中 Spring Web MVC 更新如下。

- 改进后提供更加人性化和紧凑的 DEBUG 和 TRACE 日志，通过 DispatcherServlet 中的 enableLoggingRequestDetails 属性控制潜在敏感数据的 DEBUG 记录。
- 更新了 Web 区域表示。CookieLocaleResolver 将发送符合 RFC6265 标准时区的 cookie。

- 对缺少请求头、cookie 和路径等异常定制了 MVC 异常，允许对异常进行区分和对状态代码进行区分。
- 通过 ForwardedHeaderFilter 集中处理"转发"类型头部。
- 除了 GZip 之外，还支持 Brotli 预编码静态资源。

4.2.5　Spring WebFlux 更新

Spring WebFlux 更新如下。

- 使用 Reactor Netty 0.8 运行时服务器端支持 HTTP/2。
- 改进后更加人性化和紧凑的 DEBUG 和 TRACE 日志。
- HTTP 请求和 WebSocket 会话的相关日志记录。
- 控制潜在敏感数据的 DEBUG 记录。通过 CodecConfigurer 的 defaultCodecs 属性控制。
- 会话 cookie 已具有 SameSite=Lax 功能，可以防止 CSRF 攻击。
- 支持 Protobuf 序列化，包括消息流。
- 支持 Jetty 响应式 HTTP 客户端的 WebClient 连接器。
- 支持 WebSocketSession 属性设置。
- 改进有关反应式 WebSocket API 文档。

4.2.6　Spring Messaging 更新

Spring Messaging 更新如下。

- 在"@MessageMapping"方法中支持响应式客户端，并支持 Reactor 和 RxJava 返回值的开箱即用。
- 提供选项以保留 STOMP 代理的消息发布顺序。
- "@SendTo"和"@SendToUser"都可以用于控制器方法。
- 改进了有关处理消息和消息订阅的文档。

4.2.7　Spring ORM 更新

Spring ORM 更新如下。

- 支持 Hibernate ORM 5.3：Bean 容器与 Hibernate 的新 SPI 集成。
- LocalSessionFactoryBean 和 HibernateTransactionManager 支持 JPA 交互，在同一事务中允许原生 Hibernate 和 JPA 共同访问。
- 只读事务不再在内存中保留 Hibernate 实体快照。

4.2.8　测试更新

- WebTestClient 中的 Hamcrest 和 XML 断言更新。
- 可以使用固定的 WebSession 配置 MockServerWebExchange。

第 5 章

Java 8 新特性概述

由于 Spring 5 是基于 Java 8 开发的，Spring 5 中使用了很多 Java 8 新特性。因此本书在介绍 Spring 5 应用之前，首先介绍 Java 8 的新特性。

5.1 Lambda 表达式

5.1.1 Lambda 表达式初探

Lambda 表达式，也可称为闭包，是 Java 8 最重要的新特性之一。Lambda 允许把函数作为一个方法的参数使用。使用 Lambda 表达式可以使代码变得更加简洁紧凑。其实 Lambda 表达式的本质只是一个"语法糖"，由编译器推断并转换为常规的代码，因此可以使用更少的代码来实现同样的功能。

Lambda 表达式的语法格式如下：

```
(parameters) -> expression
(parameters) ->{ statements; }
```

Lambda 表达式的一些重要特征如下。

- 可选类型声明：不需要声明参数类型，编译器可以统一识别参数值。
- 可选的参数圆括号：当只有一个参数时，无须定义圆括号，但如果有多个参数则需要定义圆括号。
- 可选的大括号：如果主体包含了一个语句，就不需要使用大括号。
- 可选的返回关键字：如果主体只有一个表达式返回值则编译器会自动返回值。

下面是 Lambda 表达式常见的书写方式：

```
// 不需要参数,返回值为 5
() -> 5
// 接收一个参数(数值类型),返回其 2 倍的值
x -> 2 * x
// 接受两个参数(数值类型),并返回两者的差值
(x, y) -> x - y
// 接收两个 int 型参数,返回两者的和
(int x, int y) -> x + y
// 接受一个 string 对象,并在控制台打印,不返回任何值
(String s) -> System.out.print(s)
```

下面将通过案例展示 Lambda 表达式的使用:

```
/**
 * @Author zhouguanya
 * @Date 2018/10/11
 * @Description LambdaTest
 */
public class Java8LambdaDemo {

    public static void main(String[] args){
        Java8LambdaDemo java8LambdaDemo = new Java8LambdaDemo();
        // 类型声明
        Calculator addition = (int a, int b) -> a + b;
        // 不用类型声明
        Calculator subtraction = (a, b) -> a - b;
        // 大括号中的返回语句
        Calculator multiplication = (int a, int b) -> { return a * b; };
        // 没有大括号及返回语句
        Calculator division = (int a, int b) -> a / b;
        // 测试用例
        System.out.println("100 + 50 = " + java8LambdaDemo.operate(100, 50, addition));
        System.out.println("100 - 50 = " + java8LambdaDemo.operate(100, 50, subtraction));
        System.out.println("100 x 50 = " + java8LambdaDemo.operate(100, 50, multiplication));
        System.out.println("100 / 50 = " + java8LambdaDemo.operate(100, 50, division));
    }

    /**
     * 算术运算接口
     */
    interface Calculator {
        /**
         * 数学运算操作
```

```
     * @param a 第一个操作数
     * @param b 第二个操作数
     * @return int
     */
    int calculate(int a, int b);
}

/**
 * 操作方法
 * @param a
 * @param b
 * @param calculator MathOperation 对象
 * @return int
 */
private int operate(int a, int b, Calculator calculator){
    return calculator.calculate(a, b);
}
}
```

运行测试代码得到如下测试结果：

```
100 + 50 = 150
100 - 50 = 50
100 x 50 = 5000
100 / 50 = 2
```

5.1.2 Lambda 表达式作用域

可以直接在 Lambda 表达式中访问外层的局部变量，但在 Lambda 表达式内部不能修改定义在 Lambda 表达式外部的局部变量,否则会编译错误。Lambda 表达式的局部变量可以不用声明为 final，但是必须不可被后面的代码修改（即隐性的具有 final 的语义）。在 Lambda 表达式当中不允许声明一个与外部变量同名的参数或者局部变量。

```
/**
 * @Author zhouguanya
 * @Date 2018/10/11
 * @Description Lambda 作用域测试
 */
public class Java8LambdaScopeDemo {

    public static void main(String[] args) {
        final String salutation = "Hello ";
        String myName = "I am Lambda ~";
        String today = "2018/10/11";
        SayHello greetingService = message -> {
            System.out.println(salutation + message + myName);
```

```
            //此处修改 today 将会出现编译错误
            //today = "2018/10/12";
            //此处定义局部变量 myName 将会出现编译错误
            //String myName = "Java";
        };
        greetingService.say("World！");
    }

    /**
     * 打招呼接口
     */
    interface SayHello {
        /**
         * 打招呼方法
         * @param message
         */
        void say(String message);
    }
}
```

如果将 today = "2018/10/12"这一行代码取消注释，将会出现 Variable used in lambda expression should be final or effectively final 编译错误。如果将 String myName = "Java"这一行代码取消注释，也会出现编译错误。

执行以上代码，得到执行结果如下：

```
Hello World！
```

5.1.3 在线程中使用 Lambda 表达式

下面通过案例代码将不使用 Lambda 表达式和使用 Lambda 表达式时的简洁代码进行对比：

```
/**
 * @Author zhouguanya
 * @Date 2018/10/12
 * @Description 测试 Lambda 表达式配合线程使用
 */
public class Java8LambdaInThreadDemo {

    public static void main(String[] args) throws Exception {
        // 不使用 Lambda 表达式，使用匿名类
        // 或者定义一个类实现 Runnable 接口
        new Thread(new Runnable() {
            @Override
            public void run () {
                System.out.println("线程1");
```

```
            }
        }).start();

        // 使用 lambda 表达式
        new Thread(() ->System.out.println("线程2")).start();

    }
}
```

执行测试代码，得到执行结果如下：

```
线程1
线程2
```

5.1.4　在集合中使用 Lambda 表达式

经常需要在集合中对集合中的元素进行排序。下面使用一个集合元素排序的案例对 Lambda 表达式的使用进行阐释。首先定义一个 Person 类，其中包含两个属性，姓名 name 和年龄 age。

```
/**
 * @Author zhouguanya
 * @Date 2018/10/12
 * @Description
 */
public class Person {
    /**
     * 姓名
     */
    public String name;
    /**
     * 年龄
     */
    public int age;

    /**
     * 构造器
     */
    public Person(String name, int age) {
        this.name = name;
        this.age = age;
    }

    @Override
    public String toString() {
        return this.name + ":" + this.age;
    }
}
```

在测试代码中，对不使用 Lambda 表达式和使用 Lambda 表达式的场景进行对比。

```java
/**
 * @Author zhouguanya
 * @Date 2018/10/12
 * @Description 集合元素排序
 */
public class Java8LambdaInCollectionDemo {

    public static void main(String[] args) {
        List<Person> personList = new ArrayList<>();
        Person personLi = new Person("李四", 22);
        Person personZhang = new Person("张三",20);
        Person personWang = new Person("王五", 26);
        personList.add(personLi);
        personList.add(personZhang);
        personList.add(personWang);
        //按年龄从小到大排序
        Collections.sort(personList, new Comparator<Person>() {
            @Override
            public int compare(Person o1, Person o2) {
                return o1.age - o2.age;
            }
        });
        System.out.println(personList);
        //使用 Lambda 表达式，按年龄从大到小排序
        Collections.sort(personList, (o1, o2) -> o2.age - o1.age);
        System.out.println(personList);
        //使用 Lambda 表达式，按年龄从小到大排序
        Collections.sort(personList, Comparator.comparingInt(o -> o.age));
        System.out.println(personList);
    }
}
```

执行测试代码，得到如下测试结果：

```
[张三:20, 李四:22, 王五:26]
[王五:26, 李四:22, 张三:20]
[张三:20, 李四:22, 王五:26]
```

5.1.5　在 Stream 中使用 Lambda 表达式

Stream 是对集合的包装，通常和 Lambda 表达式一起使用。使用 Lambda 表达式可以支持许多操作，如 map、filter、limit、sorted、count、min、max、sum 和 collect 等等。在接下来的案例中，将使用 Lambda 表达式和 Stream 对 Person 集合进行排序。

```java
/**
 * @Author zhouguanya
 * @Date 2018/10/12
 * @Description Stream 中使用 Lambda
 */
public class Java8LambdaInStreamDemo {
    public static void main(String[] args) {
        List<Person> personList = new ArrayList<Person>() {
            {
                add(new Person("张三", 24));
                add(new Person("李四", 32));
                add(new Person("王五", 28));
                add(new Person("赵六", 26));
                add(new Person("赵七", 30));
            }
        };
        //使用 stream 和 Lambda 对 personList 进行排序
        List<Person> sortedList = personList.stream()
                .sorted(Comparator.comparingInt(p -> p.age))
                .limit(5).collect(Collectors.toList());
        System.out.println(sortedList);
    }
}
```

在测试案例中，将 personList 对象转换为 Stream 对象，并配合 Lambda 表达式对其进行排序。执行代码，可得到如下测试结果：

[张三:24, 赵六:26, 王五:28, 赵七:30, 李四:32]

5.2　接口默认方法

在 Java 8 之前，interface 之中可以定义变量和方法，接口中的变量必须是被 public static final 修饰的，接口中的方法必须是被 public abstract 修饰的。由于这些修饰符都是默认的，所以在 Java 8 之前，以下的写法都是等价的：

```java
public interface Jdk8PreInterface {
    // field1 和 field2 都是 public static final 修饰的
    public static final int field1 = 0;
    int field2 = 0;
    // method1 和 method2 都是 public static final 修饰的
    public abstract void method1(int a);
    void method2(int a);
}
```

在 Java 8 之前的版本中，接口是一柄双刃剑，优点是接口是面向抽象而不是面向具体编程的；缺陷是当需要修改接口时，需要修改全部实现该接口的类，修改成本高。

Java 8 及以上版本中，Java 允许在接口中定义 static 方法和 default 方法。Java 8 通过默认方法解决了这个旧接口升级带来的成本过高的问题，在 Java 8 接口中可以添加新的方法，却不会破坏已有的接口实现，这个特性为旧接口升级提供了兼容性。

下面通过一个简单的案例阐述抽象方法的使用，案例中定义了一个 Vehicle 接口，其中包含一个抽象方法 drive()方法和默认方法 print()，接口如下：

```java
/**
 * @Author: zhouguanya
 * @Date: 2018/12/23
 * @Description:
 */
public interface Vehicle {
    /**
     * 默认方法
     */
    default void print() {
        System.out.println("我是一辆车");
    }

    /**
     * 抽象方法
     */
    void drive();
}
```

Car 实现了 Vehicle 接口：

```java
/**
 * @Author: zhouguanya
 * @Date: 2018/12/23
 * @Description: Vehicle 接口实现类 Car
 */
public class Car implements Vehicle {
    /**
     * 抽象方法
     */
    @Override
    public void drive() {
        System.out.println("开一辆轿车");
    }
}
```

下面测试类 DefaultMethodDemo 中，通过 Car 对象分别调用两个方法：

```java
/**
 * @Author: zhouguanya
 * @Date: 2018/12/23
 * @Description:
 */
public class DefaultMethodDemo {
    public static void main(String[] args) {
        Vehicle car = new Car();
        // 调用 Vehicle 接口默认方法
        car.print();
        // 调用 Car 中重写的 drive 方法
        car.drive();
    }
}
```

当两个接口中有两个相同的默认方法时,子类如果同时实现这两个接口,将会出现编译错误,需要在子类中重写默认方法。

Java 8 的接口除了可以声明默认方法,还可以声明并且实现静态方法。

在下面的案例代码中创建了 Whistle 接口并声明默认方法 print() 和静态方法 horn(),Whistle 代码如下:

```java
/**
 * @Author: zhouguanya
 * @Date: 2018/12/23
 * @Description:
 */
public interface Whistle {
    /**
     * 默认方法
     */
    default void print() {
        System.out.println("我要鸣笛");
    }

    /**
     * 静态方法
     */
    static void horn(){
        System.out.println("按喇叭~");
    }
}
```

创建 Bus 类实现 Vehicle 接口和 Whistle 接口,Bus 代码如下:

```java
/**
 * @Author: zhouguanya
```

```
 * @Date: 2018/12/23
 * @Description:
 */
public class Bus implements Vehicle, Whistle {
    /**
     * 同时实现Vehicle、Whistle接口,需要重写默认方法
     */
    public void print() {
        System.out.println("我是一辆巴士");
    }

    /**
     * 实现抽象方法
     */
    @Override
    public void drive() {
        System.out.println("开一辆巴士");
    }
}
```

测试代码中创建 Bus 对象,并调用各个方法:

```
Bus bus = new Bus();
bus.print();
bus.drive();
Whistle.horn();
```

执行测试代码,得到如下输出:

```
我是一辆巴士
开一辆巴士
按喇叭~
```

5.3 小　　结

本章主要介绍了 Java 8 重要的新特性,我们在使用 Spring 5 编程和创建项目都会用到,希望大家了解和掌握。

第 6 章

Spring WebFlux 响应式编程

Spring WebFlux 是随 Spring 5 推出的响应式 Web 框架，本章将详细介绍 WebFlux 的功能和使用。

6.1 传统的编程模型

传统的编程模型采用的是每条指令依次执行的方式，如果上一条指令没有执行完，当前线程将先等待，无论如何提升机器性能或者优化代码，都不能改变要得到响应结果需要等待的本质，即便是使用 Java 多线程编程，每个线程也是按照代码编写的先后顺序执行的。

```java
/**
 * @Author zhouguanya
 * @Date 2018/10/10
 * @Description 传统编程模型，暂不考虑重排序
 */
public class Test {
    public static void main(String[] args) {
        int a = 1;
        int b = 2;
        //c和d有依赖关系
        int c = a + b;
        //如果c没有执行完，d就不能执行
        //对应到企业开发场景中，如果c是一个远程调用，d是对远程调用结果进行分析
        //那么d只能等待c的结果，造成d后的程序都必须同步等待
        int d = c / 3;
        System.out.println(c);
```

```
    System.out.println(d);
  }
}
```

在本例中,对 c 进行计算后,得到 c 的结果才能继续执行对 d 的计算,如果 c 长时间得不到结果(比如 c 是一个 HTTP/RPC 请求的响应结果),那么 d 就会被阻塞。

在执行程序时,为了提供性能,处理器和编译器常常会对指令进行重排序。重排序分为编译器重排序和处理器重排序两种。

- 编译器重排序:编译器保证不改变单线程执行结果的前提下,可以调整多线程语句执行顺序。
- 处理器重排序:如果不存在数据依赖性,处理器可以改变语句对应机器指令的执行顺序。

若要实现快速的响应,就得把程序执行指令的方式换一换——将同步方式改成异步方式,方法执行改成消息发送,因此诞生了响应式编程模型。

6.2 响应式编程模型

响应式编程(Reactive Programming)就是与异步数据流交互的一种编程范式。就 6.1 节案例而言,在传统编程模型中,执行完 int c = a + b;这一行代码后,c 的值等于 3,如果后续修改 a = 2,b = 3,c 是感知不到 a、b 的变化的,因此 c 仍然等于 3。但在响应式编程模型中,c 是可以感知到 a、b 的变化的,因此对修改 a = 2,b = 3 后,c 的值变成 5。

常用的 Excel 表格就是一个响应式编程的例子,假设为单元格添加了类似 "=B1+C1" 的公式,那么当前单元格的值会随着 B1 和 C1 单元格中值的变化而变化。

在企业开发场景中,可以将所有的业务场景都以数据流的形式进行建模——普通的内存计算、数据库操作或者远程调用等。这种数据流动可以归纳为以下形式。

```
Command → CommandHandler → Event → EventHandler → ......
```

根据 CQRS(Command Query Responsibility Segregation,命令查询的责任分离)模式的思想,任何业务都可以分解为两种基本的消息形式,Query 和 Command。Query 模型相对比较简单,其本质上是一个没有副作用的制度操作,Command 模型是状态变更的一种封装,开发人员可以使用事件记录每次状态变更。熟悉 git 的读者会发现,git 的版本管理与此建模思想不谋而合。无论是新增、修改还是删除代码,都可以视为是一次全新的提交。

当将编程范式切换到"流(Stream)"时,普通的数据流编程范式并不能满足"响应式 Reactive"的定义。想要实现迅速响应,如何才能做到?那就是要做到没有阻塞,这就是通常所说的异步工作方式。

响应式编程的设计原则如下。

- 保持数据的不变性。
- 没有共享。
- 阻塞是有害的。

6.3 Reactor

Reactor 是第四代 Reactive 库，基于 Reactive Streams 规范在 JVM 上构建非阻塞应用程序。Reactor 侧重于服务器端响应式编程，是一个基于 Java 8 实现的响应式流规范（Reactive Streams specification）响应式库。

作为 Reactive Engine/SPI，Reactor Core 和 IO 模块都为重点使用场景提供了响应流构造，最终与 Spring、RxJava、Akka Streams 和 Ratpack 等框架结合使用，作为 Reactive API，Reactor 框架模块提供了丰富的消费功能，如组合和发布订阅事件。

本节对 Reactor 的介绍以基本的概念和简单使用为主，更多 Reactor 高级特性可参考 Reactor 官网：http://projectreactor.io/。

6.3.1 Flux 与 Mono

在 Reactor 中，数据流发布者（Publisher）由 Flux 和 Mono 两个类表示，它们都提供了丰富的操作符（operator）。一个 Flux 对象代表一个包含 0 个或多个（0..N）元素的响应式序列，而一个 Mono 对象代表一个包含 0 或一个（0..1）元素的结果。

作为数据流发布者，Flux 和 Mono 都可以发出三种数据信号，元素值、错误信号和完成信号。错误信号和完成信号都是终止信号。完成信号用来告知下游订阅者，数据流是正常结束的。错误信号在终止数据流的同时将错误信息传递给下游订阅者。这三种信号不是一定要完全具备的。

图 6-1 所示是一个 Flux 类型的数据流，横坐标是时间轴，⑥后的黑色竖线是完成信号。连续发出 1~6 共 6 个元素值，以及一个完成信号，完成信号告知订阅者数据流已经结束。

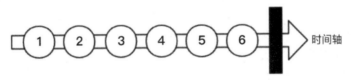

图 6-1　Flux 类型的数据流图

图 6-2 是一个 Mono 类型的数据流，其发出一个元素值后，立刻发出一个完成信号。

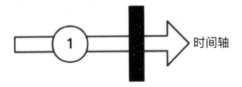

图 6-2　Mono 类型的数据流图

下面通过案例分析 Reactor 的使用。

首先创建一个 maven 项目，然后在 pom.xml 中加入对 maven 的依赖。可以到 maven 仓库 https://mvnrepository.com/ 查询最新版本的 Reactor。截止本书出版，Reactor 最新的版本是 3.2.0.RELEASE。

```xml
<dependency>
    <groupId>io.projectreactor</groupId>
    <artifactId>reactor-core</artifactId>
    <version>3.2.0.RELEASE</version>
</dependency>
```

为了方便测试，还需要添加对 reactor-test 的依赖和 Junit 的依赖。

```xml
<dependency>
    <groupId>io.projectreactor</groupId>
    <artifactId>reactor-test</artifactId>
    <version>3.2.0.RELEASE</version>
    <scope>test</scope>
</dependency>
<dependency>
  <groupId>junit</groupId>
  <artifactId>junit</artifactId>
  <version>4.12</version>
  <scope>test</scope>
</dependency>
```

下面就可以开始用 Reactor 进行编码了。

首先使用代码声明图 6-1 和图 6-2 中的 Flux 和 Mono，代码如下：

```java
Flux.just(1, 2, 3, 4, 5, 6);
Mono.just(1);
```

Flux 和 Mono 提供了多种创建数据流的方法，just 是一种比较直接的声明数据流的方式，其参数就是数据元素。

对于图 6-1 中的场景，还可以使用如下多种声明方式。

基于数组的声明方式：

```java
Integer[] array = new Integer[]{1,2,3,4,5,6};
Flux.fromArray(array);
```

基于集合的声明方式：

```java
List<Integer> list = Arrays.asList(array);
Flux.fromIterable(list);
```

基于 Stream 的声明方式：

```java
Stream<Integer> stream = list.stream();
Flux.fromStream(stream);
```

上文中提到元素值、错误信号和完成信号三者并不是要完全具备的，下面就给出几种情况：

```java
// 只有完成信号的空数据流
Flux.just();
Flux.empty();
```

```
Mono.empty();
Mono.justOrEmpty(Optional.empty());
// 只有错误信号的数据流
Flux.error(new Exception("some error"));
Mono.error(new Exception("some error"));
```

6.3.2 subscribe()

subscribe()方法表示对数据流的订阅动作,subscribe()方法有多个重载的方法,下面介绍几种常见的subscribe()方法:

```
// 订阅并触发数据流
public final Disposable subscribe();
// 订阅并对正常数据元素进行处理
public final Disposable subscribe(Consumer<? super T> consumer);
// 订阅并分别对正常数据元素和异常信号进行处理
public final Disposable subscribe(@Nullable Consumer<? super T> consumer,
Consumer<? super Throwable> errorConsumer);
// 订阅并分别对正常数据元素、错误信号和完成信号进行处理
public final Disposable subscribe(
@Nullable Consumer<? super T> consumer,
@Nullable Consumer<? super Throwable> errorConsumer,
@Nullable Runnable completeConsumer);
// 订阅并分别对正常数据元素、错误信号、完成信号和订阅发生时进行处理
public final Disposable subscribe(
@Nullable Consumer<? super T> consumer,
@Nullable Consumer<? super Throwable> errorConsumer,
@Nullable Runnable completeConsumer,
@Nullable Consumer<? super Subscription> subscriptionConsumer);
```

下面通过一个案例验证 Flux、Mono 和几种常见的 subscribe()方法的使用。案例代码如下:

```
/**
 * @Author zhouguanya
 * @Date 2018/10/22
 * @Description 第一个 Reactor 程序
 */
public class FirstReactorDemo {
public static void main(String[] args) {
// 测试 Flux
Flux.just(1, 2, 3, 4, 5, 6).subscribe(System.out::print);
System.out.println("\n--------------------------");
// 测试 Mono
Mono.just(1).subscribe(System.out::println);
System.out.println("--------------------------");
// 测试两个参数的 subscribe 方法
Flux.just(1, 2, 3, 4, 5, 6)
```

```java
.subscribe(System.out::print, System.err::println);
System.out.println("\n---------------------------");
// 测试三个参数的 subscribe 方法
Flux.just(1, 2, 3, 4, 5, 6)
    .subscribe(System.out::print, System.err::println,
        () -> System.out.println("\ncomplete"));
System.out.println("---------------------------");
// 测试四个参数的 subscribe 方法
Flux.just(1, 2, 3, 4, 5, 6)
    .subscribe(System.out::print, System.err::println,
        () -> System.out.println("\ncomplete"), subscription -> {
            System.out.println("订阅发生了");
            subscription.request(10);
        });
    }
}
```

运行案例代码，得到如下运行结果：

```
123456
---------------------------
1
---------------------------
123456
---------------------------
123456
complete
---------------------------
订阅发生了
123456
complete
```

在命令式或同步式编程世界中，调试通常都是非常直观的——直接看 stack trace 就可以找到问题出现的位置以及异常信息等。

当切换到响应式的异步代码，事情就变得复杂多了。先了解一个基本的单元测试工具——StepVerifier。当测试关注点是每个数据元素的时候，就与 StepVerifier 的使用场景非常贴切。例如期望的数据或信号是什么，是否使用 Flux 发出某个特殊值，接下来 100ms 做什么，这些场景都可以使用 StepVerifier API 表示。

下面分别使用 StepVerifier 测试 Flux 和 Mono，测试代码如下：

```java
/**
 * @Author zhouguanya
 * @Date 2018/10/25
 * @Description StepVerifier 测试案例
 */
public class StepVerifierDemo {
```

```java
public static void main(String[] args) {
    Flux flux = Flux.just(1, 2, 3, 4, 5, 6);
    // 使用StepVerifier测试Flux，应该正常
    StepVerifier.create(flux)
    //测试下一个期望的数据元素
    .expectNext(1, 2, 3, 4, 5, 6)
    //测试下一个元素是否为完成信号
    .expectComplete()
    .verify();

    Mono mono = Mono.error(new Exception("some error"));
    // 使用StepVerifier测试Mono，应该会出现异常
    StepVerifier.create(mono)
    //测试下一个元素是否为完成信号
    .expectComplete()
    .verify();
}
```

运行测试代码，测试结果如下：

```
Exception in thread "main" java.lang.AssertionError: expectation "expectComplete" fail
    ed (expected: onComplete(); actual: onError(java.lang.Exception: some error))
    at reactor.test.ErrorFormatter.assertionError(ErrorFormatter.java:105)
    at reactor.test.ErrorFormatter.failPrefix(ErrorFormatter.java:94)
    at reactor.test.ErrorFormatter.fail(ErrorFormatter.java:64)
    at reactor.test.ErrorFormatter.failOptional(ErrorFormatter.java:79)
    at reactor.test.DefaultStepVerifierBuilder.lambda$expectComplete$4(DefaultStepVerifierBuilder.java:322)
```

6.3.3 操作符（Operator）

本节介绍 Reactor 一些常用的操作符。

1. map

map 可以将数据元素转换成映射表，得到一个新的元素。map 操作符示意图如图 6-3 所示。

图中上方的箭头是原始序列的时间轴，下方的箭头是经过 map 处理后的数据序列时间轴。map 接受一个 Function 函数式接口，该接口用于定义转换操作的策略：

```java
public final <V> Flux<V> map(Function<? super T,? extends V> mapper)
public final <R> Mono<R> map(Function<? super T, ? extends R> mapper)
```

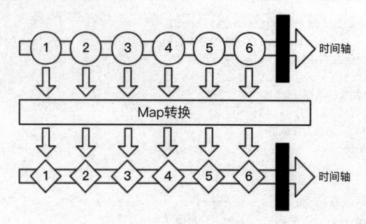

图 6-3 map 操作示意图

下面使用案例阐述 map 操作符的用法。案例代码如下：

```java
/**
 * @Author zhouguanya
 * @Date 2018/10/30
 * @Description map 操作符测试
 */
public class MapOperatorDemo {
    public static void main(String[] args) {
        // 生成从1开始，步长为1的6个整型数据
        StepVerifier.create(Flux.range(1, 6)
                // 将元素进行立方操作
                .map(i -> i * i * i))
                // 期望值
                .expectNext(1, 8, 27, 64, 125, 216)
                // 异常情况模拟
                //.expectNext(10, 8, 27, 64, 125, 216)
                // 完成信号
                .expectComplete()
                .verify();
    }
}
```

执行案例代码发现控制台无异常输出。如果修改立方后的数据为 expectNext(10, 8, 27, 64, 125, 216)将会出现如下异常：

```
Exception in thread "main" java.lang.AssertionError: expectation
"expectNext(10)" failed (expected value: 10; actual value: 1)
```

2. flatMap

flatMap 操作可以将每个数据元素转换/映射为各个流，然后将每个流合并为一个大的数据流。flatMap 操作符示意图如图 6-4 所示。

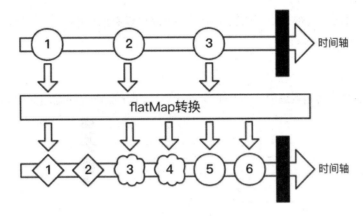

图 6-4 flatMap 操作示意图

flatMap 接收一个 Function 函数式接口为参数,这个函数式的输入为一个 T 类型数据值,输出可以是 Flux 或 Mono:

```
public final <R> Flux<R>
    flatMap(Function<? super T, ? extends Publisher<? extends R>> mapper)
public final <R> Mono<R>
    flatMap(Function<? super T, ? extends Mono<? extends R>> transformer)
```

下面使用案例阐述 flatMap 操作符的用法。案例代码如下:

```
/**
 * @Author zhouguanya
 * @Date 2018/10/31
 * @Description flatMap 操作符测试
 */
public class FlatMapOperationDemo {
    public static void main(String[] args) {
        StepVerifier.create(
        Flux.just("flux", "mono")
            // 将每个字符串拆分为包含一个字符串的字节流
            .flatMap(s -> Flux.fromArray(s.split("\\s*")))
            //对每个元素延迟 100ms
            .delayElements(Duration.ofMillis(1000)))
            // 对每个元素进行打印,doOnNext 不会消费数据流
            .doOnNext(System.out::print))
            //验证是否发出了 8 个元素
            .expectNextCount(8)
            .verifyComplete();
    }
}
```

执行案例代码,得到类似如下结果:

```
fmlounxo
```

多次执行案例代码，会得到不同的输出结果。由此可以看出，流的合并是异步的，先来先到，并非是严格按照原始序列的顺序。

3. filter

filter 操作可以对数据元素过滤，得到剩余的元素。filter 操作符示意图如图 6-5 所示。

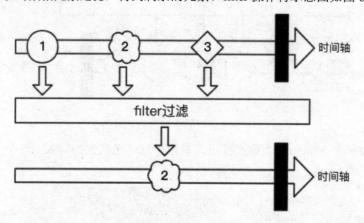

图 6-5　filter 操作示意图

filter 接受一个 Predicate 的函数式接口为参数，这个函数式接口的作用是进行判断并返回 boolean 值：

```
public final Flux<T> filter(Predicate<? super T> tester)
public final Mono<T> filter(Predicate<? super T> tester)
```

下面使用案例阐述 filter 操作符的用法。案例代码如下：

```
/**
 * @Author zhouguanya
 * @Date 2018/10/31
 * @Description filter 操作符测试
 */
public class FilterOperationDemo {
    public static void main(String[] args) {
        StepVerifier.create(Flux.range(1, 6)
            // 过滤奇数
            .filter(i -> i % 2 == 1)
            // 过滤后的元素进行立方操作
            .map(i -> i * i * i))
            // 期望的结果
            .expectNext(1, 27, 125)
            // 异常情况模拟
            //.expectNext(1, 127, 125)
            .verifyComplete();
    }
}
```

执行案例代码发现控制台无异常输出。如果修改立方后的数据为 expectNext(1, 127, 125)将会出现如下异常：

```
Exception in thread "main" java.lang.AssertionError: expectation
"expectNext(127)" fai
    led (expected value: 127; actual value: 27)
    at reactor.test.ErrorFormatter.assertionError(ErrorFormatter.java:105)
    at reactor.test.ErrorFormatter.failPrefix(ErrorFormatter.java:94)
    at reactor.test.ErrorFormatter.fail(ErrorFormatter.java:64)
    at reactor.test.ErrorFormatter.failOptional(ErrorFormatter.java:79)
```

4. zip

zip 能够将多个流一对一的合并起来。zip 有多个方法变体，这里只介绍一个最常见的二合一的场景。zip 操作符示意图如图 6-6 所示。

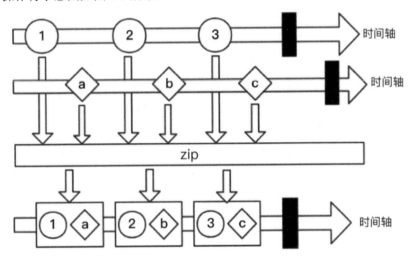

图 6-6　zip 操作示意图

zip 可以从两个 Flux/Mono 流中，每次各取一个元素，组成一个二元组：

```
public static <T1,T2> Flux<Tuple2<T1,T2>>
    zip(Publisher<? extends T1> source1, Publisher<? extends T2> source2)
public static <T1, T2> Mono<Tuple2<T1, T2>>
    zip(Mono<? extends T1> p1, Mono<? extends T2> p2)
```

下面使用案例阐述 zip 操作符的用法。案例代码如下：

```
/**
 * @Author zhouguanya
 * @Date 2018/10/31
 * @Description zip 操作符测试
 */
public class ZipOperationDemo {
    public static void main(String[] args) {
```

```
String desc = "I am Reactor";
StepVerifier.create(
    // 将字符串拆分为一个一个的单词并以每 500ms/个的速度发出
    Flux.zip(Flux.fromArray(desc.split("\\s+"))
        , Flux.interval(Duration.ofMillis(500)))
    // 打印
    .doOnNext(System.out::print))
    // 验证发出 3 个元素
    .expectNextCount(3)
    .verifyComplete();
    }
}
```

执行案例代码，得到如下结果：

```
[I,0][am,1][Reactor,2]
```

5. 更多

（1）除了以上几个常见的操作符意外，Reactor 中提供了非常丰富的操作符。

（2）用于编程方式自定义生成数据流的 create 和 generate 等及其变体方法。

（3）用于"无副作用的 peek"场景的 doOnNext、doOnError、doOncomplete、doOnSubscribe、doOnCancel 等及其变体方法。

（4）用于数据流转换的 when、and/or、merge、concat、collect、count、repeat 等及其变体方法。

（5）用于过滤/拣选的 take、first、last、sample、skip、limitRequest 等及其变体方法。

（6）用于错误处理的 timeout、onErrorReturn、onErrorResume、doFinally、retryWhen 等及其变体方法。

（7）用于分批的 window、buffer、group 等及其变体方法。

（8）用于线程调度的 publishOn 和 subscribeOn 方法。

更多操作请见官方文档：https://projectreactor.io/docs/core/release/api/reactor/core/publisher/Flux.html

6.3.4 线程模型

JDK 提供的多线程工具类 Executors 提供了多种线程池，使开发人员可以方便地定义线程池进行多线程开发。Reactor 使多线程编程更加容易，Schedulers 类提供的静态方法可以更快创建以下几种多线程环境。

- 获取当前线程环境 Schedulers.immediate()。
- 获取可重用的单线程环境 Schedulers.single()。
- 获取弹性线程池环境 Schedulers.elastic()。
- 获取固定大小线程池环境 Schedulers.parallel()。
- 获取自定义线程池环境 Schedulers.fromExecutorService(ExecutorService)。

下面通过案例对比单线程同步阻塞和使用 Schedulers 异步非阻塞的场景。案例中分别有两个方法，hello()方法同步阻塞 2s 后，返回字符串"Hello, Reactor!"，helloAsync()方法使用 Schedulers 改进为异步非阻塞方式。案例代码如下：

```java
/**
 * @Author zhouguanya
 * @Date 2018/11/2
 * @Description Schedulers demo
 */
public class SchedulerOperationDemo {

    public static void main(String[] args) throws InterruptedException {
        // 同步阻塞场景
        System.out.println(hello());
        System.out.println("--------同步阻塞场景执行结束--------");
        helloAsync();
        System.out.println("-------异步非阻塞场景执行结束-------");
        Thread.sleep(3000);
    }

    /**
     * 休眠 2s 后返回 Hello, Reactor!
     */
    private static String hello() {
        try {
            TimeUnit.SECONDS.sleep(2);
        } catch (InterruptedException e) {
            e.printStackTrace();
        }
        return "Hello, Reactor!";
    }

    /**
     * Schedulers 异步非阻塞执行
     */
    private static void helloAsync() {
        // Callable 调用同步 hello 方法
        Mono.fromCallable(() -> hello())
                // 弹性线程池执行
                .subscribeOn(Schedulers.elastic())
                // 打印结果
                .subscribe(System.out::println, System.err::println);
    }
}
```

执行案例代码，得到如下执行结果：

```
Hello, Reactor!
--------同步阻塞场景执行结束--------
--------异步非阻塞场景执行结束--------
Hello, Reactor!
```

观察执行结果，可以发现 hello() 方法是同步阻塞输出"Hello, Reactor!"，helloAsync() 方法是异步非阻塞输出"Hello, Reactor!"的。

6.4 Spring WebFlux

6.4.1 基于注解的 WebFlux 开发方式

有了本章前几节的基础知识后，本节将开始进入 WebFlux 实战。首先是相关环境的搭建。在项目中使用 Spring WebFlux 需要引入以下依赖：

```xml
<dependency>
<groupId>org.springframework</groupId>
<artifactId>spring-webflux</artifactId>
<version>${spring.framework.version}</version>
</dependency>
```

如果是使用 Spring MVC 项目，那么也可以直接升级到 Spring WebFlux，需要修改 web.xml 中的 DispatcherServlet，新增以下属性：

```xml
<async-supported>true</async-supported>
```

Controller 中处理请求的返回类型采用响应式类型：

```java
/**
 * @Author zhouguanya
 * @Date 2018/11/2
 * @Description webflux hello world
 */
@RestController
public class HelloController {

    @GetMapping("/helloflux")
    public Mono<String> helloFlux() {
        return Mono.just("welcome to webflux world ~");
    }

}
```

启动应用程序，在浏览器中访问 http://localhost:8080/helloflux 得到如下输出：

```
welcome to webflux world ~
```

从上边这个简单的例子中可以看出，Spring 5 用心良苦，WebFlux 提供了与之前 WebMVC 相同的一套注解来定义请求的处理，使得 Spring MVC 使用者迁移到 Spring WebFlux 的过程变得更加轻松。

6.4.2 基于函数式的 WebFlux 开发方式

既然 WebFlux 是响应式编程，因此应该使用统一的函数式编程风格。WebFlux 提供了一套函数式编程接口，可以用来实现类似 MVC 的效果。在传统的 Spring MVC 中，主要由以下两个注解来配合工作。

（1）@Controller：定义处理逻辑。

（2）@RequestMapping：定义方法对特定 URL 进行响应。

在 WebFlux 的函数式开发模式中，提供了类似 HandlerFunction 和 RouterFunction 接口来实现 Spring MVC 的类似功能。

（3）HandlerFunction：相当于 Controller 中的具体处理方法，输入为请求，输出为封装在 Mono 中的响应。HandlerFunction 代码如下：

```
@FunctionalInterface
public interface HandlerFunction<T extends ServerResponse> {

   /**
    * Handle the given request.
    * @param request the request to handle
    * @return the response
    */
   Mono<T> handle(ServerRequest request);

}
```

（4）RouterFunction：相当于"@RequestMapping"，将 URL 映射到具体的 HandlerFunction，输入为请求，输出为封装在 Mono 中的 HandlerFunction。RouterFunction 部分代码如下：

```
@FunctionalInterface
public interface RouterFunction<T extends ServerResponse> {
   Mono<HandlerFunction<T>> route(ServerRequest request);

   default RouterFunction<T> and(RouterFunction<T> other) {
      return new RouterFunctions.SameComposedRouterFunction<>(this, other);
   }

   default RouterFunction<?> andOther(RouterFunction<?> other) {
      return new RouterFunctions.DifferentComposedRouterFunction(this,
other);
   }
```

```
        default RouterFunction<T> andRoute(RequestPredicate predicate,
HandlerFunction<T> handlerFunction) {
            return and(RouterFunctions.route(predicate, handlerFunction));
        }
......
```

在 WebFlux 中，请求和响应不再是 WebMVC 中的 ServletRequest 和 ServletResponse，而是 ServerRequest 和 ServerResponse。它们提供了对非阻塞和回压特性的支持，以及 Http 消息体与响应式类型 Mono 和 Flux 的转换方法。

下面用函数式的方式开发两个接口。"/user/all" 用于查询所有的用户，"/user/{id}" 查询当前指定 id 的用户信息。

使用 SpringBoot 相关的依赖可以快速搭建 WebFlux，只需以下一个依赖就可以搭建一个 WebFlux 环境。这里将使用 SpringBoot 快速搭建一个 WebFlux 环境：

```xml
<dependency>
    <groupId>org.springframework.boot</groupId>
    <artifactId>spring-boot-starter-webflux</artifactId>
</dependency>
```

创建一个 User 实体类，通过接口操作，返回这个 User 实体类的对象。User 代码如下：

```java
/**
 * @Author zhouguanya
 * @Date 2018/11/07
 * @Description user 实体
 */
public class User {
    /** id */
    private Long id;
    /** 姓名 */
    private String name;
    /** 构造器 */
    public User(Long uid, String name) {
        this.id = uid;
        this.name = name;
    }
    /** 以下是 setter 和 getter 方法 */
    public Long getId() {
        return id;
    }
    public void setId(Long id) {
        this.id = id;
    }
```

```java
    public String getName() {
        return name;
    }
    public void setName(String name) {
        this.name = name;
    }
}
```

定义一个接口 UserService，其中包含两个方法，查询所有用户信息的 queryAllUserList()方法和查询单个用户信息的 queryUserById(Long id)方法。UserService 代码定义如下：

```java
/**
 * @Author zhouguanya
 * @Date 2018/11/07
 * @Description user 接口
 */
public interface UserService {
    /**
     * 查询所有用户
     */
    Flux<User> queryAllUserList();
    /**
     * 根据 id 查询用户
     */
    Mono<User> queryUserById(Long id);
}
```

UserServiceImpl 实现 UserService 接口，重写 UserService 接口中的抽象方法。此处不涉及 DAO 操作，因此在 UserServiceImpl 中用一个静态 HashMap 保存用户信息来模拟从数据库查询用户信息。UserServiceImpl 代码如下：

```java
/**
 * @Author zhouguanya
 * @Date 2018/11/07
 * @Description user 接口实现
 */
@Service
public class UserServiceImpl implements UserService {
    /**
     * 注意：
     * 一般企业开发中需要指定容器大小，避免频繁扩容
     * 此例中 map 初始容量为 4，避免扩容（默认负载因子 0.75）
     * 如果不指定，则会使用默认大小，即 16，造成空间浪费
     */
```

```java
    private static Map<Long,User> userMap = new HashMap<>(4);
    static {
        userMap.put(1L,new User(1L,"admin"));
        userMap.put(2L,new User(2L,"admin2"));
        userMap.put(3L,new User(3L,"admin3"));
    }

    @Override
    public Flux<User> queryAllUserList() {
        return Flux.fromIterable(userMap.values());
    }

    @Override
    public Mono<User> queryUserById(Long id) {
        return Mono.just(userMap.get(id));
    }
}
```

注意：此处 userMap 是指定初始容量的，一般在企业开发中，如果能够预知需要的集合大小，可以手动指定容器的大小，避免在后续的集合操作中频繁发生容器扩容，影响容器性能。

创建一个辅助类 UserHandler，其中调用 UserService 的方法并返回 ServerResponse 对象，UserHandler 代码如下：

```java
/**
 * @Author zhouguanya
 * @Date 2018/11/07
 * @Description 辅助类
 */
@Component
public class UserHandler {
    @Autowired
    private UserService userService;

    public Mono<ServerResponse> queryAllUserList(ServerRequest serverRequest){
        Flux<User> allUser = userService.queryAllUserList();
        return ServerResponse.ok().contentType(MediaType.APPLICATION_JSON)
                .body(allUser,User.class);
    }

    public Mono<ServerResponse> queryUserById(ServerRequest serverRequest){
        //获取url上携带的参数id
        Long uid = Long.valueOf(serverRequest.pathVariable("id"));
        Mono<User> user = userService.queryUserById(uid);
```

```java
            return ServerResponse.ok().contentType
(MediaType.APPLICATION_JSON).body(user,User.class);
    }
}
```

创建一个配置类 RoutingConfiguration，配置 RouterFunction。RoutingConfiguration 代码如下：

```java
/**
 * @Author zhouguanya
 * @Date 2018/11/07
 * @Description 配置类
 */
@Configuration
public class RoutingConfiguration {

    @Bean
    public RouterFunction<ServerResponse> monoRouterFunction(UserHandler userHandler){
        return route(GET("/user/all")
                        .and(accept(MediaType.APPLICATION_JSON)),userHandler::queryAllUserList)
                        .andRoute(GET("/user/{id}")
                        .and(accept(MediaType.APPLICATION_JSON)),userHandler::queryUserById);
    }
}
```

运行主函数，启动 SpringBoot 环境，代码如下：

```java
@SpringBootApplication
public class Chapter6WebfluxApplication {

    public static void main(String[] args) {
        SpringApplication.run(Chapter6WebfluxApplication.class, args);
    }
}
```

在浏览器中输入 http://localhost:8080/user/all，可以得到所有的用户信息结果：

```
[
    {
        "id": 1,
        "name": "admin"
    },
    {
        "id": 2,
```

```
      "name": "admin2"
    },
    {
      "id": 3,
      "name": "admin3"
    }
]
```

在浏览器中再次输入 http://localhost:8080/user/1，得到用户 id 为 1 的用户信息：

```
{
    "id": 1,
    "name": "admin"
}
```

6.5 小　　结

本章讲解 Spring 5 新特性之 Spring WebFlux 响应式编程，WebFlux 可以作为 Spring MVC 的替代方案，以异步非阻塞的方式实现编程，从而提高系统性能。Spring WebFlux 依赖于 Reactor，本章 6.3 节介绍的是 Reactor 的一些入门知识，如需更多 Reactor 高级特性参考 Reactor 官网。

第 7 章

WebClient 响应式客户端

本章将要介绍与响应式编程配套的 Spring 5 客户端框架——WebClient 响应式客户端，WebClient 使响应式更加便于调试。

7.1 RestTemplate 调试 Spring MVC

Spring MVC 是目前最主流的 MVC 框架之一，可用于实现 HTTP 接口。常见的调试 HTTP 接口的方式有通过浏览器或者 postman 访问 HTTP 接口。除了以上两种方式外，还有很多可以调试 HTTP 接口的工具，如 RestTemplate 或者第三方类库（HTTPClient）等。下面将使用 Spring 提供的 RestTemplate 调试 Spring MVC 编写的 HTTP 接口。

创建一个时间接口 DateService，其中包含一个查询当前日期的方法，DateService 代码如下：

```
/**
 * @Author zhouguanya
 * @Date 2018/11/7
 * @Description 时间接口
 */
public interface DateService {
    /**
     * 当前日期
     */
    String queryCurrentDate();
}
```

创建 DateService 接口的实现类 DateServiceImpl，代码如下：

```java
/**
 * @Author zhouguanya
 * @Date 2018/11/7
 * @Description DateService 接口实现
 */
@Service
public class DateServiceImpl implements DateService {
    /**
     * 当前日期
     */
    @Override
    public String queryCurrentDate() {
        SimpleDateFormat format = new SimpleDateFormat("yyyy-MM-dd");
        return "Today is " + format.format(new Date());
    }
}
```

创建 DateController，调用 DateService 接口。DateController 代码如下：

```java
/**
 * @Author zhouguanya
 * @Date 2018/11/7
 * @Description 控制器
 */
@RestController("dateController")
public class DateController {
    @Autowired
    private DateService dateService;

    @RequestMapping("/date/currentDate")
    public String getCurrentDate() {
        return dateService.queryCurrentDate();
    }
}
```

创建一个单元测试类，单元测试中使用 RestTemplate 测试 HTTP 接口，RestTemplate 相关配置如下：

```xml
<bean id="restTemplate" class="org.springframework.web.client.RestTemplate">
    <constructor-arg ref="simpleClientHttpRequestFactory"/>
    <property name="messageConverters">
        <list>
            <bean class="org.springframework.http.converter.FormHttpMessageConverter"/>
```

```xml
<bean class="org.springframework.http.converter
            .StringHttpMessageConverter">
    <property name="supportedMediaTypes">
        <list>
            <value>text/plain;charset=UTF-8</value>
        </list>
    </property>
</bean>
        </list>
    </property>
</bean>

<bean id="simpleClientHttpRequestFactory" class="org.springframework.http.client.SimpleClientHttpRequestFactory">
    <property name="readTimeout" value="10000"/>
    <property name="connectTimeout" value="5000"/>
</bean>
```

完整的测试代码如下：

```java
@RunWith(SpringJUnit4ClassRunner.class)
//获取 Spring 上下文环境
@ContextConfiguration(locations = {
    "classpath*:chapter7.xml"})
public class RestTemplateTest {
    @Autowired
    RestTemplate restTemplate;
    /**
     * 测试 currentDate 接口
     */
    @Test
    public void testCurrentDate() {
        ResponseEntity<String> responseEntity = restTemplate.getForEntity
("http://localhost:8080/date/currentDate", String.class);
        if (responseEntity.getStatusCodeValue() == 200) {
            System.out.println(responseEntity.getBody());
        }
    }
}
```

执行单元测试代码，得到如下结果：

```
Today is 2018-11-12
```

7.2　WebClient 调试 Spring WebFlux

　　WebClient 是从 Spring WebFlux 5.0 版本开始提供的一个非阻塞的基于响应式编程的进行 Http 请求的客户端工具。它的响应式编程是基于 Reactor 的。WebClient 中提供了标准 Http 请求方式对应的 get、post、put 和 delete 等方法，可以用来发起相应 Http 的请求。下面的代码是一个简单的 WebClient 请求示例。先通过 WebClient.create()创建一个 WebClient 的实例，之后通过 get()、post() 等设置请求方法；uri() 指定需要请求的路径；retrieve() 用来发起请求并获得响应；bodyToMono(String.class)用来指定请求结果，需要处理为 String，并包装为 Reactor 的 Mono 对象。

　　创建时间接口 DateWebFluxService，获取当前时间，将返回一个 Mono 对象，时间接口代码　如下：

```java
/**
 * @Author zhouguanya
 * @Date 2018/11/7
 * @Description 时间接口
 */
public interface DateWebFluxService {
    /**
     * 当前日期
     */
    Mono<String> queryCurrentDate();
}
```

创建 DateWebFluxService 接口的实现类 DateWebFluxServiceImpl，代码如下：

```java
/**
 * @Author zhouguanya
 * @Date 2018/11/7
 * @Description DateService 接口实现
 */
@Service
public class DateWebFluxServiceImpl implements DateWebFluxService {
    /**
     * 当前日期
     */
    @Override
    public Mono<String> queryCurrentDate() {
        SimpleDateFormat format = new SimpleDateFormat("yyyy-MM-dd");
        return Mono.just("Today is " + format.format(new Date()));
    }
}
```

创建一个控制器 DateWebFluxController 调用时间服务：

```java
/**
 * @Author zhouguanya
 * @Date 2018/11/7
 * @Description 控制器
 */
@RestController("dateWebFluxController")
public class DateWebFluxController {
    @Resource
    private DateWebFluxService dateWebFluxService;

    @RequestMapping("/date/webflux/currentDate")
    public Mono<String> getCurrentDate() {
        return dateWebFluxService.queryCurrentDate();
    }

}
```

编写单元测试代码，其中使用 WebClient 发起 Http 请求，由于 WebFlux 是异步非阻塞的，因此在单元测试中使用 block()方法强制阻塞直到获取到 Http 接口返回的结果，单元测试代码如下：

```java
@RunWith(SpringJUnit4ClassRunner.class)
//获取 Spring 上下文环境
@ContextConfiguration(locations = {"classpath*:chapter7.xml"})
public class WebClientTest {
    /**
     * 测试 currentDate 接口
     */
    @Test
    public void testCurrentDate() {
        WebClient webClient = WebClient.create("http://localhost:8080");
        Mono<String> resp =
webClient.get().uri("/date/webflux/currentDate").retrieve().bodyToMono(String.class);
        System.out.println(resp.block());
    }

}
```

执行单元测试代码，得到如下结果：

```
Today is 2018-11-12
```

在应用中使用 WebClient 时也许需要访问的 URL 都来自同一个应用，只是对应不同的 URL 地址，这个时候可以把公用的部分抽出来定义为 baseUrl，然后在进行 WebClient 请求的时候只指定相对于 baseUrl 的 URL 部分即可。这样的好处是修改 baseUrl 的时候只要修改一处即可。上面的代码在创建 WebClient 时定义了 baseUrl 为 http://localhost:8080。

案例中的请求响应结果为 String，如果想要将响应结果解析为对象，则可以使用类似如下的编码方式：

```
Flux<User> userFlux =
webClient.get().uri(basePath).retrieve().bodyToFlux(User.class);
```

URL 中也可以使用路径变量，路径变量的值可以通过 uri 方法的第 2 个参数指定。下面的代码就定义了 URL 中拥有一个路径变量的 id，然后实际访问时该变量将取值 1。

```
webClient.get().uri("basePath/{id}", 1);
```

URL 中也可以使用多个路径变量，多个路径变量的赋值将依次使用 uri 方法的第 2 个、第 3 个、第 N 个参数。下面的代码中就定义了 URL 中拥有路径变量 p1 和 p2，实际访问的时候将被替换为 var1 和 var2。所以实际访问的 URL 是 basePath/var1/var2。

```
webClient.get().uri(basePath/{p1}/{p2}", "var1", "var2");
```

当传递的请求体对象是一个 MultiValueMap 对象时，WebClient 默认发起的是 Form 提交。下面的代码就通过 Form 提交模拟了用户进行登录操作，给 Form 表单传递了参数 username，值为 zhangsan，传递了参数 password，值为 123456。

```
MultiValueMap<String, String> map = new LinkedMultiValueMap<>();
map.add("username", "zhangsan");
map.add("password", "123456");
Mono<String> mono =
webClient.post().uri(path).syncBody(map).retrieve().bodyToMono(String.class);
```

前面介绍的示例都是直接获取到了响应的内容，如果想获取响应的头信息、Cookie 等，在通过 WebClient 请求时把调用 retrieve() 改为调用 exchange()，就可以访问代表响应结果的 ClientResponse 对象，通过它可以获取响应的状态码、Cookie 等信息：

```
Mono<ClientResponse> mono =
webClient.post().uri("login").syncBody(map).exchange();
ClientResponse response = mono.block();
ResponseCookie cookie= response.cookies()
```

本章只介绍 WebClient 常见用法，更多有关 WebClient 的使用信息请参考其 API 文档。

7.3 小　　结

本章介绍与 Spring WebFlux 配套使用的客户端工具 WebClient，并对比 RestTemplate 与 WebClient 的使用，介绍了使用了 WebClient 一些常见的与 Http 请求相关的方法。通过使用 WebFlux 可以更加方便地对 WebFlux 响应式编程进行运行和调试。

第 8 章

Spring 5 结合 Kotlin 编程

Spring 5 新特性中的一个重要更新是支持了 Kotlin 这种编程语言，本章将介绍 Kotlin 语言的使用和 Spring 对 Kotlin 的支持。

8.1 Kotlin 简介

Kotlin 是一种在 Java 虚拟机上运行的编程语言，被称之为 Android 世界的 Swift，是由 JetBrains 设计开发并开源的。Kotlin 可以编译成 Java 字节码，也可以编译成 JavaScript，方便在没有 JVM 的设备上运行。在 Google I/O 2017 中，Google 宣布 Kotlin 成为 Android 官方开发语言。Spring 5 对 Kotlin 有很好的支持，本节对 Kotlin 的介绍以基本的概念和简单使用为主，更多 Kotlin 高级特性请访问 Kotlin 官网。

8.1.1 Kotlin 的特性

Kotlin 语言的特点如下：

- Kotlin 完全兼容 Java。
- Kotlin 使用极少的代码量可实现功能，且代码末尾无需分号结尾。
- Kotlin 是空安全的，使用 Kotlin 语言可以有效避免空指针的出现。
- Kotlin 支持 Lambda 表达式。

与 Java 中的变量类似，Kotlin 中的变量也是有作用域的，其含有以下几种作用域。

- public: 默认作用域，表示总是可见。
- internal: 同模块可见。

- protected：类似于 Java 中的 protected，对子类也可见。
- private：只能在当前源文件内使用，常量 val 和变量 var，默认都是 private 的。

很多编程语言都有关键字，Kotlin 的关键字与 Java 中关键字略有不同，下面介绍一些 Kotlin 中常见的关键字及其含义。

- abstract：抽象声明，被标注对象默认是 open。
- annotation：注解声明。
- by：类委托、属性委托。
- class：声明类。
- companion：伴生对象声明。
- const：声明编译期常量。
- constructor：声明构造函数。
- crossinline：标记内联函数的 lambda 表达式参数，标识该 lambda 函数返回为非局部返回，不允许非局部控制流。
- data：数据类，声明的类默认实现 equals()/hashCode()/toString/copy()/componentN()。
- enum：声明枚举类。
- field：属性的幕后字段。
- fun：声明函数。
- import：导入。
- in：修饰类型参数，使其逆变——只可以被消费而不可以被生产。
- init：初始化块，相当于主构造函数的方法体。
- inline：声明内联函数。
- inner：标记嵌套类，使其成为内部类——可访问外部类的成员。
- interface：声明接口。
- internal：可见性修饰符，相同模块内可见。
- lateinit：延迟初始化，避免空检查。
- noinline：禁用内联，标记内联函数不需要内联参数。
- object：对象表达式、对象声明。
- open：允许其他类继承，kotlin 类默认都是 final，禁止继承。
- operator：标记重载操作符的函数。
- out：修饰类型参数，使其协变——只可以被生产而不可以被消费。
- override：标注复写的方法、属性。
- package：包声明。
- private：可见性修饰符，文件内可见。
- protected：可见性声明，只修饰类成员，子类中可见。
- public：kotlin 默认的可见性修饰符，随处可见。
- reified：限定类型参数，需要配合 inline 关键字使用。
- sealed：声明密封类，功能类似枚举。
- super：访问超类的方法、属性。

- suspend：声明挂起函数，该函数只能从协程和其他挂起函数中调用。
- throw：抛异常。
- typealias：声明类型别名。
- val：声明只读属性。
- var：声明可变属性。
- vararg：修饰函数参数：声明为可变数量参数。

8.1.2 Kotlin 基本数据类型

Kotlin 的基本数值类型包括 Byte、Short、Int、Long、Float 和 Double 等。表 8-1 对比了每种数据类型和位宽度。

表 8-1 Kotlin 基本数据类型

类型	位宽度
Double	64 位
Float	32 位
Long	64 位
Int	32 位
Short	16 位
Byte	8 位

8.1.3 Kotlin 开发环境搭建

Kotlin 开发环境搭建有多种方式，本书以 Maven 为例说明 Kotlin 环境的搭建。

打开 Intellij IDEA，新建一个项目，选择 Maven→选择 org.jetbrains.kotlin:kotlin-archetype-jvm 即可创建一个 Kotlin 项目，如图 8-1 所示。

创建完 Kotlin 项目后，默认会生成一个 Hello.kt 文件：

```kotlin
fun main(args: Array<String>) {
    println("Hello, World")
}
```

运行程序，控制台打印如下：

```
Hello, World
```

至此就完成了一个基本的 Kotlin 环境的搭建。下面将介绍 Kotlin 编程语言的基本使用。

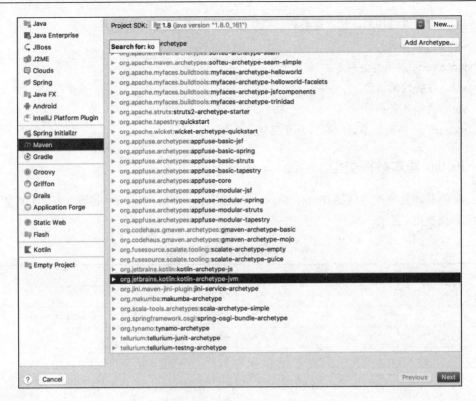

图 8-1　Kotlin 项目创建示意图

8.1.4　在 Kotlin 中定义常量与变量

可变变量定义：var 关键字，其使用语法如下：

```
var <标识符> : <类型> = <初始化值>
```

不可变变量定义：val 关键字，只能赋值一次的变量（类似 Java 中 final 修饰的变量），其使用语法如下：

```
val <标识符> : <类型> = <初始化值>
```

下面给出一些简单的 Kotlin 变量的定义示例：

```
val age: Int = 10
val num = 12            // 系统自动推断变量类型为 Int
val grade: Int          // 如果不在声明时初始化则必须提供变量类型
grade = 91              // 明确赋值
var x = 5               // 系统自动推断变量类型为 Int
x += 10                 // 变量可修改
```

8.1.5　字符串模板

字符串可以包含模板表达式，即一些小段代码，可以求值后并把结果合并到字符串中。模板表达式以美元符（$）开头，由一个简单的名字构成。

- $：表示一个变量名或者变量值。
- $varName：表示变量值。
- ${varName.fun()}：表示变量的方法返回值。

8.1.6 NULL 检查机制

Kotlin 的空安全设计可以用于声明可为空的参数，有两种处理方式，字段后加"!!"，像 Java 一样抛出空异常，另一种字段后加"?"可不做处理，返回值为 null 或配合"?:"做空判断处理。

```
//类型后面加"?"表示可为空
var age: String? = "23"
//如果 age 为空，抛出空指针异常
val ages = age!!.toInt()
//如果 age 为空，不做处理返回 null
val ages1 = age?.toInt()
//如果 age 为空，返回-1
val ages2 = age?.toInt() ?: -1
```

8.1.7 For 循环和区间

Kotlin 使用 for 关键字进行循环遍历，区间表达式由具有操作符形式".."的 rangeTo 函数辅以 in 和"!in"形成。

```
// 等同于 i >= 1 && i <= 10
for (i in 1..10) print(i)
// 什么都会不输出
for (i in 10..1) print(i)
// 使用 step 指定步长，等同于 i >= 1; i <= 10; i += 2
for (i in 1..10 step 2) print(i)
// 等同于 i <= 10; i >= 1; i -= 2
for (i in 10 downTo 1 step 2) print(i)
// 使用 until 函数排除结束元素
// i in [1, 10) 排除了 10
for (i in 1 until 10) {
    println(i)
}
```

下面通过一个案例给出 Kotlin 变量定义、字符串模板及空值处理等操作的具体使用方式，案例代码如下：

```
/**
 * 字符串模版的使用
 */
fun printAge(age: Int): String {
    // 模板中的简单名称
```

```kotlin
    val lastYear = "my age is $age last year"
    // 模板中的任意表达式:
    val thisYear = "${lastYear.replace("is", "was")}, but now I am ${age + 1}"
    return thisYear
}

/**
 * 空值处理函数
 */
fun parseInt(str: String): Int? {
    return str.toIntOrNull()
}

/**
 * 字符串乘法
 */
fun multiplicationString(arg1: String, arg2: String) {
    // 将字符串转为 Int
    val x = parseInt(arg1)
    val y = parseInt(arg2)
    // 对 x 和 y 做非空判断
    if (x != null && y != null) {
        // 打印 x * y 的值
        println(x * y)
    }
    else {
        println("参数异常，'$arg1' 或 '$arg2' 不是数字")
    }
}

/**
 * 区间操作函数
 */
fun range(): Unit {
    // 等同于 i >= 1 && i <= 10
    for (i in 1..10) print("$i\t")
    println("\n------分割线------")
    // 什么都不输出
    for (i in 10..1) print("$i\t")
    println("\n------分割线------")
    // 使用 step 指定步长, 等同于 i >= 1; i <= 10; i += 2
    for (i in 1..10 step 2) print("$i\t")
    println("\n------分割线------")
    // 等同于 i <= 10; i >= 1; i -= 2
    for (i in 10 downTo 1 step 2) print("$i\t")
    println("\n------分割线------")
    // 使用 until 函数排除结束元素
```

```kotlin
    // i in [1, 10) 排除了 10
    for (i in 1 until 10) {
        print("$i\t")
    }
    println("\n------分割线------")
}
/**
 * for 循环迭代集合
 */
fun collection(): Unit {
    val items = listOf("a", "b", "c")
    for (item in items) {
        println(item)
    }
    // 使用 in 运算符
    val fruits = setOf("apple", "banana", "pear")
    when {
        "orange" in fruits -> println("orange is yellow")
        "apple" in fruits -> println("apple is red")
    }
}
fun main(args: Array<String>) {
    println(printAge(10))
    println("------分割线------")
    multiplicationString("3", "5")
    println("------分割线------")
    multiplicationString("hello", "3")
    println("------分割线------")
    range()
    collection()
}
```

执行案例代码，得到如下输出结果：

```
my age was 10 last year, but now I am 11
------分割线------
15
------分割线------
参数异常，'hello' 或 '3' 不是数字
------分割线------
1   2   3   4   5   6   7   8   9   10
------分割线------
------分割线------
1   3   5   7   9
------分割线------
```

```
10  8  6  4  2
------分割线------
1  2  3  4  5  6  7  8  9
------分割线------
a
b
c
apple is red
```

8.1.8 定义函数

Kotlin 函数的定义需要用 fun 来定义，Kotlin 函数可以指定返回值类型，也可以使用推断返回值类型，无返回值可以使用 Unit 标示，也可以不写明返回值。

下面通过案例说明 Kotlin 函数的定义，案例代码如下：

```
/**
 * 具有两个 Int 类型的入参和 Int 返回值的 sum 方法
 */
fun addition(a: Int, b: Int): Int {
    return a + b
}
/**
 * 类型自动推断的 substract 方法
 */
fun subtraction(a: Int, b: Int) = a - b

/**
 * 无返回值的 multiplication 方法
 */
fun multiplication(a: Int, b: Int): Unit {
    println(a + b)
}
/**
 * 可变长参数函数的 variableParam 方法
 */
fun variableParam(vararg v:Int){
    for(vt in v){
        println(vt)
    }
}

fun main(args: Array<String>) {
    println("1 + 2 = " + addition(1, 2))
    println("------分割线------")
    println("1 - 2 = " + subtraction(1, 2))
    println("------分割线------")
```

```
    print("1 + 2 = ")
    multiplication(1, 2)
    println("------分割线------")
    variableParam(1, 2, 3)
    println("------分割线------")
    // lambda 匿名函数
    val division: (Int, Int) -> Int = {x, y -> x / y}
    println("4 / 2 = " + division(4, 2))
}
```

执行案例代码，得到如下测试结果：

```
1 + 2 = 3
------分割线------
1 - 2 = -1
------分割线------
1 + 2 = 3
------分割线------
1
2
3
------分割线------
4 / 2 = 2
```

8.1.9 类和对象

与 Java 类相似，Kotlin 类可以包含 Kotlin 构造函数、初始化代码块、函数、属性、内部类和对象声明等。Kotlin 类的声明也是使用 class 关键字。

Kotlin 类的属性可以用关键字 var 声明为可变的变量，也可以说是使用关键字 val 将变量声明为不可变。

Kotlin 中没有 new 关键字，因此使用构造函数创建类的实例对象可以使用如下语法：

```
val helloWorld = HelloWorld()
```

Kotlin 类中可以包含主构造器和次构造器。主构造器中不能包含任何代码，可以将初始化代码放在初始化代码段中，初始化代码段使用 init 关键字作为修饰。次构造器需要加上 constructor 作为前缀。

下面通过案例说明 Kotlin 类和对象的使用方式，其中定义了两个构造器，分别是主构造器和次构造器，通过不同的构造器创建 User 对象进行测试，案例代码如下：

```
/**
 * @Author zhouguanya
 * @Date 2018/11/23
 * @Description Kotlin 类和对象
 */
class User constructor(name: String){
```

```kotlin
        var userAge: Int = 20
        var userName = name
        // 初始化代码
        init {
            println("init code executed")
        }
        /**
         * 次构造函数
         */
        constructor (name: String, age: Int) : this(name) {
            userName = name
            userAge = age
            println("my name is $userName, i am $userAge years old.")
        }

        /**
         * sayHello方法
         */
        fun sayHello() {
            println("$userName say hello world")
        }
}
fun main(args: Array<String>) {
    val allen = User("allen")
    println(allen.userAge)
    println(allen.userName)
    allen.sayHello()
    println("------分割线------")
    val michael = User("michael", 24)
    println(michael.userAge)
    println(michael.userName)
    michael.sayHello()
}
```

执行案例代码,得到如下执行结果:

```
init code executed
20
allen
allen say hello world
------分割线------
init code executed
my name is michael, i am 24 years old.
24
michael
michael say hello world
```

8.1.10 Kotlin 与 Java 互操作

Kotlin 在设计时就考虑了与 Java 的兼容性和互操作性，这使得开发者可以方便地从 Java 代码过渡到 Kotlin 的开发。本节将介绍 Kotlin 与 Java 相互调用的一些实战操作。

下面定义一个方法，该方法的入参是 java.util.ArrayList 对象中存放 Kotlin 类型的 Int 值，在方法体内，分别使用 Kotlin 和 Java 的方式对方法入参中的 java.util.ArrayList 集合对象进行遍历，具体代码如下：

```kotlin
/**
 * @Author zhouguanya
 * @Date 2018/12/8
 * @Description Kotlin 与 Java 互操作演示案例
 */
class KotlinAndJava {
    /**
     * Kotlin 与 Java 互操作
     */
    fun kotlinJavaInteract(source: java.util.ArrayList<Int>) {
        val list = ArrayList<Int>()
        // 使用 Java 中 for 遍历集合
        for (item in source) {
            list.add(item)
        }
        System.out.println(list)
        System.out.println("------分隔符------")
        // Kotlin 操作符遍历集合
        for (i in 0 until source.size) {
            System.out.println(source[i])
        }
    }
}

fun main(args: Array<String>) {
    val city = KotlinAndJava()
    // 使用 Java 中的 ArrayList 类
    val list = java.util.ArrayList<Int>()
    list.add(1)
    list.add(2)
    list.add(3)
    list.add(4)
    list.add(5)
    city.kotlinJavaInteract(list)
}
```

执行以上代码，得到如下执行结果：

```
[1, 2, 3, 4, 5]
------分隔符------
1
2
3
4
5
```

8.2　Spring 5 集成 Kotlin

通过以上介绍，读者应该对 Kotlin 的特性有了大致的了解。Spring 5 提供了对 Kotlin 的支持，本节将介绍如何使用 Kotlin 编写 MVC 模块。

为了简便起见，这里使用 SpringBoot 搭建 Kotlin Web 环境。搭建环境需要用到如下依赖关系：

```xml
<dependency>
    <groupId>org.jetbrains.kotlin</groupId>
    <artifactId>kotlin-stdlib</artifactId>
    <version>${kotlin.version}</version>
</dependency>
<dependency>
    <groupId>org.jetbrains.kotlin</groupId>
    <artifactId>kotlin-test-junit</artifactId>
    <version>${kotlin.version}</version>
    <scope>test</scope>
</dependency>
<dependency>
    <groupId>org.jetbrains.kotlin</groupId>
    <artifactId>kotlin-stdlib-jre8</artifactId>
    <version>${kotlin.version}</version>
</dependency>
<dependency>
    <groupId>org.jetbrains.kotlin</groupId>
    <artifactId>kotlin-reflect</artifactId>
    <version>${kotlin.version}</version>
</dependency>
<dependency>
    <groupId>org.springframework.boot</groupId>
    <artifactId>spring-boot-starter-web</artifactId>
    <version>2.1.1.RELEASE</version>
</dependency>
```

下面使用 Kotlin 定义一个员工实体层 Staff，其中包含属性 id 和 name，Staff 实体类的代码如下：

```kotlin
/**
 * @Author zhouguanya
 * @Date 2018/12/8
 * @Description 员工类
 */
class Staff(
    var id: Int = -1,
    var name: String = ""
) {
    override fun toString(): String {
        return "Staff(id=$id, name='$name')"
    }
}
```

接下来定义一个员工接口 StaffService，代码如下：

```kotlin
/**
 * @Author zhouguanya
 * @Date 2018/12/8
 * @Description 员工接口
 */
interface StaffService {
    fun findByName(name: String): Staff
}
```

StaffService 接口的实现如下：

```kotlin
/**
 * @Author zhouguanya
 * @Date 2018/12/8
 * @Description 员工接口实现类
 */
@Service
class StaffServiceImpl : StaffService {
    override fun findByName(name: String): Staff {
        return Staff(100, name)
    }
}
```

然后定义一个控制器 StaffController，其中调用 StaffService 接口中的 findByName()方法，查找名叫 Michal 的员工，返回员工的编号和员工姓名。StaffController 控制的代码如下：

```kotlin
/**
 * @Author zhouguanya
```

```
 * @Date 2018/12/8
 * @Description 员工控制器
 */
@RestController
class StaffController(private val manService:StaffService) {

    @GetMapping("/staff/find")
    fun home(): String {
        val staff = manService.findByName("michael")
        return "staff id = "+ staff.id + ", staff name = " + staff.name
    }
}
```

最后一步定义一个启动类 SpringBootWithKotlinApplication，启动整个 Web 环境只需要执行 SpringBootWithKotlinApplication 类即可。SpringBootWithKotlinApplication 的代码如下：

```
/**
 * @Author zhouguanya
 * @Date 2018/12/8
 * @Description 启动类
 */
@SpringBootApplication
open class SpringBootWithKotlinApplication fun main(args: Array<String>) {

    SpringApplication.run(SpringBootWithKotlinApplication::class.java,
*args)

}
```

单击右键，执行 SpringBootWithKotlinApplication 类，即完成了一个 Kotlin 环境的创建和运行。在浏览器中输入链接 http://localhost:8080/staff/find，在浏览器上会得到如下的执行结果：

```
staff id = 100, staff name = michael
```

8.3 小　　结

Kotlin 允许开发者使用简洁而优雅的代码来实现与 Java 同样的功能，同时提供对现有的 Java 类库的互操作性。Spring 框架提供了 Kotlin 支持，使得 Java 开发可以方便地使用 Kotlin，同时也允许 Kotlin 开发者无缝使用 Spring 框架。

第 9 章

Spring 5 更多新特性

本章介绍更多有关 Spring 5 的新特性细节。

9.1　Resource 接口

Spring 5 为 org.springframework.core.io.Resource 接口新增了 isFile()方法，此方法用于判断当前的资源是否为一个文件。如果 isFile()返回 true，那么 getFile()方法将极有可能（但并不能保证）成功读取文件。isFile()方法代码如下：

```
/**
 * Determine whether this resource represents a file in a file system.
 * A value of {@code true} strongly suggests (but does not guarantee)
 * that a {@link #getFile()} call will succeed.
 * <p>This is conservatively {@code false} by default.
 * @since 5.0
 * @see #getFile()
 */
default boolean isFile() {
    return false;
}
```

除了新增 isFile()方法外，Spring 5 为 Resource 接口提供了基于 NIO 的可读通道访问器 readableChannel()方法，方法代码如下：

```
/**
 * Return a {@link ReadableByteChannel}.
```

```
 * <p>It is expected that each call creates a <i>fresh</i> channel.
 * <p>The default implementation returns {@link
Channels#newChannel(InputStream)}
 * with the result of {@link #getInputStream()}.
 * @return the byte channel for the underlying resource (must not be {@code
null})
 * @throws java.io.FileNotFoundException if the underlying resource doesn't
exist
 * @throws IOException if the content channel could not be opened
 * @since 5.0
 * @see #getInputStream()
 */
default ReadableByteChannel readableChannel() throws IOException {
    return Channels.newChannel(getInputStream());
}
```

9.2　HTTP 2

Spring 5 提供对 HTTP 2 的支持。下面将分析 HTTP 2 是如何做到提高传输性能、降低延迟，并帮助提高应用程序吞吐量的。

9.2.1　HTTP 的现状

从最早的 HTTP 0.9 开始，演变到 HTTP 1.0、HTTP 1.1 再到如今的 HTTP 2，每个版本都带来了惊人的升级体验。

HTTP 协议是一个很成功的协议，已经被广泛使用。然而，随着互联网的发展，网页变得越来越复杂，HTTP 1.1 的底层传输方式已经对应用的整体性能产生了负面影响。特别是，HTTP 1.0 在每次的 TCP 连接上只允许发送一次请求。在 HTTP 1.1 中增加了请求管线，但是这仅仅解决了部分的并发问题，阻塞的现象仍然存在，因此需要发送多个请求的 HTTP 1.0 和 HTTP 1.1 客户端就需要与服务器建立多个连接，以达到高并发低延迟的目的。

9.2.2　HTTP 2 的新特性

1. 二进制协议

HTTP 1.x 的解析是基于文本的，HTTP 2 的解析是基于二进制的，新协议称为二进制分帧层（binary framing layer），它重新设计了编码机制，更加适用于服务器间信息传输。

2. 多路复用

每个请求都有一个 ID，这样在一个连接上可以发送多个请求，并且它们在传输过程中是混杂在一起的，接收方可以根据请求的 ID 将请求再归属到不同的服务端请求里。

3. 请求优先级

在 HTTP 2 中，只有一个连接来实现连接复用，所有资源通过一个连接传输，为了避免线头堵塞（Head Of Line Block），这时资源传输的顺序就更重要了。优先加载重要资源，可以尽快渲染页面，提升用户体验。

4. 报头压缩

HTTP 2 协议拥有配套的 HPACK，HPACK 的目的是尽可能减少客户端请求与服务器响应之间的头部信息重复所导致的性能开销。报头压缩的实现方式是，要求客户端和服务器都维护之前看见的头部字段的列表，减少网络传输的内容。

5. 服务端推送

HTTP 1.x 只能是客户端主动拉取资源，HTTP 2 支持从服务器端推送资源至客户端。当服务器在处理请求的同时，可以将一些静态资源如 CSS 或 JavaScript 推送到客户端。

6. 流控制

流控制管理数据的传输，使数据发送者不会让数据接收者不堪重负。流控制允许接收者停止或减少发送的数据量。例如，一个视频网站，当观众观看一个视频流时，服务器向客户端发送数据；如果视频暂停，客户端会通知服务器停止或者减少发送视频数据，以避免客户端过载。

9.2.3 多路复用与长连接的区别

在 HTTP 1.0 中，一次请求响应就需要建立一次连接，用完后连接即关闭。每个新的请求又要重新建立一个新的连接。

在 HTTP 1.1 中，使用 Pipeling 优化了 HTTP 1.0 中的问题，即通常所说的 Keep-Alive 模式。当连接建立后，多个请求通过串行化的单线程方式进行处理，排在后面的请求必须等待前面的请求处理完才能获得执行机会。一旦有请求处理耗时较长，后续的请求都只能被迫阻塞（Head-of-Line Blocking，即常说的线头阻塞）。如图 9-1 所示。

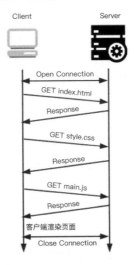

图 9-1　HTTP 1.1 长连接示意图

在 HTTP 2 的多路复用中，多个请求在同一个连接上并行执行，即使某个请求耗时严重，也不会影响到其他请求的正常执行，如图 9-2 所示。

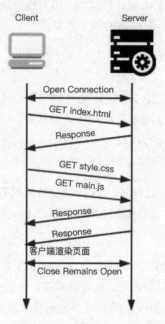

图 9-2　HTTP 2 多路复用示意图

9.3　JUnit 5

在 Spring TestContext Framework 中完全支持 JUnit 5 Jupiter 编程，本节介绍 Spring 5 如何集成 JUnit 5 进行编程。

9.3.1　JUnit 5 简介

与以前版本的 JUnit 不同，JUnit 5 由三个不同子项目中的几个不同模块组成。各个模块的功能和职责如下。

（1）JUnit Platform 是基于 JVM 的运行测试的基础框架，JUnit Platform 定义了开发运行在这个测试框架上的 TestEngine API。此外该平台提供了一个控制台启动器，可以从命令行启动平台，也可以为 Gradle 和 Maven 构建插件。

（2）JUnit Jupiter 是在 JUnit 5 中编写测试用例和扩展的新编程模型和扩展模型。Jupiter 子项目提供了一个 TestEngine 在平台上运行基于 Jupiter 的测试。

（3）JUnit Vintage 提供了一个 TestEngine 在平台上运行基于 JUnit 3 和 JUnit 4 的测试。

9.3.2　JUnit 5 快速体验

下面案例通过简单的加法计算阐述 JUnit 5 的使用方式。

1. 准备开发环境

在 maven 项目的 pom.xml 中加入对 JUnit 5 的依赖关系:

```xml
<dependency>
    <groupId>org.junit.jupiter</groupId>
    <artifactId>junit-jupiter</artifactId>
    <version>5.4.0</version>
    <scope>test</scope>
</dependency>
```

2. 创建一个计算类 Calculator

创建 Calculator 类,其中包含一个 add()方法:

```java
/**
 * @Author: zhouguanya
 * @Date: 2019/02/09
 * @Description: 计算类
 */
public class Calculator {

    public int add(int a, int b) {
        return a + b;
    }

}
```

3. 创建单元测试

创建单元测试,代码如下:

```java
/**
 * @Author: zhouguanya
 * @Date: 2019/02/09
 * @Description: JUnit 5测试用例
 */
class CalculatorTests {

    /**
     * @Test 声明一个测试用例
     * @DisplayName 为测试用例声明一个自定义的显示名称
     */
    @Test
    @DisplayName("1 + 1 = 2")
    void addsTwoNumbers() {
        Calculator calculator = new Calculator();
        assertEquals(2, calculator.add(1, 1), "1 + 1 should equal 2");
```

```
    }
    /**
     * @ParameterizedTest 参数化测试
     * @CsvSource 用 csv 文件进行测试
     */
    @DisplayName("addFromCSV")
    @ParameterizedTest(name = "{0} + {1} = {2}")
    @CsvSource({
        "0,   1,   1",
        "1,   2,   3",
        "49, 51, 100",
        "1,  100, 101"
    })
    void add(int first, int second, int expectedResult) {
        Calculator calculator = new Calculator();
        assertEquals(expectedResult, calculator.add(first, second),() -> first
+ " + " + second + " should equal " + expectedResult);
    }
}
```

执行单元测试，运行结果如图 9-3 所示。

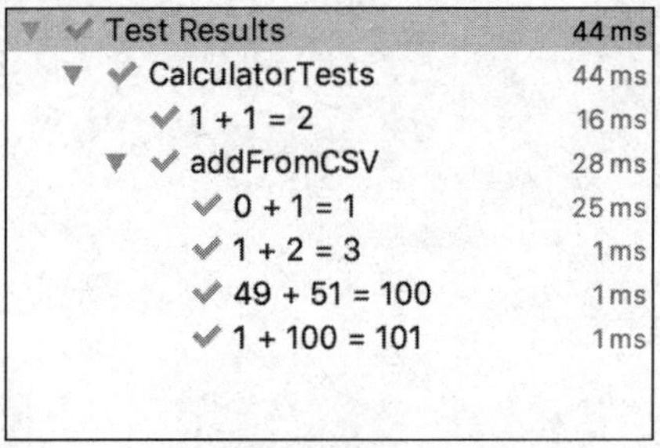

```
▼ ✓ Test Results                    44 ms
    ▼ ✓ CalculatorTests             44 ms
        ✓ 1 + 1 = 2                 16 ms
      ▼ ✓ addFromCSV                28 ms
            ✓ 0 + 1 = 1             25 ms
            ✓ 1 + 2 = 3             1 ms
            ✓ 49 + 51 = 100         1 ms
            ✓ 1 + 100 = 101         1 ms
```

图 9-3 JUnit 5 单元测试示意图

9.3.3 JUnit 5 常用注解

JUnit 5 框架常用注解如表 9-1 所示。

表 9-1 JUnit 5 常用注解

注解	描述
@Test	表示此方法是一个测试方法。与 JUnit 4 的"@Test"注解不同的是，JUnit 5 的"@Test"注解没有声明任何属性，因为 JUnit 5 中的测试扩展是基于专用注解来完成的
@ParameterizedTest	表示此方法是一个参数化测试用例，可以被继承
@RepeatedTest	表示此方法是一个可以用于重复测试的测试模板，可以被继承
@TestFactory	表示此方法是一个动态测试的测试工厂，可以被继承

（续表）

注　解	描　述
@TestTemplate	表示此方法是一个测试模板，依据注册的提供者所返回上下文的调用数量，可以被继承
@TestMethodOrder	配置测试方法的执行顺序。与 JUnit 4 中的 "@FixMethodOrder" 注解类似，此注解可以被继承
@TestInstance	用于配置测试类中的测试用例的生命周期，可以被继承
@DisplayName	为测试类或者测试方法指定一个自定义的显示名称，此注解不能被继承
@DisplayNameGeneration	为测试类指定一个自定义的显示名称生成器，可以被继承
@BeforeEach	在声明 "@Test/@RepeatedTest/@ParameterizedTest/@TestFactory" 的方法之前执行此方法。类似于 JUnit 4 中的 "@Before"，此方法可以被继承
@AfterEach	在声明 "@Test/@RepeatedTest/@ParameterizedTest/@TestFactory" 的方法之后执行此方法。类似于 JUnit 4 中的 "@After"，此方法可以被继承
@BeforeAll	在声明 "@Test/@RepeatedTest/@ParameterizedTest/@TestFactory" 的方法之前执行此方法。类似于 JUnit 4 中的 "@BeforeClass"，此方法可以被继承。与 "@BeforeEach" 相比，"@BeforeAll" 标注的方法在测试类中只执行一次
@AfterAll	在声明 "@Test/@RepeatedTest/@ParameterizedTest/@TestFactory" 的方法之后执行此方法。类似于 JUnit 4 中的 "@AfterClass"，此方法可以被继承。与 "@AfterEach" 相比，"@AfterAll" 标注的方法在测试类中只执行一次
@Nested	表示使用了该注解的类是一个内嵌、非静态的测试类。"@BeforeAll/@AfterAll" 标注的方法不能直接在 "@Nested" 测试类中使用，此注解不能被继承
@Tag	用于声明过滤测试的 tags，该注解可以用在方法或类上。类似于 TesgNG 的测试组或 JUnit 4 的分类。此注解能被继承，但仅限于类级别，而非方法级别
@Disabled	用于禁用一个测试类或测试方法。类似于 JUnit 4 的 "@Ignore"。此注解不能被继承
@ExtendWith	用于声明式的注册扩展程序，此注解可以被继承
@RegisterExtension	用于通过字段以编程方式注册扩展程序。如果这些字段不被隐藏，这些字段都可以被继承
@TempDir	用于在生命周期方法或测试方法中通过字段注入或参数注入提供临时目录

9.4　小　结

Spring 5 对 HTTP/2 的支持和对 Junit 5 的支持将带给开发者更好的用户体验，提升开发者的开发效率。

第三篇

Spring系统集成篇

本篇主要讲解 Spring 与多种第三方组件的集成和使用。

第 10 章

Spring 集成 Log4j2

　　Log4j 是 Apache 的一个开源日志项目，通过 Log4j，开发人员可以控制日志信息输送的目的地是控制台、文件、GUI 组件，甚至是远程服务器（如互联网公司常用的日志三剑客 ELK）等；开发人员也可以控制每一条日志的输出格式；通过定义每一条日志信息的级别，能够更加细致地控制日志的生成过程。最令人感兴趣的就是，这些可以通过一个配置文件来灵活地进行配置，而不需要修改应用程序代码，大大降低了对程序代码的侵入性。

　　在企业开发过程中，一般不直接使用 Log4J 的 API 进行日志输出，更加常见的是使用 SLF4J，即简单日志门面（Simple Logging Facade for Java）进行日志的输出操作。SLF4J 并不是具体的日志解决方案，它只服务于各种各样的日志系统。按照官方的说法，SLF4J 是一个用于日志系统的简单 Facade，允许终端用户在部署其应用时使用其所选择的日志系统。有关门面模式更多详情请参考本书附录。

10.1　Log4j2 配置详解

　　一般企业开发中会使用一个 XML 配置文件来对 Log4j2 进行配置，一般命名为 log4j2.xml，配置文件的获取可以从官网下载，下面将详细介绍配置文件的组成及其含义。

　　Log4j2 的配置文件结构如图 10-1 所示的树状图。

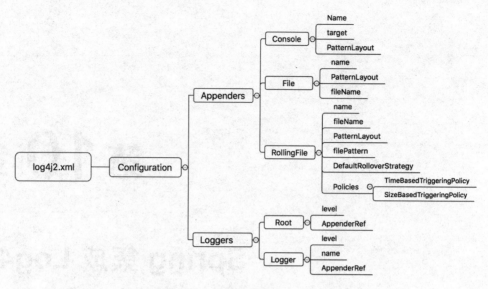

图 10-1　Log4j2 配置文件结构图

一个基本的 log4j2.xml 配置文件应该包含如下元素：

```xml
<?xml version="1.0" encoding="UTF-8"?>
 <Configuration status="WARN">
   <Appenders>
     <Console name="Console" target="SYSTEM_OUT">
        <PatternLayout pattern="%d{yyyy-MM-dd HH:mm:ss.SSS}[%t] %-5level %logger{36} - %msg%n"/>
     </Console>
   </Appenders>
   <Loggers>
     <Root level="info">
       <AppenderRef ref="Console"/>
     </Root>
   </Loggers>
</Configuration>
```

下面解析配置文件中各个元素的含义：

（1）Configuration 根节点

Configuration 元素的属性分析如下。

- status 属性：指定 Log4j2 本身的日志打印级别。
- monitorinterval 属性：用于指定 Log4j 自动重新配置的监测间隔时间，单位是秒，最小值是 5s。

（2）Appenders 元素

Appenders 元素是 Configuration 元素的子节点，可以指定日志输出的路径，常见输出路径有控制台、文件和网络 Socket 等。Appenders 元素常见子节点如下：

- Console：用来将日志输出到控制台。name 属性指定 Appender 的名字；target 属性可以为 SYSTEM_OUT 或 SYSTEM_ERR，一般使用 SYSTEM_OUT；PatternLayout 元素设置日志输出的格式。
- File：用于设置将日志输出到指定的文件中。fileName 指定输出日志的目标文件带全路径的文件名；name 和 PatternLayout 元素的作用同 Appenders 中的 PatternLayout。
- RollingFile：用于将日志输出到滚动文件中，日志输出时会判断文件是否满足封存文件的要求，若满足，则将文件封存并把日志写入到下一个滚动文件。其中的 name、fileName 和 PatternLayout 的作用同 File；filePattern 属性用于指定新建日志文件的命名格式；DefaultRolloverStrategy 用来指定同一个目录下最多有几个日志文件时开始删除最旧的日志文件，并创建新的文件（通过 max 属性）；Policies 元素指定滚动日志的策略，即何时新建文件输出日志。

Policies 节点有以下两个子节点。

- TimeBasedTriggeringPolicy：基于时间的滚动策略，interval 属性用于配置滚动一次的时间，默认是 1h；modulate=true 用于调整时间。
- SizeBasedTriggeringPolicy：基于指定文件大小的滚动策略，size 属性用来定义每个日志文件的大小。

（3）Loggers 元素

- Root：用于指定项目的根日志，如果没有单独指定 Logger，则会使用 Root 作为默认的日志输出。level 属性用于定义日志输出级别，共有 8 个，从低到高顺序为 All<Trace<Debug<Info<Warn<Error<Fatal<OFF；AppenderRef 是 Root 的子节点，用于指定将日志输出到 Appenders 元素定义的 Appenders 中。
- Logger：用于单独指定日志的形式，如为指定包下的 class 指定不同的日志级别等。level 属性用于定义 Logger 日志的输出级别，共有 8 个，从低到高顺序为 All<Trace<Debug<Info<Warn<Error<Fatal<OFF；name 属性用于指定该 Logger 所适用的类或者类所在的包全路径；AppenderRef 元素是 Logger 的子节点，用于设置将该日志输出到指定 Appender，如果没有指定，就会默认继承自 Root。如果已指定，那么在指定的 Appender 和 Root 的 Appender 中都会进行日志输出。

（4）PatternLayout

PatternLayout 是用于控制日志输出格式的，其详细配置规则如下。

PatternLayout 的属性如表 10-1 所示。

表10-1　PatternLayout的属性

属 性 名	作　　用
charset	控制输出日志的字符集
pattern	控制输出的日志格式
alwaysWriteExceptions	该属性默认为 true，表示输出异常

(续表)

属性名	作用
heade	可选项，表示包含在每个日志文件顶部的内容
footer	可选项，表示包含在每个日志文件尾部的内容

pattern 属性参数格式描述如表 10-2 所示。

表 10-2　Pattern 属性参数

参数格式	作用
%c{参数} 或 %logger{参数}	控制输出日志名称
%C{参数} 或 %class{参数}	控制输出类型
%d{参数}	控制输出时间
%F|%file	输出文件名
highlight{pattern}{style}	设置高亮显示
%l	设置输出错误的完整位置
%L	输出错误发生的行号
%m 或 %msg 或 %message	用于输出信息
%M 或 %method	输出方法名
%n	输出换行符
%level{参数 1}{参数 2}{参数 3}	控制输出日志的级别
%t 或 %thread	输出创建 logging 事件的线程名

pattern 表示日志对齐。

在任何 pattern 和 "%" 之间加入一个小数，可以是正数，也可以是负数，正数表示右对齐，负数表示左对齐，如下所示：

%20——右对齐，不足 20 个字符则在信息前面用空格补足，超过 20 个字符则保留原信息
%-20——左对齐，不足 20 个字符则在信息后面用空格补足，超过 20 个字符则保留原信息

也可以用小数的形式控制日志的对齐方式，整数位表示输出信息最小为 n 个字符，如果输出信息不够 n 个字符，将用空格补齐，小数位表示输出信息的最大字符数，如果超过 n 个字符，则只保留最后 n 个字符的信息

%.30——如果信息超过 30 个字符，则只保留最后 30 个字符
%20.30——右对齐，不足 20 个字符则在信息前面用空格补足，超过 30 个字符则只保留最后 30 个字符
%-20.30——左对齐，不足 20 个字符则在信息后面用空格补足，超过 30 个字符则只保留最后 30 个字符

10.2　Log4j2 日志级别

Log4j2 中定义了 8 种不同的日志级别，各种日志级别如下（> 表示大于）：

```
OFF > Fatal > Error > Warn > Info > Debug > Trace > All
```

各个日志级别的含义如表 10-3 所示。

表 10-3　Log4j2 的日志级别

参数格式	作　　用
OFF	最高级别，用于关闭所有日志记录
Fatal	输出将会导致应用程序退出的错误事件日志
Error	输出发送错误的日志信息
Warn	输出警告及 Warn 级别以下的日志
Info	消息在粗粒度级别上突出强调应用程序的运行过程，输出开发者感兴趣的或者重要的信息，便于监控程序运行状态
Debug	输出细粒度信息事件，对调试应用程序非常有帮助
Trace	追踪级别，不常用
All	最低级别的，用于打开所有日志记录

当开发者打印日志时需要注意，程序会打印高于或等于所设置级别的日志，设置的日志级别越高，打印出来的日志就越少；设置的日志级别越低，打印的日志就越多。选择合适的日志级别对于高并发场景下控制日志输出量很有帮助，否则可能会导致输出的日志过多或过少，不便于定位和监控系统运行情况。

10.3　Log4j2 实战演练

在本章的开始，本书已经说明企业开发过程中一般不直接使用 Log4j2 的 API 进行日志打印，而是使用 SLF4J 的 API 进行日志打印。Log4j2 的执行过程可以用图 10-2 表示。

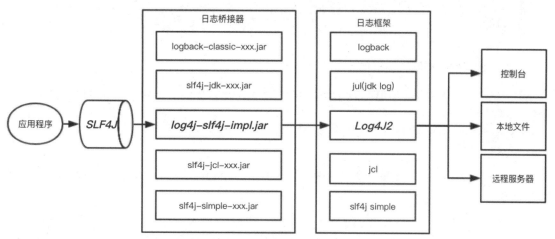

图 10-2　Log4j2 执行原理图

从图 10-2 可以看出，SLF4J 可以有不同的日志实现，Log4J2 只是其中的一种，当程序中将 SFL4J、SFL4J-Log4J2 桥接器、Log4J2 配合使用时，即完成了一个 Log4J2 环境的搭建。一般情况下，将 log4j2.xml 文件放置在 classpath 的 resources 目录下即可（约定优于配置），如果读者想要使用别的配置文件名，或者想将配置文件放置其他目录下，则需要修改相关的配置文件，如 web.xml 中加入如下配置：

```xml
<context-param>
    <param-name>log4jConfiguration</param-name>
    <!-- 日志配置文件路径，请根据具体项目自行调整 -->
    <param-value>classpath:conf/log4j2.xml</param-value>
</context-param>
```

本书案例全部基于默认的配置，即将配置文件 log4j2.xml 放置在 classpath 的 resources 目录下。下面将通过分析一些案例，使读者更容易理解 Log4J2 的使用。

在本书 10.1 节中阐述了最简单的 Log4J2 的配置方式，下面就用这种配置方式通过 Log4J2 实现日志输出。

案例代码中使用的日志操作类都是 SLF4J 的 API，但是在程序执行过程中将会使用 Log4j2 的实现类进行日志打印，案例代码如下：

```java
/**
 * @Author zhouguanya
 * @Date 2018/12/11
 * @Description 使用 Log4j2 实现的第一个 HelloWorld 程序
 */
public class HelloWorld {
    private static final Logger LOGGER = LoggerFactory.getLogger(HelloWorld.class);
    public static void main(String[] args) {
        LOGGER.info("hello world");
        LOGGER.warn("hello world");
    }
}
```

运行案例代码，控制台输出如下：

```
 2018-12-14 22:59:36.565 [main] INFO  com.test.log.helloworld.HelloWorld - hello world
 2018-12-14 22:59:36.569 [main] WARN  com.test.log.helloworld.HelloWorld - hello world
```

下面使用 Log4j2 将日志输出到文件中，修改 log4j2.xml 配置文件，在 Appenders 元素下加入如下配置：

```xml
<File name="File" fileName="log/test.log">
    <PatternLayout pattern="%d{yyyy-MM-dd HH:mm:ss.SSS}
        [%t] %-5level %logger{36} - %msg%n"/>
</File>
```

修改 Root 元素，加入如下配置：

```
<AppenderRef ref="File"/>
```

再次运行案例代码，会发现当前项目路径下多出一个 log 目录，其中包含 test.log 文件，文件内容如下：

```
    2018-12-14 23:09:11.027 [main] INFO  com.test.log.helloworld.HelloWorld - hello world
    2018-12-14 23:09:11.032 [main] WARN  com.test.log.helloworld.HelloWorld - hello world
```

接下来验证 RollingFile 的使用，将日志打印到文件中，并设置每个日志文件的大小为 1KB，当日志输出量大于 1KB 时，将超过 1KB 的历史日志文件进行存档，具体是在 Log4j2 的配置文件 Appenders 元素下新增配置如下：

```
<!-- 这个会打印出所有的 info 及以下级别的信息，每次大小超过 size,
    则这 size 大小的日志会自动存入按年份-月份建立的文件夹下面并进行压缩，作为存档-->
<RollingFile name="RollingFileInfo" fileName="log/RollingFileInfo.log"
    filePattern="log/$${date:yyyy-MM}/info-%d{yyyy-MM-dd}-%i.log">
    <!--控制台只输出 level 及以上级别的信息（onMatch），其他的直接拒绝（onMismatch）-->
    <ThresholdFilter level="info" onMatch="ACCEPT" onMismatch="DENY"/>
    <PatternLayout pattern="%d{yyyy-MM-dd HH:mm:ss.SSS} [%t] %-5level %logger{36} - %msg%n"/>
    <Policies>
    <TimeBasedTriggeringPolicy/>
    <SizeBasedTriggeringPolicy size="1 KB"/>
    </Policies>
</RollingFile>
```

修改 Root 元素，加入如下配置：

```
<AppenderRef ref="RollingFileInfo"/>
```

执行案例代码时需要注意，当第一次执行案例代码时，log 目录下将会创建一个 RollingFileInfo.log 文件，其执行效果如图 10-3 所示。

图 10-3　Log4j2 RollingFile 第一次验证效果图

RollingFileInfo.log 中记录了开发者输出的日志信息：

```
    2018-12-14 23:34:43.362 [main] INFO  com.test.log.helloworld.HelloWorld - hello world
    2018-12-14 23:34:43.367 [main] WARN  com.test.log.helloworld.HelloWorld - hello world
```

反复执行案例代码多次（大约执行 7~8 次），将会发现 log 目录下多出一个按当前月份命名的新的文件夹，其中会将历史日志文件进行存档，执行效果如图 10-3 所示。

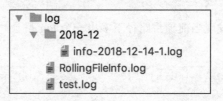

图 10-4　Log4j2 RollingFile 多次验证效果图

存档的日志文件中的记录类似如下的日志：

```
    2018-12-14 23:34:43.362 [main] INFO  com.test.log.helloworld.HelloWorld - hello world
    2018-12-14 23:34:43.367 [main] WARN  com.test.log.helloworld.HelloWorld - hello world
    2018-12-14 23:34:47.224 [main] INFO  com.test.log.helloworld.HelloWorld - hello world
    2018-12-14 23:34:47.228 [main] WARN  com.test.log.helloworld.HelloWorld - hello world
    2018-12-14 23:34:52.897 [main] INFO  com.test.log.helloworld.HelloWorld - hello world
    2018-12-14 23:34:52.902 [main] WARN  com.test.log.helloworld.HelloWorld - hello world
    2018-12-14 23:34:55.837 [main] INFO  com.test.log.helloworld.HelloWorld - hello world
    2018-12-14 23:34:55.842 [main] WARN  com.test.log.helloworld.HelloWorld - hello world
    2018-12-14 23:34:59.193 [main] INFO  com.test.log.helloworld.HelloWorld - hello world
    2018-12-14 23:34:59.198 [main] WARN  com.test.log.helloworld.HelloWorld - hello world
    2018-12-14 23:35:02.337 [main] INFO  com.test.log.helloworld.HelloWorld - hello world
    2018-12-14 23:35:02.341 [main] WARN  com.test.log.helloworld.HelloWorld - hello world
```

10.4　小　　结

本章介绍 SLF4J 与 Log4j2 配合使用的原理，并介绍了门面设计模式的相关知识，更多有关设计模式的知识请参见本书附录。

本章 10.3 节实战演练部分介绍了常见的企业开发中的 Log4j2 的配置，读者可以将其中的配置运用到自己的生产实践中，输出更丰富的系统运行中的日志，为监控系统稳定性提供更好的保障。

第 11 章

Spring 集成 Spring MVC

Spring MVC 是一个优秀的"模型—视图—控制器（MVC）"架构的 Web 框架，Spring MVC 负责发送每个请求到合适的处理程序，使用视图来最终返回响应结果。Spring MVC 是 Spring 产品组合的一部分。本章将讲解 Spring MVC 在企业开发中的实际应用。

11.1 Spring MVC 快速体验

Spring MVC 框架是一个开源的 Java Web 框架，为开发强大的基于 Java 的 Web 应用程序提供全面的架构支持。

Spring MVC 框架提供了 MVC 架构和用于开发灵活和松散耦合的 Web 应用程序的组件。模型（Model）封装了应用程序数据，通常由 POJO 类组成。视图（View）负责渲染模型数据，一般可以输出 HTML、JSP 或 Excel 等。控制器（Controller）负责处理用户请求并构建适当的模型，并将其传递给视图进行渲染。

11.1.1 web.xml 配置

Spring MVC 环境的搭建需要修改 web.xml 文件，在其中加入 DispatcherServlet，配置文件修改如下：

```xml
<servlet>
  <servlet-name>springDispatcherServlet</servlet-name>
  <servlet-class>org.springframework.web.servlet.DispatcherServlet</servlet-class>
  <!-- 配置 Spring mvc 下的配置文件的位置和名称 -->
  <init-param>
```

```xml
      <param-name>contextConfigLocation</param-name>
      <param-value>classpath:spring*.xml</param-value>
    </init-param>
    <load-on-startup>1</load-on-startup>
    <async-supported>true</async-supported>
</servlet>

<servlet-mapping>
    <servlet-name>springDispatcherServlet</servlet-name>
    <url-pattern>/</url-pattern>
</servlet-mapping>
```

11.1.2 创建 Spring MVC 的配置文件

在 11.1.1 节中，可以看到在 DispatcherServlet 中，为属性 contextConfigLocation 指定了配置文件是 classpath 下以 spring 开头的 xml 文件，因此在这一节中，创建一个 springmvc.xml 配置文件，配置文件中通过 InternalResourceViewResolver 配置视图解析器，在本例中，使用 JSP 作为视图层。配置文件内容如下：

```xml
<?xml version="1.0" encoding="UTF-8"?>
<beans xmlns="http://www.springframework.org/schema/beans"
       xmlns:xsi="http://www.w3.org/2001/XMLSchema-instance"
       xmlns:context="http://www.springframework.org/schema/context"
       xmlns:mvc="http://www.springframework.org/schema/mvc"
       xsi:schemaLocation="http://www.springframework.org/schema/beans
http://www.springframework.org/schema/beans/spring-beans.xsd
        http://www.springframework.org/schema/context
http://www.springframework.org/schema/context/spring-context-4.0.xsd
        http://www.springframework.org/schema/mvc
http://www.springframework.org/schema/mvc/spring-mvc-4.0.xsd">
    <!-- 配置自动扫描的包 -->
    <context:component-scan base-package="com.test">
</context:component-scan>
    <!-- 配置视图解析器 把handler 方法返回值解析为实际的物理视图 -->
    <bean
class="org.springframework.web.servlet.view.InternalResourceViewResolver">
        <property name = "prefix" value="/pages/"></property>
        <property name = "suffix" value = ".jsp"></property>
    </bean>
</beans>
```

11.1.3 创建 Spring MVC 的视图文件

创建视图文件 hellospringmvc.jsp，其中"${message}"是动态输出的，其代码如下：

```jsp
<%@ page contentType="text/html; charset=UTF-8" %>
<html>
    <head>
        <title>Hello World</title>
    </head>
    <body>
        <h2>Hello, ${message}</h2>
    </body>
</html>
```

11.1.4 创建控制器

创建 HelloSpringMvcController 控制器，返回字符串 hellospringmvc，其代码如下：

```java
/**
 * @Author zhouguanya
 * @Date 2018/12/15
 * @Description
 */
@Controller
@RequestMapping("/hello/springmvc")
public class HelloSpringMvcController {

    @RequestMapping(method = RequestMethod.GET)
    public String printHello(ModelMap model) {
        model.addAttribute("message", "Welcome to Spring MVC");
        return "hellospringmvc";
    }

}
```

11.1.5 测试运行

通过 tomcat 部署开发好的 Spring MVC 项目，即可验证开发的功能。在浏览器中输入 url：http://localhost:8080/hello/springmvc，执行结果如图 11-1 所示。

Hello, Welcome to Spring MVC

图 11-1　Spring MVC JSP 视图运行结果图

至此完成了一个最简单的 Spring MVC 的创建和验证过程，下一节将介绍更多有关 Spring MVC 视图层的呈现。

11.2 Spring MVC 视图呈现

Spring MVC 除了可以返回 JSP 视图，还可以返回多种视图，本节将介绍各种视图层的使用。

11.2.1 FreeMarker 视图的实现

Spring MVC 集成 FreeMarker 的环境搭建，需要使用到如下一些依赖：

```xml
<dependency>
    <groupId>org.freemarker</groupId>
    <artifactId>freemarker</artifactId>
    <version>${freemarker.version}</version>
</dependency>
<dependency>
  <groupId>org.springframework</groupId>
  <artifactId>spring-context-support</artifactId>
  <version>${spring.framework.version}</version>
</dependency>
```

修改 Spring 的配置文件，在其中加入 FreeMarker 的视图解析器：

```xml
<!-- 配置 freeMarker 的模板路径 -->
<bean class="org.springframework.web.servlet.view.freemarker.FreeMarkerConfigurer">
    <property name="templateLoaderPath" value="/template/" />
    <property name="defaultEncoding" value="UTF-8" />
</bean>
<!-- freemarker 视图解析器 -->
<bean class="org.springframework.web.servlet.view.freemarker.FreeMarkerViewResolver">
    <property name="suffix" value=".ftl" />
    <property name="contentType" value="text/html;charset=UTF-8" />
    <property name="order" value="0"></property>
</bean>
```

值得注意的是，当有多个视图解析器时，可以通过 order 属性指定各个视图解析器的优先级。
创建一个 FreeMarker 文件 helloFreeMarker.ftl，代码如下：

```html
<html>
    <body>
        <h1>Hello,${msg}</h1><br/>
    </body>
</html>
```

创建一个控制器 FreeMarkerController，返回 helloFreeMarker.ftl 视图，控制器代码如下：

```java
/**
 * @Author zhouguanya
 * @Date 2018/12/16
 * @Description
 */
@Controller
public class FreeMarkerController {
    @RequestMapping("/hello/freemarker")
    public String hello(ModelMap map){
        map.put("msg", "Welcome to FreeMarker World");
        return "helloFreeMarker";
    }
}
```

部署 tomcat 启动应用程序，用浏览器访问地址 http://localhost:8080/hello/freemarker，运行结果如图 11-2 所示。

Hello,Welcome to FreeMarker World

图 11-2　Spring MVC FreeMarker 视图运行结果图

11.2.2　XML 视图的实现

视图层不仅仅只有页面文件，也可以是数据文件，比如在多个系统间进行交互的时候，可以使用 XML 进行通信。本节讲解 Spring MVC 使用 XML 视图的实现。

首先修改 Spring 的配置文件，加入如下配置：

```
<mvc:annotation-driven/>
```

该注解会自动注册 RequestMappingHandlerMapping 与 RequestMappingHandlerAdapter 两个 Bean，这是 Spring MVC 为"@Controller"分发请求所必需的，并且提供了数据绑定支持、"@NumberFormatannotation"支持、"@DateTimeFormat"支持、"@Valid"读写 XML 的支持（JAXB）和读写 JSON 的支持（默认 Jackson）等功能。

下一步创建一个 POJO 类 User，其中包含两个属性 userName 和 userAge，User 类的代码如下：

```java
/**
 * @Author zhouguanya
 * @Date 2018/12/16
 * @Description
 */
@XmlRootElement(name = "user")
public class User implements Serializable {
    private String userName;
    private int userAge;
```

```java
    public String getUserName() {
        return userName;
    }
    @XmlElement
    public void setUserName(String userName) {
        this.userName = userName;
    }

    public int getUserAge() {
        return userAge;
    }
    @XmlElement
    public void setUserAge(int userAge) {
        this.userAge = userAge;
    }
}
```

其中"@XmlRootElement"表示 XML 文件的根节点，"@XmlElement"表示子节点。

首先创建一个 Controller，返回 User 对象，Controller 代码如下：

```java
/**
 * @Author zhouguanya
 * @Date 2018/12/16
 * @Description
 */
@Controller
public class XmlController {
    @RequestMapping("/hello/xml")
    @ResponseBody
    public User getUser() {
        User user = new User();
        user.setUserName("Michael");
        user.setUserAge(20);
        return user;
    }
}
```

其中"@ResponseBody"注解的作用是将 Controller 的方法返回的对象通过适当的转换器转换为指定的格式之后，写入到 Response 对象的 body 区，通常用来返回 JSON 数据或者是 XML 数据。

再部署项目，在浏览器中访问地址 http://localhost:8080/hello/xml，执行结果如图 11-3 所示。

```
<user>
  <userAge>20</userAge>
  <userName>Michael</userName>
</user>
```

图 11-3　Spring MVC XML 视图运行结果图

11.2.3 JSON 视图的实现

除了 XML 以外，JSON 也是常用的数据交互方式，本节介绍使用 Spring MVC 返回 JSON 视图的案例。

首先创建一个 Book 实体类，其中含有书名 name、价格 price 和作者 author 三个属性，Book 代码如下：

```java
/**
 * @Author zhouguanya
 * @Date 2018/12/16
 * @Description
 */
public class Book {
    private String name;
    private int price;
    private String author;
    public String getName() {
        return name;
    }

    public void setName(String name) {
        this.name = name;
    }

    public int getPrice() {
        return price;
    }

    public void setPrice(int price) {
        this.price = price;
    }

    public String getAuthor() {
        return author;
    }

    public void setAuthor(String author) {
        this.author = author;
    }
}
```

再创建控制器 JsonController，返回 Book 对象：

```java
/**
 * @Author zhouguanya
 * @Date 2018/12/16
```

```
 * @Description
 */
@Controller
public class JsonController {
    @RequestMapping("/hello/book")
    @ResponseBody
    public Book getBook() {
        Book book = new Book();
        book.setName("Spring 5");
        book.setPrice(50);
        book.setAuthor("Michael");
        return book;
    }
}
```

用 tomcat 部署项目，在浏览器中访问地址 http://localhost:8080/hello/book，得到的执行结果如图 11-4 所示。

```
{
    "name": "Spring 5",
    "price": 50,
    "author": "Michael"
}
```

图 11-4　Spring MVC JSON 视图运行结果图

11.3　Spring MVC 拦截器

Spring MVC 提供了拦截器功能，下面讲解 Spring MVC 拦截器使用。
首先创建一个控制器 InterceptorController，其中包含两个方法 hello()和 bye()，代码如下：

```
/**
 * @Author zhouguanya
 * @Date 2018/12/17
 * @Description SpringMVC 拦截器使用
 */
@Controller
@RequestMapping("/interceptor")
public class InterceptorController {

    @RequestMapping("/hello")
    @ResponseBody
    public String hello() {
        return "hello interceptor";
```

```java
    }

    @RequestMapping("/bye")
    @ResponseBody
    public String bye() {
        return "bye bye interceptor";
    }
}
```

创建拦截器 HelloInterceptor 类，其实现了 HandlerInterceptor 接口，拦截器实现如下：

```java
/**
 * @Author zhouguanya
 * @Date 2018/12/17
 * @Description
 */
public class HelloInterceptor implements HandlerInterceptor {
    @Override
    public boolean preHandle(HttpServletRequest request, HttpServletResponse response, Object handler) throws Exception {
        System.out.printf("进入 preHandle 方法，请求 URL=%s,请求 URI=%s%n",request.getRequestURL().toString(), request.getRequestURI());
        return true;
    }

    @Override
    public void postHandle(HttpServletRequest request, HttpServletResponse response, Object handler, ModelAndView modelAndView) throws Exception {
        System.out.printf("进入 postHandle 方法，请求 URL=%s,请求 URI=%s%n",request.getRequestURL().toString(), request.getRequestURI());
    }

    @Override
    public void afterCompletion(HttpServletRequest request, HttpServletResponse response, Object handler, Exception ex) throws Exception {
        System.out.printf("进入 afterCompletion 方法，请求 URL=%s,请求 URI=%s%n",request.getRequestURL().toString(), request.getRequestURI());
    }
}
```

修改 Spring 配置文件，加入<mvc:interceptors>标签如下：

```xml
<mvc:interceptors>
    <mvc:interceptor>
        <mvc:mapping path="/interceptor/**"/>
        <bean class="com.test.mvc.interceptor.HelloInterceptor"/>
```

```
        </mvc:interceptor>
    </mvc:interceptors>
```

部署 Spring MVC 项目，在浏览器中输入地址 http://localhost:8080/interceptor/hello，可以发现浏览器中输出如下信息：

```
hello interceptor
```

控制台，将输出如下信息：

```
    进入 preHandle 方法，请求 URL=http://localhost:8080/interceptor/hello，请求
URI=/interceptor
    /hello
    进入 postHandle 方法，请求 URL=http://localhost:8080/interceptor/hello，请求
URI=/interceptor/hello
    进入 afterCompletion 方法，请求 URL=http://localhost:8080/interceptor/hello，请
求 URI=/interceptor/hello
```

在浏览器中输入 http://localhost:8080/interceptor/bye，可以发现浏览器中输出如下信息：

```
bye bye interceptor
```

控制台将输入如下信息：

```
    进入 preHandle 方法，请求 URL=http://localhost:8080/interceptor/bye，请求
URI=/interceptor/bye
    进入 postHandle 方法，请求 URL=http://localhost:8080/interceptor/bye，请求
URI=/interceptor/bye
    进入 afterCompletion 方法，请求 URL=http://localhost:8080/interceptor/bye，请求
URI=/interceptor/bye
```

从以上结果可以发现，Spring MVC 拦截器在正常执行过程中执行了 HelloInterceptor 中的 3 个方法，即 preHandle()方法、postHandle()方法和 afterCompletion()方法。有关 Spring MVC 拦截器原理请参考图 11-6，更多有关 Spring MVC 拦截器的底层实现请参考 11.4 节代码解析部分。

11.4 Spring MVC 代码解析

从 web.xml 的配置可以看到，Spring MVC 的核心处理逻辑是从 DispatcherServlet 这个 servlet 开始的，它在容器启动时初始化参数 contextConfigLocation，本书中使用的是 classpath 下的以 spring 命名开头的 xml 配置文件。

进入到 DispatcherServlet 类的代码，查看 DispatcherServlet 类的集成结构如图 11-5 所示。

从类的集成结构上可以看出，DispatcherServlet 类是 Servlet 的子类。熟悉 Servlet 规范的读者应该都知道，Servlet 提供了一些核心方法，如 doGet、doPost、service 等。

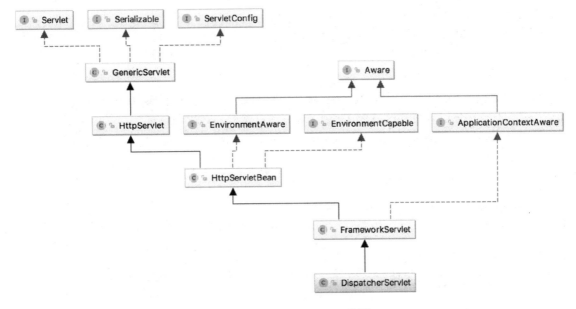

图 11-5 DispatcherServlet 类图

首先分析初始化过程。从图 11-5 可以看到，DispatcherServlet 继承了 HttpServletBean 类，其中含有初始化方法 init()，该方法的代码如下：

```java
@Override
public final void init() throws ServletException {

    // Set bean properties from init parameters.
    PropertyValues pvs = new ServletConfigPropertyValues(getServletConfig(), this.requiredProperties);
    if (!pvs.isEmpty()) {
        try {
            BeanWrapper bw = PropertyAccessorFactory.forBeanPropertyAccess(this);
            ResourceLoader resourceLoader = new
                ServletContextResourceLoader(getServletContext());
            bw.registerCustomEditor(Resource.class,
                new ResourceEditor(resourceLoader, getEnvironment()));
            initBeanWrapper(bw);
            bw.setPropertyValues(pvs, true);
        }
        catch (BeansException ex) {
            if (logger.isErrorEnabled()) {
                logger.error("Failed to set bean properties on servlet '" +
                    getServletName() + "'", ex);
            }
            throw ex;
        }
    }
```

```
    // Let subclasses do whatever initialization they like.
    initServletBean();
}
```

该方法是获取 web.xml 中配置的属性参数的，并将这些属性设置到 DispatcherServlet 中；init() 方法中还包含一个模板方法 initServletBean()，该方法需要其子类去实现。

从图 11-5 中可以看到，HttpServletBean 类的子类是 FrameworkServlet，这里实现了模板方法 initServletBean()，代码如下：

```
@Override
protected final void initServletBean() throws ServletException {
    getServletContext().log("Initializing Spring " + getClass().getSimpleName() + " '" + getServletName() + "'");
    if (logger.isInfoEnabled()) {
        logger.info("Initializing Servlet '" + getServletName() + "'");
    }
    long startTime = System.currentTimeMillis();

    try {
        this.webApplicationContext = initWebApplicationContext();
        initFrameworkServlet();
    }
    catch (ServletException | RuntimeException ex) {
        logger.error("Context initialization failed", ex);
        throw ex;
    }

    if (logger.isDebugEnabled()) {
        String value = this.enableLoggingRequestDetails ?"shown which may lead to unsafe logging of potentially sensitive data" :"masked to prevent unsafe logging of potentially sensitive data";
        logger.debug("enableLoggingRequestDetails='" + this.enableLoggingRequestDetails + "': request parameters and headers will be " + value);
    }

    if (logger.isInfoEnabled()) {
        logger.info("Completed initialization in " + (System.currentTimeMillis() - startTime) + " ms");
    }
}
```

可以看到 initServletBean() 方法的核心是初始化 WebApplicationContext 对象，即是 initWebApplicationContext()方法：

```java
    protected WebApplicationContext initWebApplicationContext() {
        WebApplicationContext rootContext =
WebApplicationContextUtils.getWebApplicationContext (getServletContext());
        WebApplicationContext wac = null;

        if (this.webApplicationContext != null) {
           wac = this.webApplicationContext;
           if (wac instanceof ConfigurableWebApplicationContext) {
              ConfigurableWebApplicationContext cwac =
(ConfigurableWebApplicationContext) wac;
              if (!cwac.isActive()) {
                 if (cwac.getParent() == null) {
                    // The context instance was injected without an explicit parent -> set
                    // the root application context (if any; may be null) as the parent
                    cwac.setParent(rootContext);
                 }
                 configureAndRefreshWebApplicationContext(cwac);
              }
           }
        }
        if (wac == null) {
           wac = findWebApplicationContext();
        }
        if (wac == null) {
           // No context instance is defined for this servlet -> create a local one
           wac = createWebApplicationContext(rootContext);
        }

        if (!this.refreshEventReceived) {
            synchronized (this.onRefreshMonitor) {
               onRefresh(wac);
            }
        }

        if (this.publishContext) {
           // Publish the context as a servlet context attribute.
           String attrName = getServletContextAttributeName();
           getServletContext().setAttribute(attrName, wac);
        }

        return wac;
     }
```

这个方法比较长，这里将分析最重要的 createWebApplicationContext() 方法。createWebApplicationContext()方法是创建 WebApplicationContext 对象的，进入该方法，发现其调用了 FrameworkServlet 的重载方法 createWebApplicationContext(ApplicationContext parent)，重载方法的代码如下：

```java
protected WebApplicationContext createWebApplicationContext(@Nullable
ApplicationContext parent) {
    Class<?> contextClass = getContextClass();
    if (!ConfigurableWebApplicationContext.class.isAssignableFrom
(contextClass)) {
        throw new ApplicationContextException("Fatal initialization error in
servlet with name '" + getServletName() +"': custom WebApplicationContext class
[" + contextClass.getName() +"] is not of type ConfigurableWebApplicationContext");
    }
    ConfigurableWebApplicationContext wac =(ConfigurableWebApplicationContext)
BeanUtils.instantiateClass (contextClass);

    wac.setEnvironment(getEnvironment());
    wac.setParent(parent);
    String configLocation = getContextConfigLocation();
    if (configLocation != null) {
      wac.setConfigLocation(configLocation);
    }
    configureAndRefreshWebApplicationContext(wac);

    return wac;
}
```

在该方法中使用 BeanUtils 类的 instantiateClass()方法通过反射创建了 web 上下文对象，即 ConfigurableWebApplicationContext 对象，并且将 web.xml 中配置的 contextConfigLocation 元素设置到 ConfigurableWebApplicationContext 对象中。最后执行刷新 web 应用上下文的方法 configureAndRefreshWebApplicationContext()：

```java
protected void
configureAndRefreshWebApplicationContext(ConfigurableWebApplicationCon
text wac) {
    if (ObjectUtils.identityToString(wac).equals(wac.getId())) {
      // The application context id is still set to its original default value
      // -> assign a more useful id based on available information
      if (this.contextId != null) {
        wac.setId(this.contextId);
      }
      else {
        // Generate default id...
```

```
            wac.setId(ConfigurableWebApplicationContext.
APPLICATION_CONTEXT_ID_PREFIX + ObjectUtils.getDisplayString
(getServletContext().getContextPath()) + '/' + getServletName());
        }
    }

    wac.setServletContext(getServletContext());
    wac.setServletConfig(getServletConfig());
    wac.setNamespace(getNamespace());
    wac.addApplicationListener(new SourceFilteringListener(wac, new
ContextRefreshListener()));
    ConfigurableEnvironment env = wac.getEnvironment();
    if (env instanceof ConfigurableWebEnvironment) {
    ((ConfigurableWebEnvironment)env).initPropertySources(getServletContext(),
getServletConfig());
    }

    postProcessWebApplicationContext(wac);
    applyInitializers(wac);
    wac.refresh();
}
```

该方法的最后一行 ConfigurableWebApplicationContext.refresh()方法会进入到 IoC 容器的初始化过程，具体细节请参考本书第 2 章 IoC 容器的代码解析，此处不再赘述。

回到 FrameworkServlet 的 initWebApplicationContext()方法，当 web 应用上下文对象创建完后，将会继续执行 onRefresh()方法。onRefresh()方法由 DispatcherServlet 类实现：

```
@Override
protected void onRefresh(ApplicationContext context) {
    initStrategies(context);
}
```

onRefresh()方法中只有一行代码，调用 initStrategies()方法：

```
protected void initStrategies(ApplicationContext context) {
    // 初始化文件上传处理器
    initMultipartResolver(context);
    // 初始化本地化处理器
    initLocaleResolver(context);
    // 初始化主题处理器
    initThemeResolver(context);
    // 初始化处理器映射器(用来保存 Controller 中配置的 RequestMapping 与 Method 映射
关系)
    initHandlerMappings(context);
    // 初始化处理器适配器(用来动态匹配 Method 参数，包括类转换、动态赋值)
    initHandlerAdapters(context);
```

```
    // 初始化异常处理器
    initHandlerExceptionResolvers(context);
    // 初始化请求至视图名转换
    initRequestToViewNameTranslator(context);
    // 初始化视图解析器
    initViewResolvers(context);
    // 初始化 flash 映射管理器
    initFlashMapManager(context);
}
```

初始化策略的方法 initStrategies()在不指定个性化配置文件的情况下，会使用默认的配置进行初始化，默认配置位于 DispatcherServlet.properties 中：

```
org.springframework.web.servlet.LocaleResolver=
    org.springframework.web.servlet.i18n.AcceptHeaderLocaleResolver

    org.springframework.web.servlet.ThemeResolver=
    org.springframework.web.servlet.theme.FixedThemeResolver

    org.springframework.web.servlet.HandlerMapping=
    org.springframework.web.servlet.handler.BeanNameUrlHandlerMapping,\
        org.springframework.web.servlet.mvc.method.annotation.
RequestMappingHandlerMapping

    org.springframework.web.servlet.HandlerAdapter=
    org.springframework.web.servlet.mvc.HttpRequestHandlerAdapter,\
    org.springframework.web.servlet.mvc.
SimpleControllerHandlerAdapter,\
        org.springframework.web.servlet.mvc.method.annotation.
RequestMappingHandlerAdapter

    org.springframework.web.servlet.HandlerExceptionResolver=
    org.springframework.web.servlet.mvc.method.annotation.
ExceptionHandlerExceptionResolver,\
        org.springframework.web.servlet.mvc.annotation.
ResponseStatusExceptionResolver,\ org.springframework.web.servlet.mvc.support.
DefaultHandlerExceptionResolver

    org.springframework.web.servlet.RequestToViewNameTranslator=
    org.springframework.web.servlet.view.
DefaultRequestToViewNameTranslator

    org.springframework.web.servlet.ViewResolver=
    org.springframework.web.servlet.view.InternalResourceViewResolver

    org.springframework.web.servlet.FlashMapManager=
    org.springframework.web.servlet.support.SessionFlashMapManager
```

从 DispatcherServlet.properties 默认配置策略中可以看到，其中对每种处理器配置了默认的实现，即本地化处理器 LocaleResolver 默认使用的是 AcceptHeaderLocaleResolver。初始化各种处理器的功能已经通过注释的形式 initStrategies() 方法中，如初始化处理器映射器的方法 initHandlerMappings()是用来处理 Controller 中 RequestMapping 和 Method 之间的映射关系的，initHandlerMappings()方法的实现如下：

```java
    private void initHandlerMappings(ApplicationContext context) {
        this.handlerMappings = null;

        if (this.detectAllHandlerMappings) {
            // Find all HandlerMappings in the ApplicationContext, including ancestor contexts.
            Map<String, HandlerMapping> matchingBeans =
                    BeanFactoryUtils.beansOfTypeIncludingAncestors(context, HandlerMapping.class, true, false);
            if (!matchingBeans.isEmpty()) {
                this.handlerMappings = new ArrayList<>(matchingBeans.values());
                // We keep HandlerMappings in sorted order.
                AnnotationAwareOrderComparator.sort(this.handlerMappings);
            }
        }
        else {
            try {
                HandlerMapping hm = context.getBean(HANDLER_MAPPING_BEAN_NAME, HandlerMapping.class);
                this.handlerMappings = Collections.singletonList(hm);
            }
            catch (NoSuchBeanDefinitionException ex) {
                // Ignore, we'll add a default HandlerMapping later.
            }
        }

        // Ensure we have at least one HandlerMapping, by registering
        // a default HandlerMapping if no other mappings are found.
        if (this.handlerMappings == null) {
            this.handlerMappings = getDefaultStrategies(context, HandlerMapping.class);
            if (logger.isTraceEnabled()) {
                logger.trace("No HandlerMappings declared for servlet '" + getServletName() +"': using default strategies from DispatcherServlet.properties");
            }
        }
    }
```

分析完初始化过程后，下面继续分析 DispatcherServlet 的执行过程。

熟悉 Servlet 规范的读者都应该知道，在图 11-5 所示的类继承结构中，最顶层的接口 Servlet 中含有 service()方法，该方法是处理请求的，该方法的具体实现是图 11-5 所示类图的 HttpServlet

中。细心的读者应该会发现 HttpServlet 类并不是 Spring 中的类,因为该类位于 javax.servlet.http 包下,HttpServlet 是 JavaEE 扩展中的类。

再次回到图 11-5 所示的类图中,FrameworkServlet 继承了 HttpServlet,并重写了其中的 service()方法:

```
@Override
protected void service(HttpServletRequest request, HttpServletResponse response)
    throws ServletException, IOException {

  HttpMethod httpMethod = HttpMethod.resolve(request.getMethod());
  if (httpMethod == HttpMethod.PATCH || httpMethod == null) {
    processRequest(request, response);
  }
  else {
    super.service(request, response);
  }
}
```

FrameworkServlet 是 Spring 中的类,即 FrameworkServlet 扩展了 Java EE 中 HttpServlet 的功能,新增了 processRequest()方法。

再次回到 HttpServlet 类中,发现除了处理 service()方法以外,还有一些处理 Http 请求的方法,如 doGet()、doPost()和 doDelete()等。

```
protected void doGet(HttpServletRequest req, HttpServletResponse resp)
    throws ServletException, IOException
{
  String protocol = req.getProtocol();
  String msg = lStrings.getString("http.method_get_not_supported");
  if (protocol.endsWith("1.1")) {
    resp.sendError(HttpServletResponse.SC_METHOD_NOT_ALLOWED, msg);
  } else {
    resp.sendError(HttpServletResponse.SC_BAD_REQUEST, msg);
  }
}
```

回到 FrameworkServlet 类中,发现该类也重写了 HttpServlet 类中的有关处理 HTTP 请求的方法:

```
@Override
protected final void doGet(HttpServletRequest request, HttpServletResponse response)
    throws ServletException, IOException {

  processRequest(request, response);
}
```

在 FrameworkServlet 重写的众多 HttpServlet 的方法中,可以发现重写方法中都使用到了

processRequest()方法,因此可以证明 processRequest()方法是 FrameworkServlet 类中最核心的处理请求的方法。下面开始分析 processRequest()方法:

```
    protected final void processRequest(HttpServletRequest request,
HttpServletResponse re sponse)
        throws ServletException, IOException {
    long startTime = System.currentTimeMillis();
    Throwable failureCause = null;

    LocaleContext previousLocaleContext = LocaleContextHolder.
getLocaleContext();
    LocaleContext localeContext = buildLocaleContext(request);

    RequestAttributes previousAttributes = RequestContextHolder.
getRequestAttributes();
    ServletRequestAttributes requestAttributes = buildRequestAttributes
(request, response, previousAttributes);

    WebAsyncManager asyncManager = WebAsyncUtils.getAsyncManager(request);
    asyncManager.registerCallableInterceptor
(FrameworkServlet.class.getName(), new RequestBindingInterceptor());

    initContextHolders(request, localeContext, requestAttributes);

    try {
        doService(request, response);
    }
    catch (ServletException | IOException ex) {
        failureCause = ex;
        throw ex;
    }
    catch (Throwable ex) {
        failureCause = ex;
        throw new NestedServletException("Request processing failed", ex);
    }

    finally {
        resetContextHolders(request, previousLocaleContext,
previousAttributes);
        if (requestAttributes != null) {
            requestAttributes.requestCompleted();
        }
        logResult(request, response, failureCause, asyncManager);
        publishRequestHandledEvent(request, response, startTime, failureCause);
    }
}
```

processRequest()方法比较复杂,挑选其中最核心的 doService()方法进行分析。从

FrameworkServlet 代码可以发现 doService()方法是一个抽象方法,并没有任何的实现逻辑,需要FrameworkServlet 子类 DispatcherServlet 实现:

```java
@Override
protected void doService(HttpServletRequest request, HttpServletResponse response) throws Exception {
    logRequest(request);
    Map<String, Object> attributesSnapshot = null;
    if (WebUtils.isIncludeRequest(request)) {
        attributesSnapshot = new HashMap<>();
        Enumeration<?> attrNames = request.getAttributeNames();
        while (attrNames.hasMoreElements()) {
            String attrName = (String) attrNames.nextElement();
            if (this.cleanupAfterInclude || attrName.startsWith(DEFAULT_STRATEGIES_PREFIX)) {
                attributesSnapshot.put(attrName, request.getAttribute(attrName));
            }
        }
    }

    // Make framework objects available to handlers and view objects.
    request.setAttribute(WEB_APPLICATION_CONTEXT_ATTRIBUTE, getWebApplicationContext());
    request.setAttribute(LOCALE_RESOLVER_ATTRIBUTE, this.localeResolver);
    request.setAttribute(THEME_RESOLVER_ATTRIBUTE, this.themeResolver);
    request.setAttribute(THEME_SOURCE_ATTRIBUTE, getThemeSource());

    if (this.flashMapManager != null) {
        FlashMap inputFlashMap = this.flashMapManager.retrieveAndUpdate(request, response);
        if (inputFlashMap != null) {
            request.setAttribute(INPUT_FLASH_MAP_ATTRIBUTE, Collections.unmodifiableMap(inputFlashMap));
        }
        request.setAttribute(OUTPUT_FLASH_MAP_ATTRIBUTE, new FlashMap());
        request.setAttribute(FLASH_MAP_MANAGER_ATTRIBUTE, this.flashMapManager);
    }

    try {
        doDispatch(request, response);
    }
    finally {
        if (!WebAsyncUtils.getAsyncManager(request).isConcurrentHandlingStarted()) {
```

```
            // Restore the original attribute snapshot, in case of an include.
            if (attributesSnapshot != null) {
                restoreAttributesAfterInclude(request, attributesSnapshot);
            }
        }
    }
}
```

FrameworkServlet 类中的 doService()方法调用了该类中的 doDispatch()方法，该方法将寻找合适的处理方法执行请求。doDispatch()方法比较复杂，挑选重要的部分代码：

```
......省略代码......
    WebAsyncManager asyncManager = WebAsyncUtils.getAsyncManager(request);

    try {
        ModelAndView mv = null;
        Exception dispatchException = null;

        try {
            processedRequest = checkMultipart(request);
            multipartRequestParsed = (processedRequest != request);

            // 寻找请求对应的 handler（即 Controller 中的处理方法）
            mappedHandler = getHandler(processedRequest);
            if (mappedHandler == null) {
                noHandlerFound(processedRequest, response);
                return;
            }

            // 根据处理器查找 handerler 适配器
            HandlerAdapter ha = getHandlerAdapter(mappedHandler.getHandler());

            // Process last-modified header, if supported by the handler.
            String method = request.getMethod();
            boolean isGet = "GET".equals(method);
            if (isGet || "HEAD".equals(method)) {
                long lastModified = ha.getLastModified(request, mappedHandler.getHandler());
                if (new ServletWebRequest(request, response).checkNotModified(lastModified) && isGet) {
                    return;
                }
            }

            if (!mappedHandler.applyPreHandle(processedRequest, response)) {
                return;
```

```
        }

        // 处理请求,并返回相应的视图
        mv = ha.handle(processedRequest, response, mappedHandler.
getHandler());

        if (asyncManager.isConcurrentHandlingStarted()) {
            return;
        }

        applyDefaultViewName(processedRequest, mv);
        mappedHandler.applyPostHandle(processedRequest, response, mv);
    }
    ......省略代码......
}
```

下面分析 getHandler()方法如何找到请求对应的处理程序,这里涉及一个设计模式——拦截过滤器模式。由于本节关注的重点是 Spring MVC 的代码解析,因此这里只简单介绍拦截过滤器模式,更多有关拦截过滤器模式的介绍请参考本书附录部分。

假设有如下的开发场景,对用户提交的数据进行加密处理,防止明文传输造成数据泄露。用户 A 发起了一次请求,请求中携带身份证 userIdNo、phone 和 password 等敏感信息,假设请求参数是 User 对象:

```
{
    "userIdNo": "1234567890",
    "phone": 18219021754,
    "password": "124"
}
```

一般情况下,开发人员拿到这样的需求都是直接把每个参数值取出后,通过算法将每个参数进行加密操作:

```
//明文 userIdNo
String userIdNo = user.getUserIdNo();
//密文 userIdNo
String encryptionUserIdNo = EncryptionUtils.encryption(userIdNo);
//修改 user 对象 userIdNo 为密文
user.setUserIdNo(encryptionUserIdNo);
```

当处理完后,需要调用加解密工具类中的解密方法,将数据解析出后供用户或第三方使用。对于 phone 和 password 这两个参数也要进行类似的处理。这样的做法主要缺点是正常的业务流程中加入了与业务处理无关的加解密操作,代码可读性降低;当有更多的用户信息需要加密时(如对用户地址 userAddress 进行加密),必须对核心代码逻辑进行修改,不易于维护。

拦截过滤器模式"优雅地"解决了这个问题,拦截过滤器模式原理如图 11-6 所示。

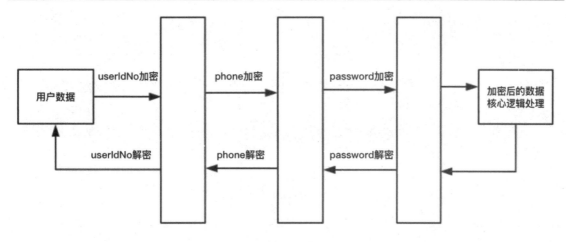

图 11-6 拦截过滤器模式示意图

从图 11-6 可以看到，拦截过滤器模式对核心的代码没有侵入性，当需要增加对用户地址 userAddress 进行加密时，只需要在核心逻辑之前加入对 userAddress 进行加解密的拦截器即可。

回到 DispatcherServlet 类的 getHandler()方法，方法实现如下：

```
protected HandlerExecutionChain getHandler(HttpServletRequest request) throws
  Exception {
  if (this.handlerMappings != null) {
    for (HandlerMapping mapping : this.handlerMappings) {
      HandlerExecutionChain handler = mapping.getHandler(request);
      if (handler != null) {
        return handler;
      }
    }
  }
  return null;
}
```

getHandler()会从 List<HandlerMapping> handlerMappings 中查找对应的 HandlerMapping 对象，并由 HandlerMapping 对象创建 HandlerExecutionChain 对象。细心的读者应该可以发现，List<HandlerMapping> handlerMappings 是初始化时在 initStrategies()方法中完成的。

从 DispatcherServlet 类的 getHandler()方法代码中可以看到，在 getHandler()方法内部调用了 HandlerMapping 类的 getHandler()方法。

HandlerMapping 相关类图如图 11-7 所示。

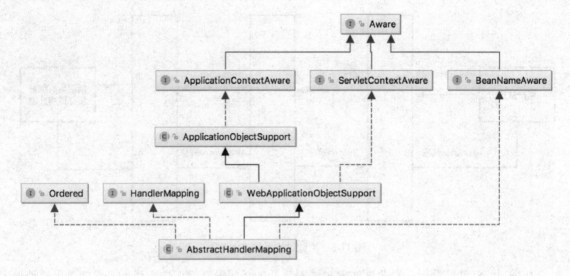

图 11-7　HandlerMapping 类图

从图 11-7 可知，DispatcherServlet 中的 getHandler()方法其实是调用了 AbstractHandlerMapping 的 getHandler()方法，该方法实现如下：

```
    public final HandlerExecutionChain getHandler(HttpServletRequest request) throws Excep
    tion {
      Object handler = getHandlerInternal(request);
      if (handler == null) {
        handler = getDefaultHandler();
      }
      if (handler == null) {
        return null;
      }
      // Bean name or resolved handler?
      if (handler instanceof String) {
        String handlerName = (String) handler;
        handler = obtainApplicationContext().getBean(handlerName);
      }

      HandlerExecutionChain executionChain = getHandlerExecutionChain(handler, request);

      if (logger.isTraceEnabled()) {
         logger.trace("Mapped to " + handler);
      }
      else if (logger.isDebugEnabled() && !request.getDispatcherType().equals(DispatcherType.ASYNC)) {
         logger.debug("Mapped to " + executionChain.getHandler());
      }
```

```
        if (CorsUtils.isCorsRequest(request)) {
            CorsConfiguration globalConfig = this.corsConfigurationSource.getCorsConfiguration(request);
            CorsConfiguration handlerConfig = getCorsConfiguration(handler, request);
            CorsConfiguration config = (globalConfig != null ? globalConfig.combine(handlerConfig) : handlerConfig);
            executionChain = getCorsHandlerExecutionChain(request, executionChain, config);
        }

        return executionChain;
    }
```

在上述代码中 getHandlerInternal()方法会根据请求找到对应的请求处理器 handler，如果没有找到，则会使用 getDefaultHandler()方法获取默认的处理器。

getHandler()方法中最重要的是 getHandlerExecutionChain()方法，因为整个 getHandler()方法就是为了获得一个 HandlerExecutionChain 对象。getHandlerExecutionChain()方法的实现如下：

```
    protected HandlerExecutionChain getHandlerExecutionChain(Object handler,HttpServletRequest request) {
        HandlerExecutionChain chain = (handler instanceof HandlerExecutionChain ?(HandlerExecutionChain) handler : new HandlerExecutionChain(handler));

        String lookupPath = this.urlPathHelper.getLookupPathForRequest(request);
        for (HandlerInterceptor interceptor : this.adaptedInterceptors) {
          if (interceptor instanceof MappedInterceptor) {
            MappedInterceptor mappedInterceptor = (MappedInterceptor) interceptor;
            if (mappedInterceptor.matches(lookupPath, this.pathMatcher)) {
              chain.addInterceptor(mappedInterceptor.getInterceptor());
            }
          }
          else {
            chain.addInterceptor(interceptor);
          }
        }
        return chain;
    }
```

从 getHandlerExecutionChain()方法实现可以看到，首先会判断根据请求获取的处理器是不是 HandlerExecutionChain 对象。如果是，则直接使用；如果不是，则通过请求处理器 Handler 创建一个 HandlerExecutionChain 对象，然后将多个拦截器对象保存到 HandlerExecutionChain 对象中的

List<HandlerInterceptor> interceptorList 属性中。因此 HandlerExecutionChain 封装了请求的处理程序（即 Controller 中的处理方法）和相关的拦截器，正如该类代码中的注释所述：

```
Handler execution chain, consisting of handler object and any handler
interceptors.
```

HandlerExecutionChain 类部分代码如下：

```java
public class HandlerExecutionChain {

    private static final Log logger =
LogFactory.getLog(HandlerExecutionChain.class);

    private final Object handler;

    @Nullable
    private HandlerInterceptor[] interceptors;

    @Nullable
    private List<HandlerInterceptor> interceptorList;
```

返回到 DispatcherServlet 类的 doDispatch() 方法，执行完 getHandler() 方法以后，得到的 HandlerExecutionChain 对象如果为空，则会执行 noHandlerFound() 方法，执行完后直接返回，流程结束。noHandlerFound() 方法的代码如下：

```java
    protected void noHandlerFound(HttpServletRequest request,
HttpServletResponse response) throws Exception {
        if (pageNotFoundLogger.isWarnEnabled()) {
            pageNotFoundLogger.warn("No mapping for " + request.getMethod() + " " +
getRequestUri(request));
        }
        if (this.throwExceptionIfNoHandlerFound) {
            throw new NoHandlerFoundException(request.getMethod(),
getRequestUri(request),
                new ServletServerHttpRequest(request).getHeaders());
        }
        else {
            response.sendError(HttpServletResponse.SC_NOT_FOUND);
        }
    }
```

一般开发过程中不会将 private boolean throwExceptionIfNoHandlerFound = false 属性修改为 true，因此当 HandlerExecutionChain 为空时不会抛出 NoHandlerFoundException 异常，而是会进入 HttpServletResponse 的 sendError 方法。SC_NOT_FOUND 响应码可能不太容易辨识，进入其代码：

```java
    public static final int SC_NOT_FOUND = 404;
```

这个 404 的返回码可能各位读者都很熟悉，这就是常见的 HTTP 请求找不到资源的响应码。

上面介绍完 HandlerExecutionChain 为空的情况，下面将分析 HandlerExecutionChain 非空的情况。

如果执行完 getHandler()方法后得到的 HandlerExecutionChain 对象非空，将会执行 getHandlerAdapter()方法，该方法的作用是根据请求处理器，获取执行操作的请求适配器：

```
protected HandlerAdapter getHandlerAdapter(Object handler) throws
ServletException {
    if (this.handlerAdapters != null) {
      for (HandlerAdapter adapter : this.handlerAdapters) {
        if (adapter.supports(handler)) {
          return adapter;
        }
      }
    }
    throw new ServletException("No adapter for handler [" + handler + "]:The
DispatcherServlet configuration needs to include a HandlerAdapter that supports
this handler");
}
```

即从 private List<HandlerAdapter> handlerAdapters 属性中查找能够支持请求入参中的 handler 执行的 HandlerAdapter 对象。handlerAdapters 对象也是在初始化过程中完成初始化的。

获取到 HandlerAdapter 对象后将会执行以下代码段：

```
String method = request.getMethod();
boolean isGet = "GET".equals(method);
if (isGet || "HEAD".equals(method)) {
    // lastModified 属性可返回文档最后被修改的日期和时间
    long lastModified = ha.getLastModified(request,
mappedHandler.getHandler());
    // checkNotModified 逻辑对比当前 lastModfied 值和 http header 的上次缓存值
    // 如果还没有过期就设置 304 响应头并且返回并结束整个请求流程。否则继续。
    if (new ServletWebRequest(request, response).
checkNotModified(lastModified)
            && isGet) {
      return;
    }
}
```

这段代码主要用来返回 HTTP 304 状态码。

HTTP 304 状态码作用是，客户端有缓存的文档并发出了一个条件性的请求后，服务器会通知客户端，原来缓存的文档对应的服务器端资源并未被修改，客户端还可以继续使用缓存文件。

分析完 doDispatch()有关 HTTP 304 的代码段后，下面将要分析 applyPreHandle()方法，该方法是 HandlerExecutionChain 类中的：

```
boolean applyPreHandle(HttpServletRequest request, HttpServletResponse
response)
```

```
      throws Exception {
    HandlerInterceptor[] interceptors = getInterceptors();
    if (!ObjectUtils.isEmpty(interceptors)) {
      for (int i = 0; i < interceptors.length; i++) {
        HandlerInterceptor interceptor = interceptors[i];
        if (!interceptor.preHandle(request, response, this.handler)) {
          triggerAfterCompletion(request, response, null);
          return false;
        }
        this.interceptorIndex = i;
      }
    }
    return true;
}
```

getInterceptors()方法的作用是将 List<HandlerInterceptor> interceptorList 对象转换为数组对象：

```
public HandlerInterceptor[] getInterceptors() {

    if (this.interceptors == null && this.interceptorList != null) {
      this.interceptors = this.interceptorList.toArray(new HandlerInterceptor[0]);
    }
    return this.interceptors;
}
```

applyPreHandle()方法在获取到所有拦截器后，通过 for 循环调用其中每个拦截器 HandlerInterceptor 的 preHandle()方法，此过程可以类比图 11-6 对用户数据进行加密的过程。

下面将介绍最核心的处理请求的 HandlerAdapter.handle()方法，此方法通过调用对应的处理器对请求做出处理，并返回一个 ModelAndView 对象。HandlerAdapter 是一个接口，因此 handle()方法需要其子类实现。HandlerAdapter 的类图如图 11-8 所示。

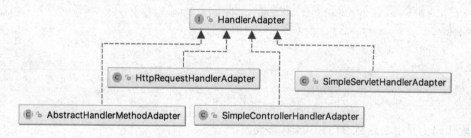

图 11-8 HandlerAdapter 类图

挑选其中的 AbstractHandlerMethodAdapter 分析，handle()方法的实现如下：

```
public final ModelAndView handle(HttpServletRequest request,
HttpServletResponse response, Object handler)
    throws Exception {
```

```
    return handleInternal(request, response, (HandlerMethod) handler);
}
```

AbstractHandlerMethodAdapter 中的 handle()方法是调用 RequestMappingHandlerAdapter 类中的 handleInternal()方法完成功能的。入参中的 handler 其实就是 Controller 中自定义的方法。handleInternal()方法实现如下：

```
protected ModelAndView handleInternal(HttpServletRequest request,HttpServletResponse response, HandlerMethod handlerMethod) throws Exception {

    ModelAndView mav;
    checkRequest(request);

    // Execute invokeHandlerMethod in synchronized block if required.
    if (this.synchronizeOnSession) {
      HttpSession session = request.getSession(false);
      if (session != null) {
        Object mutex = WebUtils.getSessionMutex(session);
        synchronized (mutex) {
           mav = invokeHandlerMethod(request, response, handlerMethod);
        }
      }
      else {
        // No HttpSession available -> no mutex necessary
        mav = invokeHandlerMethod(request, response, handlerMethod);
      }
    }
    else {
      // No synchronization on session demanded at all...
      mav = invokeHandlerMethod(request, response, handlerMethod);
    }

    if (!response.containsHeader(HEADER_CACHE_CONTROL)) {
      if (getSessionAttributesHandler(handlerMethod).hasSessionAttributes()){
        applyCacheSeconds(response, this.cacheSecondsForSessionAttributeHandlers);
      }
      else {
        prepareResponse(response);
      }
    }

    return mav;
}
```

从 invokeHandlerMethod()方法的代码可以看到，其会调用 invokeHandlerMethod()方法并返回一个 ModelAndView 对象。invokeHandlerMethod()方法部分实现如下：

```
protected ModelAndView invokeHandlerMethod(HttpServletRequest
request,HttpServletResponse response, HandlerMethod handlerMethod) throws
Exception {
    ServletWebRequest webRequest = new ServletWebRequest(request, response);
    try {
        ......省略代码......
        ServletInvocableHandlerMethod invocableMethod =
createInvocableHandlerMethod(handlerMethod);
        ......省略代码......
        if (asyncManager.hasConcurrentResult()) {
            Object result = asyncManager.getConcurrentResult();
            mavContainer = (ModelAndViewContainer)
asyncManager.getConcurrentResultContext()[0];
            asyncManager.clearConcurrentResult();
            LogFormatUtils.traceDebug(logger, traceOn -> {
                String formatted =LogFormatUtils.formatValue(result, !traceOn);
                return "Resume with async result [" + formatted + "]";
            });
            invocableMethod = invocableMethod.wrapConcurrentResult(result);
        }
        invocableMethod.invokeAndHandle(webRequest, mavContainer);
        if (asyncManager.isConcurrentHandlingStarted()) {
            return null;
        }
        return getModelAndView(mavContainer, modelFactory, webRequest);
    }
    finally {
        webRequest.requestCompleted();
    }
}
```

invokeHandlerMethod()方法主要将请求参数和处理方法进行封装，并通过封装后的方法对象 ServletInvocableHandlerMethod 进行方法调用：

```
public void invokeAndHandle(ServletWebRequest webRequest,
ModelAndViewContainer mavContainer,Object... providedArgs) throws Exception {
    Object returnValue = invokeForRequest(webRequest, mavContainer,
providedArgs);
    setResponseStatus(webRequest);
    ......省略代码......
    mavContainer.setRequestHandled(false);
```

```
        Assert.state(this.returnValueHandlers != null, "No return value handlers");
        try {
            this.returnValueHandlers.handleReturnValue(returnValue,
getReturnValueType(returnValue), mavContainer, webRequest);
        }
        catch (Exception ex) {
            if (logger.isTraceEnabled()) {
                logger.trace(formatErrorForReturnValue(returnValue), ex);
            }
            throw ex;
        }
    }
```

这里要注意,invokeForRequest()方法调用了 InvocableHandlerMethod 类中的 invokeForRequest()方法,invokeForRequest()方法实现如下:

```
    public Object invokeForRequest(NativeWebRequest request, @Nullable
ModelAndViewContainer mavContainer,Object... providedArgs) throws Exception {
        Object[] args = getMethodArgumentValues(request, mavContainer,
providedArgs);
        if (logger.isTraceEnabled()) {
            logger.trace("Arguments: " + Arrays.toString(args));
        }
        return doInvoke(args);
    }
```

从方法实现可以看出,invokeForRequest()方法会首先获取请求中的处理方法执行所需要的参数,然后调用 doInvoke()方法如下:

```
    protected Object doInvoke(Object... args) throws Exception {
        ReflectionUtils.makeAccessible(getBridgedMethod());
        try {
            return getBridgedMethod().invoke(getBean(), args);
        }
        ......省略代码......
```

doInvoke 方法会调用 getBridgedMethod()方法获取到 Controller 中开发者自定义的处理方法,并获得方法的 Method 对象,然后通过反射调用 Method.invoke()方法完成对 Controller 中自定义的调用。

回到 invokeAndHandle()方法,调用自定义方法后,将返回值通过 handleReturnValue()方法封装到 ModelAndViewContainer 对象中。

再回到 invokeHandlerMethod()方法中,执行完 invokeAndHandle()方法后,将会调用 getModelAndView()方法并返回 ModelAndView 对象。getModelAndView()方法如下:

```
    private ModelAndView getModelAndView(ModelAndViewContainer mavContainer,
ModelFactory modelFactory, NativeWebRequest webRequest) throws Exception{
        modelFactory.updateModel(webRequest, mavContainer);
```

```
        if (mavContainer.isRequestHandled()) {
            return null;
        }
        ModelMap model = mavContainer.getModel();
        ModelAndView mav = new ModelAndView(mavContainer.getViewName(), model, mavContainer.getStatus());
        if (!mavContainer.isViewReference()) {
            mav.setView((View) mavContainer.getView());
        }
        if (model instanceof RedirectAttributes) {
            Map<String, ?> flashAttributes = ((RedirectAttributes) model).getFlashAttributes();
            HttpServletRequest request = webRequest.getNativeRequest(HttpServletRequest.class);
            if (request != null) {
         RequestContextUtils.getOutputFlashMap(request).putAll(flashAttributes);
            }
        }
        return mav;
    }
```

ModelAndView 类中封装了 Controller 中自定义方法执行后的返回值和视图对象。回到 DispatcherServlet 中，至此 doDispatch()方法中的 handle()方法分析完毕。此时得到了 ModelAndView 对象。

接下来分析 DispatcherServlet 中 doDispatch()方法调用的 applyDefaultViewName()方法，该方法的作用是对默认视图的处理：

```
    private void applyDefaultViewName(HttpServletRequest request, @Nullable ModelAndView mv) throws Exception {
        if (mv != null && !mv.hasView()) {
            String defaultViewName = getDefaultViewName(request);
            if (defaultViewName != null) {
                mv.setViewName(defaultViewName);
            }
        }
    }
```

当获得的 ModelAndView 对象视图为空时，将默认视图封装到 ModelAndView 对象中。

下面分析 DispatcherServlet 中 doDispatch()方法中调用的 applyPostHandle()方法，该方法类似 applyPreHandle()方法，不同的是，两者的执行顺序不同：

```
    void applyPostHandle(HttpServletRequest request, HttpServletResponse response, @Nullable ModelAndView mv)
            throws Exception {
        HandlerInterceptor[] interceptors = getInterceptors();
        if (!ObjectUtils.isEmpty(interceptors)) {
            for (int i = interceptors.length - 1; i >= 0; i--) {
```

```
            HandlerInterceptor interceptor = interceptors[i];
            interceptor.postHandle(request, response, this.handler, mv);
        }
    }
}
```

applyPostHandle()方法获取到所有的拦截器,并通过 for 循环从最后一个拦截器开始,调用拦截器的 postHandle()方法,直到调用到第一个拦截器,即退出循环。此过程可以类比图 11-6 所示的用户信息解密过程。

接下来分析 DispatcherServlet 中的 doDispatch()方法调用的 processDispatchResult()方法,此方法是处理最终结果的,包括异常处理、渲染页面和发出完成通知触发拦截器的 afterCompletion()方法执行等,processDispatchResult()方法代码如下:

```
    private void processDispatchResult(HttpServletRequest request,
HttpServletResponse response,@Nullable HandlerExecutionChain mappedHandler,
@Nullable ModelAndView mv,@Nullable Exception exception) throws Exception {

        boolean errorView = false;

        if (exception != null) {
            if (exception instanceof ModelAndViewDefiningException) {
                logger.debug("ModelAndViewDefiningException encountered",
exception);
                mv = ((ModelAndViewDefiningException) exception).getModelAndView();
            }
            else {
                Object handler = (mappedHandler != null ? mappedHandler.getHandler() :
null);
                mv = processHandlerException(request, response, handler, exception);
                errorView = (mv != null);
            }
        }

        // Did the handler return a view to render?
        if (mv != null && !mv.wasCleared()) {
          render(mv, request, response);
           if (errorView) {
              WebUtils.clearErrorRequestAttributes(request);
           }
        }
        else {
           if (logger.isTraceEnabled()) {
              logger.trace("No view rendering, null ModelAndView returned.");
           }
        }

        if (WebAsyncUtils.getAsyncManager(request).isConcurrentHandlingStarted()){
```

```
            // Concurrent handling started during a forward
            return;
        }

        if (mappedHandler != null) {
            mappedHandler.triggerAfterCompletion(request, response, null);
        }
    }
```

processDispatchResult()方法调用 render()方法完成的对视图层的渲染，render()方法代码如下：

```
    protected void render(ModelAndView mv, HttpServletRequest request, 
HttpServletResponse response) throws Exception {
        // Determine locale for request and apply it to the response.
        Locale locale = (this.localeResolver != null ? this.localeResolver.
resolveLocale(request) : request.getLocale());
        response.setLocale(locale);

        View view;
        String viewName = mv.getViewName();
        if (viewName != null) {
            // We need to resolve the view name.
            view = resolveViewName(viewName, mv.getModelInternal(), locale, 
request);
            if (view == null) {
                throw new ServletException("Could not resolve view with name '" + 
mv.getViewName() +"' in servlet with name '" + getServletName() + "'");
            }
        }
        else {
            // No need to lookup: the ModelAndView object contains the actual View 
object.
            view = mv.getView();
            if (view == null) {
                throw new ServletException("ModelAndView [" + mv + "] neither contains 
a view name nor a " +"View object in servlet with name '" + getServletName() + "'");
            }
        }

        // Delegate to the View object for rendering.
        if (logger.isTraceEnabled()) {
            logger.trace("Rendering view [" + view + "] ");
        }
        try {
            if (mv.getStatus() != null) {
                response.setStatus(mv.getStatus().value());
            }
```

```
        view.render(mv.getModelInternal(), request, response);
    }
    catch (Exception ex) {
        if (logger.isDebugEnabled()) {
            logger.debug("Error rendering view [" + view + "]", ex);
        }
        throw ex;
    }
}
```

此方法通过调用 resolveViewName()方法获取视图名,并生成 View 对象,然后通过 render()方法实现对视图的渲染。resolveViewName()方法如下:

```
protected View resolveViewName(String viewName, @Nullable Map<String, Object>
model, Locale locale, HttpServletRequest request) throws Exception {
    if (this.viewResolvers != null) {
        for (ViewResolver viewResolver : this.viewResolvers) {
            View view = viewResolver.resolveViewName(viewName, locale);
            if (view != null) {
                return view;
            }
        }
    }
    return null;
}
```

resolveViewName()方法是从 List<ViewResolver> viewResolvers 多个视图解析器中解析视图名,并返回 View 对象。通过图 11-9 所示的类图可知,11.2 节中使用的两个视图解析器 FreeMarkerViewResolver 和 InternalResourceViewResolver 应该都保存在 viewResolvers 中。

图 11-9 ViewResolver 类图

通过 debug 进入 resolveViewName()方法,观察运行状态如图 11-10 所示。可以发现 List<ViewResolver> viewResolvers 确实保存了 FreeMarkerViewResolver 和 InternalResourceViewResolver,并且两者是按照配置文件中指定的顺序保存的。

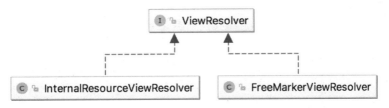

图 11-10 运行中 viewResolvers 状态

回到 DispatcherServlet 中的 render()方法，这里的 render()方法是一个抽象方法，需要其子类实现。和图 10-11 所示。

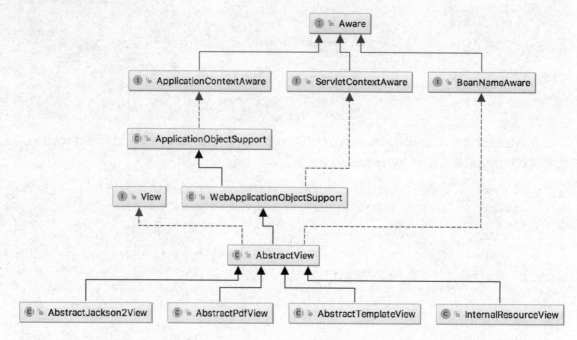

图 11-11　View 类图

通过图 11-11 所示的类图可知，AbstractView 是 View 的子类，其中实现了 render()方法：

```
public void render(@Nullable Map<String, ?> model, HttpServletRequest request,
    HttpServletResponse response) throws Exception {

  if (logger.isDebugEnabled()) {
    logger.debug("View " + formatViewName() +", model " + (model != null ?
model : Collections.emptyMap()) + (this.staticAttributes.isEmpty() ? "" : ", static
attributes " + this.staticAttributes));
  }

  Map<String, Object> mergedModel = createMergedOutputModel(model, request,
response);
  prepareResponse(request, response);
  renderMergedOutputModel(mergedModel, getRequestToExpose(request),
response);
}
```

AbstractView 类中的 render()方法调用的 renderMergedOutputModel()是抽象方法，该方法由不同的子类实现并创建不同的视图，AbstractView 部分子类如图 11-11 所示。

11.5 小　　结

Spring MVC 是企业开发过程中应用最多的 Web 层框架，Spring MVC 在面试中也经常被问到的，并且大部分面试题的侧重点是有关 Spring MVC 底层原理和对 Spring MVC 代码的学习。因此本章对 Spring MVC 代码的解析是十分有必要的。

为了便于读者理解整个 Spring MVC 框架的原理，将 Spring MVC 初始化过程和运行过程用图 11-12 表示。

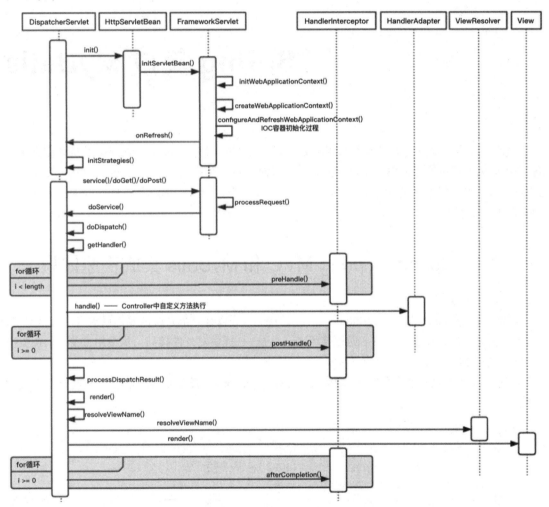

图 11-12　Spring MVC 原理图

第 12 章

Spring 集成 MyBatis

Java 企业级开发中对数据库的操作是非常重要的技术。MyBatis 是一款常见的持久层框架。MyBatis 避免了直接使用 JDBC 编写 SQL，手动设置参数以及获取结果集。MyBatis 可以使用简单的 XML 或注解来配置和映射原生信息，将接口和 Java 的 POJOs（Plain Old Java Objects）映射成数据库中对应的记录。

12.1 Spring、Spring MVC 和 MyBatis 集成快速体验

本节介绍 Spring MVC 与 MyBatis 结合使用的场景。假设有一个系统需要提供对客户信息的增、删、改、查等功能，在这个场景下通过使用 Spring MVC 提供对外的接口，供调用方使用，在数据库访问层使用 MyBatis 作为持久化方案。

首先，因此场景需要存储数据，因此需要设计 customer 表存储客户信息，customer 表结构如下：

```sql
CREATE TABLE `customer` (
  `id` int(11) unsigned NOT NULL AUTO_INCREMENT COMMENT '客户id',
  `name` varchar(20) DEFAULT '' COMMENT '客户姓名',
  `phone` varchar(11) DEFAULT '' COMMENT '客户手机号',
  `adddate` timestamp NOT NULL DEFAULT CURRENT_TIMESTAMP COMMENT '添加时间',
  `updatedate` timestamp NOT NULL DEFAULT CURRENT_TIMESTAMP ON UPDATE CURRENT_TIMESTAMP COMMENT '修改时间',
  PRIMARY KEY (`id`)
) ENGINE=InnoDB DEFAULT CHARSET=utf8
```

在 Java 实体层创建 Customer 类，其中的字段与 customer 表中的各个字段一一对应，并为各个字段生成 set()和 get()方法，Customer 实体类如下：

```java
/**
 * @Author: zhouguanya
 * @Date: 2018/12/24
 * @Description: 客户实体类
 */
public class Customer {
    /**
     * 客户号
     */
    private int id;
    /**
     * 客户姓名
     */
    private String name;
    /**
     * 客户手机号
     */
    private String phone;
    /**
     * 添加时间
     */
    private Date addDate;
    /**
     * 修改时间
     */
    private Date updateDate;
    public int getId() {
        return id;
    }
    public void setId(int id) {
        this.id = id;
    }
    public String getName() {
        return name;
    }
    public void setName(String name) {
        this.name = name;
    }
    public String getPhone() {
        return phone;
    }
    public void setPhone(String phone) {
```

```java
        this.phone = phone;
    }
    public Date getAddDate() {
        return addDate;
    }
    public void setAddDate(Date addDate) {
        this.addDate = addDate;
    }
    public Date getUpdateDate() {
        return updateDate;
    }
    public void setUpdateDate(Date updateDate) {
        this.updateDate = updateDate;
    }
}
```

接下来,设计数据库访问层(DAO)层代码,数据库访问层涉及 4 个接口,分别是保存客户信息 save()方法、查询客户信息 query()方法、更新客户信息 update()方法和删除客户信息 delete()方法:

```java
/**
 * @Author: zhouguanya
 * @Date: 2018/12/24
 * @Description: 数据库访问层
 */
public interface CustomerDao {
    /**
     * 保存客户信息
     */
    int save(Customer customer);

    /**
     * 更新用户信息
     */
    int update(Customer customer);

    /**
     * 查询用户信息
     */
    Customer query(int id);

    /**
     * 删除用户信息
     */
    int delete(int id);
}
```

有了数据库 customer 表、Customer 实体类和 CustomerDao 接口，就可以使用 MyBatis 进行数据库操作。

在 jdbc.properties 文件中配置数据库连接的相关参数，以便 MyBatis 框架能连接到数据库：

```
driver=com.mysql.jdbc.Driver
#jdbc 连接
url=jdbc:mysql://127.0.0.1:3306/test
#MySQL 用户名
username=root
#MySQL 密码
password=123456
```

在 springmybatis.xml 文件中配置数据源、MyBatis 映射文件等信息：

```xml
<!-- 引入 jdbc 配置文件 -->
<bean id="propertyConfigurer" class="org.springframework.beans.factory.config.PropertyPlaceholderConfigurer">
    <property name="location" value="classpath:jdbc.properties" />
</bean>

<bean id="dataSource" class="org.springframework.jdbc.datasource.DriverManagerDataSource">
    <property name="driverClassName" value="${driver}" />
    <property name="url" value="${url}" />
    <property name="username" value="${username}" />
    <property name="password" value="${password}" />
</bean>

<!-- spring 和 MyBatis 整合 -->
<bean id="sqlSessionFactory" class="org.mybatis.spring.SqlSessionFactoryBean">
    <property name="dataSource" ref="dataSource" />
    <!-- 自动扫描 mapping.xml 文件，**表示迭代查找 -->
    <property name="mapperLocations" value="classpath:mybatis-customer-mapper.xml" />
</bean>

<!-- DAO 接口所在包名，Spring 会自动查找其下的类，包下的类需要使用@MapperScan注解,否则容器注入会失败 -->
<bean class="org.mybatis.spring.mapper.MapperScannerConfigurer">
    <property name="basePackage" value="com.test.mybatis.dao" />
    <property name="sqlSessionFactoryBeanName" value="sqlSessionFactory" />
</bean>

<!-- 事务管理 -->
<bean id="transactionManager" class="org.springframework.jdbc.datasource.DataSourceTransactionManager">
```

```xml
        <property name="dataSource" ref="dataSource" />
    </bean>
```

以上配置文件中使用到了 mybatis-customer-mapper.xml 文件，其实这个文件就是数据库操作相关的 SQL 语句存放的地方，MyBatis 会将 SQL 中需要的动态参数注入到 SQL 中：

```xml
<?xml version="1.0" encoding="UTF-8"?>
<!DOCTYPE mapper PUBLIC "-//mybatis.org//DTD Mapper 3.0//EN"
"http://mybatis.org/dtd/mybatis-3-mapper.dtd">
<mapper namespace="com.test.mybatis.dao.CustomerDao">
    <resultMap id="BaseResultMap" type="com.test.mybatis.entity.Customer">
        <id column="id" jdbcType="INTEGER" property="id" />
        <result column="name" jdbcType="VARCHAR" property="name" />
        <result column="phone" jdbcType="VARCHAR" property="phone" />
        <result column="adddate" jdbcType="TIMESTAMP" property="addDate" />
        <result column="updatedate" jdbcType="TIMESTAMP" property="updateDate" />
    </resultMap>
    <select id="query" parameterType="java.lang.Integer" resultMap="BaseResultMap">
        select
        *
        from customer
        where id = #{id,jdbcType=BIGINT}
    </select>
    <delete id="delete" parameterType="java.lang.Integer">
        delete from customer
        where id = #{id,jdbcType=BIGINT}
    </delete>
    <insert id="save" parameterType="com.test.mybatis.entity.Customer">
        insert into customer (name, phone)
        values (#{name,jdbcType=VARCHAR}, #{phone,jdbcType=VARCHAR})
    </insert>
    <update id="update" parameterType="com.test.mybatis.entity.Customer">
        update customer set name = #{name}, phone = #{phone} where id = #{id}
    </update>
</mapper>
```

以上几个步骤完成了对数据库访问层相关的准备工作，下面就需要将数据库访问层的操作通过接口暴露给应用层服务调用，本例中使用 CustomerService 服务层接口调用 DAO 操作，创建服务层接口 CustomerService，包含 save()、query()、update() 和 delete() 方法，供 Controller 调用。

```java
/**
 * @Author: zhouguanya
 * @Date: 2018/12/24
 * @Description: service 接口
```

```java
 */
public interface CustomerService {
    /**
     * 保存客户信息
     */
    int save(Customer customer);

    /**
     * 更新用户信息
     */
    int update(Customer customer);

    /**
     * 查询用户信息
     */
    Customer query(int customerId);

    /**
     * 删除用户信息
     */
    int delete(int customerId);
}
```

创建 CustomerService 接口的实现类 CustomerServiceImpl，分别实现 CustomerService 中的抽象方法，CustomerServiceImpl 中会调用 CustomerDao 中的方法进行数据库操作。

```java
/**
 * @Author: zhouguanya
 * @Date: 2018/12/24
 * @Description:
 */
@Service
public class CustomerServiceImpl implements CustomerService {
    @Autowired
    private CustomerDao customerDao;
    /**
     * 保存客户信息
     */
    @Override
    public int save(Customer customer) {
        return customerDao.save(customer);
    }

    /**
     * 更新用户信息
     */
    @Override
```

```java
    public int update(Customer customer) {
        return customerDao.update(customer);
    }

    /**
     * 查询用户信息
     */
    @Override
    public Customer query(int customerId) {
        return customerDao.query(customerId);
    }

    /**
     * 删除用户信息
     */
    @Override
    public int delete(int customerId) {
        return customerDao.delete(customerId);
    }
}
```

创建 CustomerController，其中调用 CustomerService 接口中的方法，并对外提供 HTTP 接口。

```java
/**
 * @Author: zhouguanya
 * @Date: 2018/12/24
 * @Description: 控制层
 */
@RestController
@RequestMapping("/customer")
public class CustomerController {
    @Autowired
    private CustomerService customerService;

    /**
     * 新增客户
     */
    @RequestMapping("/save")
    public int save(Customer customer) {
        return customerService.save(customer);
    }
    /**
     * 更新客户
     */
    @RequestMapping("/update")
    public int update(Customer customer) {
        return customerService.update(customer);
    }
```

```java
/**
 * 查询客户
 */
@RequestMapping("/query")
public Customer query(int customerId) {
    return customerService.query(customerId);
}
/**
 * 删除客户
 */
@RequestMapping("/delete")
public int delete(int customerId) {
    return customerService.delete(customerId);
}
}
```

Spring MVC 集成 MyBatis 实现对客户信息的增、删、改、查的案例搭建完毕，使用 Tomcat 部署案例代码。

下面使用 postman（一款模拟 HTTP 请求的工具）发起 HTTP 请求，模拟对客户信息进行增、删、改、查操作。

首先测试新增客户信息，在 postman 中输入地址 http://localhost:8080/customer/save，模拟 post 请求，发送 name=michael，phone=12345678901，单击发送，将会发起一个 HTTP 请求，执行结果如图 12-1 所示。

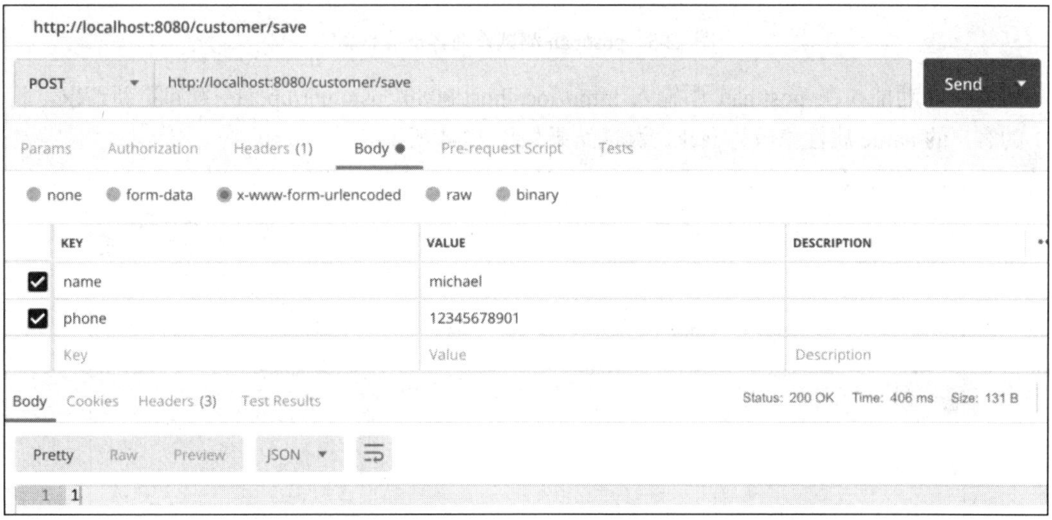

图 12-1　postman 模拟 HTTP 请求新增客户信息图

观察数据中的记录，客户信息已保存成功，如图 12-2 所示。

测试查询功能，在 postman 中输入 http://localhost:8080/customer/query?customerId=1 发起查询请求，测试结果如图 12-3 所示。

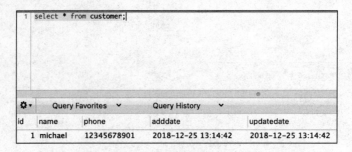

图 12-2 验证新增客户信息图

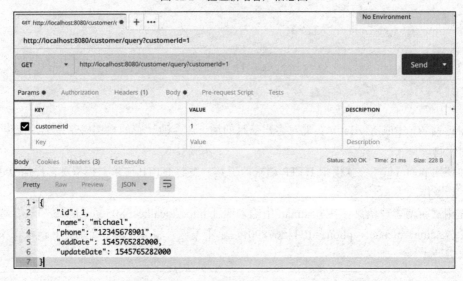

图 12-3 postman 测试查询客户信息图

测试更新功能,在 postman 中输入 http://localhost:8080/customer/update 发起更新请求,将 id 为 1 的客户的 name 属性修改为 jack,测试结果如果 12-4 所示。

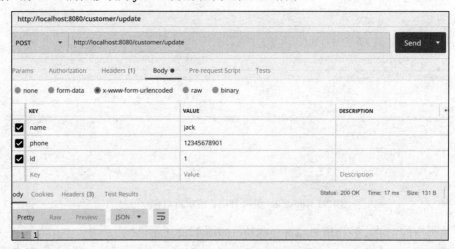

图 12-4 postman 测试更新客户信息图

查询数据库中的记录,验证更新客户信息成功,如图 12-5 所示。

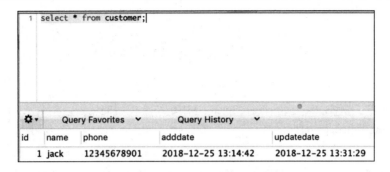

图 12-5　验证更新客户信息图

测试删除功能，在 postman 中输入 http://localhost:8080/customer/delete?customerId=1 发起删除请求，将 id 为 1 的客户信息删除，测试结果如果 12-6 所示。

图 12-6　测试删除客户信息

查询数据库中的记录，验证删除客户信息已成功，可以发现执行删除后，数据库中没有任何客户信息，如图 12-7 所示。

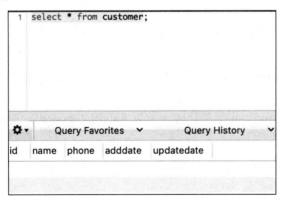

图 12-7　验证删除客户信息

本节到此阐述了 Spring 与 MyBatis 的集成使用，并通过 Spring MVC 结合 MyBatis 使用场景验证 MyBatis 对数据库操作的有效性。

12.2 MyBatis 代码解析

12.1 节中描述了 MyBatis 与 Spring、Spring MVC 结合使用的场景，本节从 MyBatis 代码角度分析 MyBatis 的原理和运行方式。

从 MyBatis 的配置文件 springmybatis.xml 中可以发现，在配置文件包含了有关 JDBC 数据源 DriverManagerDataSource 的配置和有关 SqlSessionFactoryBean 的配置。下面将从这两个类开始着手进行 MyBatis 代码分析。

DriverManagerDataSource 封装了 JDBC 连接相关的参数，如 JDBC 驱动程序、JDBC 连接、数据库用户名和数据库密码等信息。然后将数据源注入给 SqlSessionFactoryBean，下面将分析 SqlSessionFactoryBean 类。

SqlSessionFactoryBean 是什么？SqlSessionFactoryBean 是用来创建 MyBatis SqlSessionFactory 对象的。SqlSessionFactory 是用于创建 SqlSession 对象的，SqlSession 对象是 MyBatis 基本的接口，通过 SqlSession 对象可以执行 SQL 和控制事物。

进入 SqlSessionFactoryBean 类的代码发现其中含有 afterPropertiesSet()方法。根据本书第 2 章 2.4 节中有关 Bean 生命周期的介绍，此方法会在 IoC 容器启动过程中，在 Bean 的构造器执行完后执行。

```
public void afterPropertiesSet() throws Exception {
  notNull(dataSource, "Property 'dataSource' is required");
  notNull(sqlSessionFactoryBuilder, "Property 'sqlSessionFactoryBuilder' is required");
    state((configuration == null && configLocation == null)
    || !(configuration != null && configLocation != null),
      "Property 'configuration' and 'configLocation' can not specified with together");
    this.sqlSessionFactory = buildSqlSessionFactory();
}
```

可以发现 afterPropertiesSet()方法调用了 buildSqlSessionFactory()方法，此方法是生成 SqlSession 工厂——SqlSessionFactory 的地方。

进入 SqlSessionFactoryBean 类的代码，发现 buildSqlSessionFactory()方法，此方法很长，这里只挑选最重要的部分代码：

```
protected SqlSessionFactory buildSqlSessionFactory() throws IOException {
......省略代码......
return this.sqlSessionFactoryBuilder.build(configuration);
}
```

从以上 buildSqlSessionFactory()方法部分代码片段中可以看出，此方法最终会通过调用 SqlSessionFactoryBuilder 对象的 build()方法返回 SqlSessionFactory 对象。

SqlSessionFactoryBuilder 的功能是解析配置文件创建 SqlSessionFactory 对象，其内部的 build()方法如下：

```
public SqlSessionFactory build(Configuration config) {
  return new DefaultSqlSessionFactory(config);
}
```

可以发现 SqlSessionFactoryBuilder.build()方法会创建一个新的 DefaultSqlSessionFactory 对象。

DefaultSqlSessionFactory 类是创建 DefaultSqlSession 对象的地方，其中含有很多重载的 openSession()方法，这些重载方法都返回 SqlSession 对象，其中几个重载方法如下：

```
@Override
public SqlSession openSession() {
  return openSessionFromDataSource(configuration.getDefaultExecutorType(), null, false);
}

@Override
public SqlSession openSession(boolean autoCommit) {
    return openSessionFromDataSource
(configuration.getDefaultExecutorType(), null, autoCommit);
}

@Override
public SqlSession openSession(ExecutorType execType) {
    return openSessionFromDataSource(execType, null, false);
}
```

可以发现这些重载的 openSession()方法都会调用 openSessionFromConnection()方法，此方法会返回 DefaultSqlSession 对象。

```
private SqlSession openSessionFromDataSource(ExecutorType execType, TransactionIsolationLevel level, boolean autoCommit) {
  Transaction tx = null;
  try {
    final Environment environment = configuration.getEnvironment();
    final TransactionFactory transactionFactory = getTransactionFactoryFromEnvironment(environment);
    tx = transactionFactory.newTransaction(environment.getDataSource(), level, autoCommit);
    final Executor executor = configuration.newExecutor(tx, execType);
    return new DefaultSqlSession(configuration, executor, autoCommit);
  } catch (Exception e) {
    closeTransaction(tx); // may have fetched a connection so lets call close()
    throw ExceptionFactory.wrapException("Error opening session.  Cause: " + e, e);
  } finally {
```

```
      ErrorContext.instance().reset();
   }
}
```

下面总结一下 SqlSession 对象的创建过程，如图 12-8 所示。

在 MyBatis 中，数据库访问层对象（DAO）是通过 MapperProxy 对象进行代理的，即在调用 customerDao 中的方法时，是在执行 MapperProxy 代理对象中的方法。下面分析 MapperProxy 对象的获取过程。

MapperFactoryBean 类中 getObject() 方法将会返回代理对象：

```
public T getObject() throws Exception {
   return getSqlSession().getMapper(this.mapperInterface);
}
```

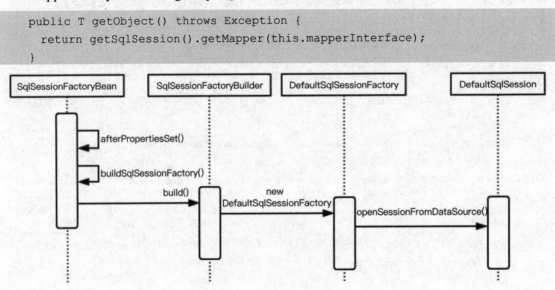

图 12-8　MyBatis SqlSession 创建过程

getObject() 会调用 SqlSessionTemplate 类中的 getMapper() 方法，并返回代理对象：

```
public <T> T getMapper(Class<T> type) {
   return getConfiguration().getMapper(type, this);
}
```

SqlSessionTemplate 类的 getMapper() 方法会调用 Configuration() 类的 getMapper() 方法：

```
public <T> T getMapper(Class<T> type, SqlSession sqlSession) {
   return mapperRegistry.getMapper(type, sqlSession);
}
```

Configuration 类的 getMapper() 方法会调用 MapperRegistry 类中的 getMapper() 方法：

```
public <T> T getMapper(Class<T> type, SqlSession sqlSession) {
   final MapperProxyFactory<T> mapperProxyFactory = (MapperProxyFactory<T>) knownMappers.get(type);
   if (mapperProxyFactory == null) {
     throw new BindingException("Type " + type + " is not known to the MapperRegistry.");
```

```
    }
    try {
      return mapperProxyFactory.newInstance(sqlSession);
    } catch (Exception e) {
      throw new BindingException("Error getting mapper instance. Cause: " + e, e);
    }
  }
}
```

MapperRegistry 类的 getMapper()方法会调用 MapperProxyFactory 类的 newInstance()方法：

```
public T newInstance(SqlSession sqlSession) {
    final MapperProxy<T> mapperProxy = new MapperProxy<T>(sqlSession, mapperInterface, methodCache);
    return newInstance(mapperProxy);
}
```

MapperProxyFactory 类的 newInstance()方法创建了 MapperProxy 对象，并将这个对象作为入参调用重载 newInstance()方法，MapperProxyFactory 定义如下：

```
public class MapperProxy<T> implements InvocationHandler, Serializable
```

MapperProxyFactory 类的 newInstance()方法会调用 MapperProxyFactory 类中重载的 newInstance()方法，这个重载的 newInstance()方法将会通过动态代理创建代理对象，并返回此代理对象：

```
protected T newInstance(MapperProxy<T> mapperProxy) {
    return (T) Proxy.newProxyInstance(mapperInterface.getClassLoader(), new Class[] { mapperInterface }, mapperProxy);
}
```

由 3.1 节中知识可知，MapperProxy 类实现了 InvocationHandler 接口。在 newInstance()方法中通过 JDK 的动态代理生成了数据库访问层（DAO）接口的代理类。

下面总结一下 MapperProxy 对象的创建过程，如图 12-9 所示。

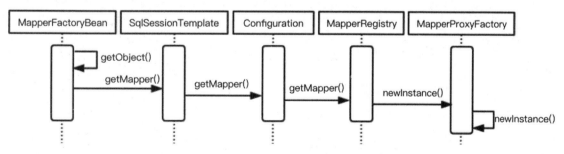

图 12-9　MyBatis MapperProxy 代理对象创建过程

获取到 MapperProxy 对象后，下面将分析 MapperProxy 是如何运行的。因为 MapperProxy 实现了 InvocationHandler 接口，因此其执行逻辑封装在 invoke()方法中：

```java
    public Object invoke(Object proxy, Method method, Object[] args) throws Throwable {
        try {
            if (Object.class.equals(method.getDeclaringClass())) {
                return method.invoke(this, args);
            } else if (isDefaultMethod(method)) {
                return invokeDefaultMethod(proxy, method, args);
            }
        } catch (Throwable t) {
            throw ExceptionUtil.unwrapThrowable(t);
        }
        final MapperMethod mapperMethod = cachedMapperMethod(method);
        return mapperMethod.execute(sqlSession, args);
    }
```

从 MapperProxy 类的 invoke()方法可以知道，方法执行调用了 MapperMethod 类的 execute()方法，MapperMethod 类的 execute()方法如下：

```java
    public Object execute(SqlSession sqlSession, Object[] args) {
        Object result;
        switch (command.getType()) {
            case INSERT: {
            Object param = method.convertArgsToSqlCommandParam(args);
                result = rowCountResult(sqlSession.insert(command.getName(), param));
                break;
            }
            case UPDATE: {
                Object param = method.convertArgsToSqlCommandParam(args);
                result = rowCountResult(sqlSession.update(command.getName(), param));
                break;
            }
            case DELETE: {
                Object param = method.convertArgsToSqlCommandParam(args);
                result = rowCountResult(sqlSession.delete(command.getName(), param));
                break;
            }
            case SELECT:
                if (method.returnsVoid() && method.hasResultHandler()) {
                    executeWithResultHandler(sqlSession, args);
                    result = null;
                } else if (method.returnsMany()) {
                    result = executeForMany(sqlSession, args);
                } else if (method.returnsMap()) {
                    result = executeForMap(sqlSession, args);
                } else if (method.returnsCursor()) {
                    result = executeForCursor(sqlSession, args);
```

```
      } else {
        Object param = method.convertArgsToSqlCommandParam(args);
        result = sqlSession.selectOne(command.getName(), param);
      }
      break;
    case FLUSH:
      result = sqlSession.flushStatements();
      break;
    default:
      throw new BindingException("Unknown execution method for: " + command.getName());
    }
    if (result == null && method.getReturnType().isPrimitive() && !method.returnsVoid()) {
      throw new BindingException("Mapper method '" + command.getName()
        + " attempted to return null from a method with a primitive return type (" + method.getReturnType() + ").");
    }
    return result;
  }
```

可以看到 MapperMethod 类的 execute()方法对数据库增、删、改、查操作做了不同的处理，对每种类型的数据库操作，都会对 SqlSession 类中对应的方法进行处理，下面就以数据库新增操作为例，分析执行过程。DefaultSqlSession 类的 insert()方法如下：

```
public int insert(String statement, Object parameter) {
  return update(statement, parameter);
}
```

由 DefaultSqlSession 类的 insert()方法可知，insert()方法会调用 update()方法：

```
public int update(String statement, Object parameter) {
  try {
    dirty = true;
    MappedStatement ms = configuration.getMappedStatement(statement);
    return executor.update(ms, wrapCollection(parameter));
  } catch (Exception e) {
    throw ExceptionFactory.wrapException("Error updating database. Cause: " + e, e);
  } finally {
    ErrorContext.instance().reset();
  }
}
```

DefaultSqlSession 类的 update()方法是通过调用 Executor 类的 update()方法运行的，以 BaseExecutor 类的 update()方法为例：

```
    public int update(MappedStatement ms, Object parameter) throws SQLException{
      ErrorContext.instance().resource(ms.getResource()).activity("executing an
update").object(ms.getId());
      if (closed) {
        throw new ExecutorException("Executor was closed.");
      }
      clearLocalCache();
      return doUpdate(ms, parameter);
    }
```

BaseExecutor 类的 update()方法会调用 doUpdate()方法，此方法由其子类实现，以其子类 SimpleExecutor 为例：

```
    public int doUpdate(MappedStatement ms, Object parameter) throws SQLException{
      Statement stmt = null;
      try {
        Configuration configuration = ms.getConfiguration();
        StatementHandler handler = configuration.newStatementHandler(this, ms,
parameter, RowBounds.DEFAULT, null, null);
        stmt = prepareStatement(handler, ms.getStatementLog());
        return handler.update(stmt);
      } finally {
        closeStatement(stmt);
      }
    }
```

在 SimpleExecutor 类的 doUpdate()方法调用 StatementHandler 类的 update()方法，这里 StatementHandler 子类较多，以 PreparedStatementHandler 这个子类为例：

```
    public int update(Statement statement) throws SQLException {
      PreparedStatement ps = (PreparedStatement) statement;
      ps.execute();
      int rows = ps.getUpdateCount();
      Object parameterObject = boundSql.getParameterObject();
      KeyGenerator keyGenerator = mappedStatement.getKeyGenerator();
      keyGenerator.processAfter(executor, mappedStatement, ps, parameterObject);
      return rows;
    }
```

到此各位读者应该比较熟悉了，此处是使用原生的 JDBC 的 PreparedStatement 进行数据库操作的地方。

下面总结 MyBatis SQL 执行流程，如图 12-10 所示。

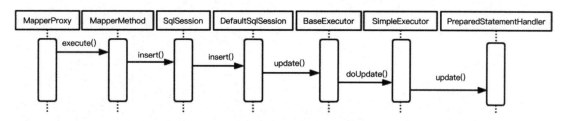

图 12-10　MyBatis SQL 执行过程

12.3　小　　结

　　MyBatis 是企业开发中最常用的 ORM 框架之一，本章通过 MyBatis 与 Spring、Spring MVC 集成阐述了一个常见的企业开发中使用 MyBatis 的场景；并通过对 MyBatis 的底层代码分析阐述了 MyBatis 的运行原理。

第 13 章

Spring 事务管理

数据库事务（Database Transaction）是指将一系列数据库操作当作一个逻辑处理单元的操作，这个单元中的数据库操作要么完全执行，要么完全不执行。通过将一组相关操作组合为一个逻辑处理单元，可以简化错误恢复，并使应用程序更加可靠。

13.1 事务的特性

一个逻辑处理单元要成为事务，必须满足 ACID（原子性、一致性、隔离性和持久性）属性。所谓的 ACID 含义如下。

- 原子性（Atomicity）：一个事务内的操作，要么全部执行成功，要么全部不成功。
- 一致性（Consistency）：事务执行后，数据库状态与其他业务规则保持一致。如转账业务，无论事务执行成功与否，参与转账的两个账号余额之和应该是不变的。
- 隔离性（Isolation）：每个事务独立运行。在并发环境中，并发的事务是互相隔离的，互不影响。
- 持久性（Durability）：事务一旦提交后，数据库中的数据必须被永久地保存下来。

13.2 事务的隔离级别

数据库事物的隔离级别分为 4 种，下面将通过一个转账业务场景对这 4 种隔离级别分别做分析。

13.2.1 READ_UNCOMMITTED

已读但未提交，即一个事务读取到了另一个事务未提交的数据。在这种隔离级别下，会造成"脏读"的情况。

假设有如下场景，有 A、B 两个事务同时对一个账户进行存款和取款操作，A 事务向账户存款 10 元，B 事务从账户取款 10 元。

在 READ_UNCOMMITTED 隔离级别下，可能会出现如表 13-1 所示场景。

表 13-1 READ_UNCOMMITTED 隔离级别演示

时 间 轴	事物 A 存款	事物 B 取款
T1	事物 A 开始	——
T2	——	事物 B 开始
T3	——	事务 B 查询余额（余额为 10 元）
T4	——	事务 B 取出 10 元（余额为 0 元）
T5	事务 A 查询余额（余额为 0 元）	——
T6	——	事务 B 撤销（余额为 10 元）
T7	事务 A 存入 10 元	——
T8	事务 A 提交（余额更新为 10 元）	——

正常情况下，A、B 两事务执行完以后，账户余额应为 20 元，但是在时刻 T5 时，事务 A 查询到的余额为 0 元，这是因为读取到了事务 B 未提交的数据，即读到了"脏"数据。这个场景就是典型的在 READ_UNCOMMITTED 隔离级别下常见的"脏读"问题。

13.2.2 READ_COMMITTED

在这个隔离级别下，可以有效避免"脏读"情况的发生。虽然解决了不可重复读的问题，但是在这个隔离级别下无法避免不可重复读取的问题。在 READ_COMMITTED 隔离级别下，可能会出现如表 13-2 所示的场景。

表 13-2 READ_COMMITTED 隔离级别演示

时 间 轴	事物 A 查询	事物 B 取款
T1	事物 A 开始	——
T2	——	事物B 开始
T3	——	事务 B 查询余额（余额为 10 元）
T4	事务 A 查询余额（余额为 10 元）	——
T5	——	事务B 取出 10 元（余额为 0 元）
T6	——	事务 B 提交
T7	事务 A 查询余额（0 元）	——
T8	事务 A 提交	——

在表 13-2 所示的场景中，事务 A 执行了两次余额查询，但第一次查询得到的余额是 10 元，第二次查询得到的余额为 0 元，这就是不可重复读取的问题。

13.2.3　REPEATABLE_READ

可重复读级别是保证在事务处理过程中多次读取同一个数据时的值始终是一致的。可重复读取是通过在事务开启后不允许其他事务对当前记录进行修改操作实现的。

这个隔离级别避免了"脏读"和不可重复读的问题，但是有可能会出现"幻读"。"幻读"场景的出现如表 13-3 所示。

表 13-3　REPEATABLE_READ 隔离级别演示

时间轴	事物 A 查询记录	事物 B 存款
T1	事物 A 开始	——
T2	——	事物 B 开始
T3	查询交易记录	——
T4	——	事务 B 存入 10 元
T5	——	事务 B 提交
T6	查询交易记录	——
T7	提交事务	——

在事务 A 中，同一个事务多次获取交易记录，发现第二次获取交易记录的结果中多出了一笔存款记录——事务 B 发生的存款操作，对事务 A 来说，好像是出现了幻觉一样，即"幻读"。

13.2.4　SERIALIZABLE

顺序读是最严格的事务隔离级别。它要求所有的事务排队依序执行，即事务只能一个接一个地处理，不能并发执行。

SERIALIZABLE 隔离级别如表 13-4 所示。

表 13-4 SERIALIZABLE 隔离级别演示

时间轴	事物 A 存款	事物 B 取款
T1	事物 A 开始	——
T2	事务 A 查询余额（余额为 0 元）	——
T3	事务 A 存入 10 元	——
T4	事务 A 提交	——
T5	——	事务 B 开始
T6	——	事务 B 查询余额（余额为 10 元）
T7	——	事务 B 取出 10 元
T8	——	事务 B 提交

针对以上 4 种隔离级别以及每种隔离级别下产生的各种问题进行总结和归纳，如表 13-5 所示。

表 13-5　各种隔离级别及产生的问题

事务隔离级别	脏读	不可重复读	幻读
READ_UNCOMMITTED	允许	允许	允许
READ_COMMITTED	禁止	允许	允许
REPEATABLE_READ	禁止	禁止	允许
SERIALIZABLE	禁止	禁止	禁止

4 种事务隔离级别从上往下，级别越高，并发性越差，但安全性越来越高。

13.3　JDBC 方式使用事务

在 JDBC 编程中，事务的管理需要结合 Connection 来实现。Connection 中与事务有关的方法如下：

```
/** 设置事务自动提交 */
void setAutoCommit(boolean autoCommit) throws SQLException;
/** 提交事务 */
void commit() throws SQLException;
/** 回滚事务 */
void rollback() throws SQLException;
```

使用 JDBC 处理事务的代码如下：

```
public class AccountDao {
    /*
     * 修改指定用户的余额
     * */
    public void updateBalance(Connection con, String name,double balance) {
        try {
            String sql = "UPDATE account SET balance=balance+? WHERE name=?";
            PreparedStatement pstmt = con.prepareStatement(sql);
            pstmt.setDouble(1,balance);
            pstmt.setString(2,name);
            pstmt.executeUpdate();
        }catch (Exception e) {
            throw new RuntimeException(e);
        }
    }
}
```

下面是使用 Connection 对象控制事务的提交和回滚。

```
public void transferAccounts(String from,String to,double money) {
    //对事务的操作
```

```
            Connection con = null;
            try{
                con = JdbcUtils.getConnection();
                con.setAutoCommit(false);
                AccountDao dao = new AccountDao();
                //更新账户余额
                dao.updateBalance(con,from,-money);
                //提交事务
                con.commit();
            } catch (Exception e) {
                try {
                    con.rollback();
                } catch (SQLException e1) {
                    e.printStackTrace();
                }
                throw new RuntimeException(e);
            }
        }
```

直接使用 JDBC 的编程方式管理事务，虽然可以完成对应的功能，但这种编程方式对代码的复用性不高。

为了解决通过直接适应 JDBC 的方式对事务进行控制，提高代码复用性的问题，Spring 也提供了对事务进行控制的相关 API，下面将介绍使用 Spring 的方式进行事务管理。

13.4　Spring 事务管理快速体验

本节针对一个对用户账户进行数据库操作的场景，对比在不使用事务的场景下和使用事务的场景下，对数据库造成的不同影响。

创建账户余额表 account_balance，建表语句如下：

```sql
CREATE TABLE `account_balance` (
  `id` int(11) unsigned NOT NULL AUTO_INCREMENT COMMENT '主键',
  `customerId` int(11) NOT NULL COMMENT '客户号',
  `balance` decimal(10,0) DEFAULT NULL COMMENT '账户余额',
  `adddate` timestamp NOT NULL DEFAULT CURRENT_TIMESTAMP COMMENT '添加时间',
  `updatedate` timestamp NOT NULL DEFAULT CURRENT_TIMESTAMP ON UPDATE CURRENT_TIMESTAMP COMMENT '修改时间',
  PRIMARY KEY (`id`),
  KEY (`customerId`)
) ENGINE=InnoDB DEFAULT CHARSET=utf8
```

创建实体类 AccountBalance，与数据库中的字段一一对应，AccountBalance 类的代码如下：

```java
/**
 * @Author: zhouguanya
 * @Date: 2018/12/27
 * @Description: 账户余额
 */
public class AccountBalance {
    /**
     * 主键
     */
    private int id;
    /**
     * 客户号
     */
    private int customerId;
    /**
     * 账户余额
     */
    private BigDecimal balance;
    /**
     * 创建时间
     */
    private Date addDate;
    /**
     * 修改时间
     */
    private Date updateDate;

    public int getId() {
        return id;
    }

    public void setId(int id) {
        this.id = id;
    }

    public int getCustomerId() {
        return customerId;
    }

    public void setCustomerId(int customerId) {
        this.customerId = customerId;
    }

    public BigDecimal getBalance() {
        return balance;
    }

    public void setBalance(BigDecimal balance) {
```

```java
        this.balance = balance;
    }

    public Date getAddDate() {
        return addDate;
    }

    public void setAddDate(Date addDate) {
        this.addDate = addDate;
    }

    public Date getUpdateDate() {
        return updateDate;
    }

    public void setUpdateDate(Date updateDate) {
        this.updateDate = updateDate;
    }
}
```

创建 DAO 接口 AccountBalanceDao，接口中定义了对账户余额的保存、查询以及更新操作：

```java
/**
 * @Author: zhouguanya
 * @Date: 2018/12/27
 * @Description: MyBatis 会通过动态代理注入
 */
public interface AccountBalanceDao {
    /**
     * 查询账户余额
     */
    int queryAccountByCustomerId(int id);

    /**
     * 保存账户余额
     */
    int saveAccountBalance(AccountBalance accountBalance);

    /**
     * 更新账户余额
     */
    int updateAccountBalance(AccountBalance accountBalance);
}
```

这个案例中依旧使用 MyBatis 作为持久层框架，Spring 集成 MyBatis 的配置如下：

```xml
<bean id="propertyConfigurer" class="org.springframework.beans.factory.config.PropertyPlaceholderConfigurer">
    <property name="location" value="classpath:jdbc.properties" />
</bean>
```

```xml
    <bean id="dataSource" class="org.springframework.jdbc.datasource.
DriverManagerDataSource">
        <property name="driverClassName" value="${driver}" />
        <property name="url" value="${url}" />
        <property name="username" value="${username}" />
        <property name="password" value="${password}" />
    </bean>

    <!-- spring 和 MyBatis 整合-->
    <bean id="sqlSessionFactory" class="org.mybatis.spring.
SqlSessionFactoryBean">
        <property name="dataSource" ref="dataSource" />
        <!-- 自动扫描 mapping.xml 文件,**表示迭代查找 -->
        <property name="mapperLocations">
            <array>
                <value>classpath:mybatis-accountbalance-mapper.xml</value>
            </array>
        </property>
    </bean>

    <!-- DAO 接口所在包名,Spring 会自动查找其下的类,包下的类需要使用@MapperScan 注解,否则
容器注入会失败 -->
    <bean class="org.mybatis.spring.mapper.MapperScannerConfigurer">
        <property name="basePackage" value="com.test.transaction.dao" />
        <property name="sqlSessionFactoryBeanName" value="sqlSessionFactory" />
    </bean>

    <!-- 事务管理 -->
    <bean id="transactionManager" class="org.springframework.jdbc.datasource.
DataSourceTransactionManager">
        <property name="dataSource" ref="dataSource" />
    </bean>
    <bean id="transactionInterceptor" class="org.springframework.transaction.
interceptor.TransactionInterceptor">
        <property name="transactionManager" ref="transactionManager"/>
        <property name="transactionAttributes">
            <props>
                <prop key="delete*">PROPAGATION_REQUIRED</prop>
                <prop key="add*">PROPAGATION_REQUIRED</prop>
                <prop key="update*">PROPAGATION_REQUIRED</prop>
                <prop key="save*">PROPAGATION_REQUIRED</prop>
                <prop key="find*">PROPAGATION_REQUIRED,readOnly</prop>
            </props>
        </property>
    </bean>
```

配置文件中配置了 DataSourceTransactionManager 类用于管理事务。

环境搭建好后，此处使用 JUnit 单元测试框架测试数据库访问操作，本例中创建一个单元测试类 AccountTransactionTest，测试代码中包含对账户余额的保存和更新的测试。其中更新账户余额的测试方法 testUpdateWithoutTransaction()不使用 Spring 事务管理，另一个更新账户余额的测试方法 testUpdateWithTransaction()通过 "@Transactional" 使用 Spring 事务管理，测试代码如下：

```java
@RunWith(SpringJUnit4ClassRunner.class)
@ContextConfiguration("classpath:spring-transaction-mybatis.xml")
public class AccountTransactionTest {
    @Autowired
    private AccountBalanceDao accountBalanceDao;

    /**
     * 插入一条测试记录
     */
    @Test
    public void testSave() {
        AccountBalance accountBalance = new AccountBalance();
        accountBalance.setCustomerId(1);
        accountBalance.setBalance(new BigDecimal(10));
        accountBalanceDao.saveAccountBalance(accountBalance);
    }
    /**
     * 测试不使用事务
     */
    @Test
    public void testUpdateWithoutTransaction() {
        AccountBalance accountBalance = new AccountBalance();
        accountBalance.setCustomerId(1);
        accountBalance.setBalance(new BigDecimal(20));
        accountBalanceDao.updateAccountBalance(accountBalance);
        //模拟异常
        int x = 1 / 0;
        accountBalance.setBalance(new BigDecimal(50));
        accountBalanceDao.updateAccountBalance(accountBalance);
    }
    /**
     * 测试使用事务
     */
    @Test
    @Transactional
    public void testUpdateWithTransaction() {
        AccountBalance accountBalance = new AccountBalance();
        accountBalance.setCustomerId(1);
        accountBalance.setBalance(new BigDecimal(50));
```

```
        accountBalanceDao.updateAccountBalance(accountBalance);
        //模拟异常
        int x = 1 / 0;
        accountBalance.setBalance(new BigDecimal(100));
        accountBalanceDao.updateAccountBalance(accountBalance);
    }
}
```

下面开始执行单元测试，首先测试对账户余额进行新增操作 testSave()方法，执行完 testSave()方法后，查询数据库中账户余额表的记录，如图 13-1 所示。

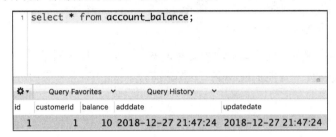

图 13-1　保存账户余额

接下来执行 testUpdateWithoutTransaction()方法，此方法中有如下代码：

```
int x = 1 / 0;
```

以上代码是用于模拟一个运行时的异常，方便观察测试方法 testUpdateWithoutTransaction()对数据库的影响。执行 testUpdateWithoutTransaction()方法，控制台输出如下：

```
java.lang.ArithmeticException: / by zero

    at com.test.transaction.AccountTransactionTest.
testUpdateWithoutTransaction(AccountTransactionTest.java:45)
    at sun.reflect.NativeMethodAccessorImpl.invoke0(Native Method)
    at sun.reflect.NativeMethodAccessorImpl.invoke
(NativeMethodAccessorImpl.java:62)
```

可以看到 testUpdateWithoutTransaction()方法执行过程中发生了异常，根据 13.1 节中事物的特性，这里 testUpdateWithoutTransaction()方法应该回滚，不会将数据持久化到账户余额表中，观察账户余额表的数据，如图 13-2 所示。

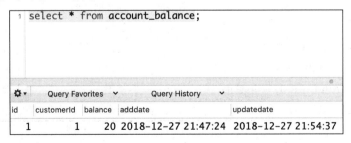

图 13-2　不使用事务测试更新账户余额

从图 13-2 所示可以发现，testUpdateWithoutTransaction()执行过程中发生了异常，但是事务发生回滚，账户余额从 10 被更新为 20，这显然是有问题的。

下面测试 testUpdateWithTransaction()方法，这个方法使用了"@Transactional"注解，执行此方法后，控制台输入如下：

```
java.lang.ArithmeticException: / by zero
    at com.test.transaction.AccountTransactionTest.
testUpdateWithoutTransaction(AccountTransactionTest.java:45)
    at sun.reflect.NativeMethodAccessorImpl.invoke0(Native Method)
    at sun.reflect.NativeMethodAccessorImpl.invoke
(NativeMethodAccessorImpl.java:62)
```

此方法依然会抛出异常。观察数据库中记录的变化，如图 13-3 所示。

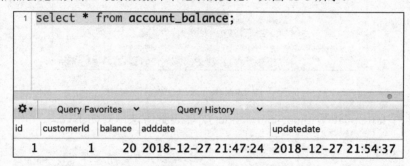

图 13-3　使用事务测试更新账户余额

从图 13-3 所示可以发现，testUpdateWithTransaction()方法执行后，虽然发生了异常，但是并未将异常发生之前的更新持久化到数据库中，这是期望得到的正确结果。

13.5　Spring 事务隔离级别

Spring 对事务的支持提供了 5 种隔离级别，各种隔离级别的含义如表 13-6 所示。

表13-6　Spring事务隔离级别

事务隔离级别	脏读	不可重复读	幻读
ISOLATION_DEFAULT	同数据库事务隔离级别	同数据库事务隔离级别	同数据库事务隔离级别
ISOLATION_READ_UNCOMMITTED	允许	允许	允许
ISOLATION_READ_COMMITTED	禁止	允许	允许
ISOLATION_REPEATABLE_READ	禁止	禁止	允许
ISOLATION_SERIALIZABLE	禁止	禁止	禁止

13.6 Spring 事务传播行为

事务传播行为是用来描述由某一个事务传播行为修饰的方法被嵌套进另一个方法的时候，事务的传播特性。Spring 中定义了 7 种事务传播行为，如表 13-7 所示。

表13-7 Spring事务隔离级别

事务传播行为类型	说 明
PROPAGATION_REQUIRED	如果当前没有事务，就新建一个事务。 如果已经存在一个事务中，加入到这个事务中
PROPAGATION_SUPPORTS	支持当前事务。如果当前没有事务，就以非事务方式执行
PROPAGATION_MANDATORY	使用当前的事务。如果当前没有事务，就抛出异常
PROPAGATION_REQUIRES_NEW	新建事务。如果当前存在事务，把当前事务挂起
PROPAGATION_NOT_SUPPORTED	以非事务方式执行操作。如果当前存在事务，就把当前事务挂起
PROPAGATION_NEVER	以非事务方式执行。如果当前存在事务，则抛出异常
PROPAGATION_NESTED	如果当前存在事务，则在嵌套事务内执行。 如果当前没有事务,则执行与 PROPAGATION_REQUIRED 类似的操作。 与 PROPAGATION_REQUIRES_NEW 的差别是 PROPAGATION_REQUIRES_NEW 另起一个事务，将会与其父事务相互独立。PROPAGATION_NESTED 事务和其父事务是相依的，其要等父事务一起提交

13.7 Spring 事务代码分析

在 13.4 节的案例中，在配置文件中配置了 TransactionInterceptor 类，TransactionInterceptor 接口实现了 AOP 联盟中的 MethodInterceptor 接口，这个接口在本书的第 3 章中已讲解过，此接口可以对目标对象进行拦截。TransactionInterceptor 的类图如图 13-4 所示。

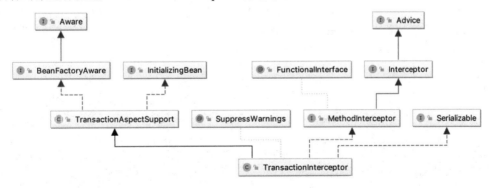

图 13-4 TransactionInterceptor 类图

分析 TransactionInterceptor 类的 invoke()方法：

```java
public Object invoke(MethodInvocation invocation) throws Throwable {
    // Work out the target class: may be {@code null}.
    // The TransactionAttributeSource should be passed the target class
    // as well as the method, which may be from an interface.
    Class<?> targetClass = (invocation.getThis() != null ?
AopUtils.getTargetClass(invocation.getThis()) : null);

    // Adapt to TransactionAspectSupport's invokeWithinTransaction...
    return invokeWithinTransaction(invocation.getMethod(),
targetClass, invocation::proceed);
}
```

TransactionInterceptor 类的 invoke()方法会调用其父类 TransactionAspectSupport 中的 invokeWithinTransaction()方法，方法实现如下：

```java
protected Object invokeWithinTransaction(Method method, @Nullable Class<?>
targetClass,final InvocationCallback invocation) throws Throwable {
    TransactionAttributeSource tas = getTransactionAttributeSource();
    final TransactionAttribute txAttr = (tas != null ?
        tas.getTransactionAttribute(method, targetClass) : null);
        final PlatformTransactionManager tm =
determineTransactionManager(txAttr);
        final String joinpointIdentification =
methodIdentification(method, targetClass, txAttr);
        if (txAttr == null || !(tm instanceof
CallbackPreferringPlatformTransactionManager)) {
            // 判断是否需要开启事务
            TransactionInfo txInfo = createTransactionIfNecessary(tm,
txAttr, joinpointIdentification);
            Object retVal = null;
            try {
                // 执行回调，如果没有后续拦截器，就进入事务方法了
                retVal = invocation.proceedWithInvocation();
            }
            catch (Throwable ex) {
                // 事务发生异常
                completeTransactionAfterThrowing(txInfo, ex);
                throw ex;
            }
            finally {
                cleanupTransactionInfo(txInfo);
            }
            // 事务未发生异常。
            commitTransactionAfterReturning(txInfo);
```

```
        return retVal;
    }
......省略代码......
```

重点分析 createTransactionIfNecessary()方法，根据事务的传播属性做出不同的处理，核心是通过 TransactionStatus 来判断事务的属性。createTransactionIfNecessary()方法如下：

```
protected TransactionInfo createTransactionIfNecessary(@Nullable
PlatformTransactionManager tm, @Nullable TransactionAttribute txAttr, final String
joinpointIdentification) {

    // If no name specified, apply method identification as transaction name.
    if (txAttr != null && txAttr.getName() == null) {
      txAttr = new DelegatingTransactionAttribute(txAttr) {
        @Override
        public String getName() {
          return joinpointIdentification;
        }
      };
    }

    TransactionStatus status = null;
    if (txAttr != null) {
      if (tm != null) {
        status = tm.getTransaction(txAttr);
      }
      else {
        if (logger.isDebugEnabled()) {
          logger.debug("Skipping transactional joinpoint [" +
joinpointIdentification + "] because no transaction manager has been configured");
        }
      }
    }
    return prepareTransactionInfo(tm, txAttr, joinpointIdentification,
status);
}
```

createTransactionIfNecessary()中会调用 PlatformTransactionManager 接口的 getTransaction()方法，这里会调用 PlatformTransactionManager 接口的子类中的 getTransaction()方法，接口的子类是 AbstractPlatformTransactionManager，getTransaction()方法实现如下：

```
@Override
public final TransactionStatus getTransaction(@Nullable TransactionDefinition
definition) throws TransactionException {
    //调用 DataSourceTransactionManager.doGetTransaction()方法
    Object transaction = doGetTransaction();
```

```java
      boolean debugEnabled = logger.isDebugEnabled();
      if (definition == null) {
         definition = new DefaultTransactionDefinition();
      }
      // 是否已经存在的一个transaction
      if (isExistingTransaction(transaction)) {
         return handleExistingTransaction(definition, transaction, debugEnabled);
      }
   ......省略代码......
      //如果是PROPAGATION_REQUIRED, PROPAGATION_REQUIRES_NEW, PROPAGATION_NESTED
      //这三种类型将开启一个新的事务
      else if (definition.getPropagationBehavior() ==
            TransactionDefinition.PROPAGATION_REQUIRED ||
               definition.getPropagationBehavior() ==
            TransactionDefinition.PROPAGATION_REQUIRES_NEW ||
               definition.getPropagationBehavior() ==
            TransactionDefinition.PROPAGATION_NESTED) {
               SuspendedResourcesHolder suspendedResources = suspend(null);
         ......省略代码......
               try {
                  boolean newSynchronization =
(getTransactionSynchronization() != SYNCHRONIZATION_NEVER);
                  DefaultTransactionStatus status = newTransactionStatus
(definition, transaction, true, newSynchronization, debugEnabled,
suspendedResources);
                  //开启新事物
                  doBegin(transaction, definition);
                  prepareSynchronization(status, definition);
                  return status;
               }
               catch (RuntimeException | Error ex) {
                  resume(null, suspendedResources);
                  throw ex;
               }
            }
   ......省略代码......
      return prepareTransactionStatus(definition, null, true, newSynchronization,
debugEnabled, null);
            }
      }
```

getTransaction()方法会调用子类 DataSourceTransactionManager 中的 doGetTransaction()方法，doGetTransaction()方法代码如下：

```java
   @Override
   protected Object doGetTransaction() {
```

```
    DataSourceTransactionObject txObject = new DataSourceTransactionObject();
    txObject.setSavepointAllowed(isNestedTransactionAllowed());
    ConnectionHolder conHolder =(ConnectionHolder)
TransactionSynchronizationManager.getResource(obtainDataSource());
    txObject.setConnectionHolder(conHolder, false);
    return txObject;
}
```

doGetTransaction()方法根据 DataSource 数据源获取 DataSourceTransactionObject 对象。接下来分析 doBegin()方法，doBegin()方式的实现在 DataSourceTransactionManager 中，doBegin()方法如下：

```
@Override
protected void doBegin(Object transaction, TransactionDefinition definition){
    DataSourceTransactionObject txObject = (DataSourceTransactionObject)
transaction;
    Connection con = null;

    try {
      if (!txObject.hasConnectionHolder() ||
          txObject.getConnectionHolder().isSynchronizedWithTransaction()) {
        Connection newCon = obtainDataSource().getConnection();
        if (logger.isDebugEnabled()) {
          logger.debug("Acquired Connection [" + newCon + "] for JDBC
transaction");
        }
        txObject.setConnectionHolder(new ConnectionHolder(newCon), true);
      }

      txObject.getConnectionHolder().setSynchronizedWithTransaction(true);
      con = txObject.getConnectionHolder().getConnection();

      Integer previousIsolationLevel = DataSourceUtils.
prepareConnectionForTransaction(con, definition);
      txObject.setPreviousIsolationLevel(previousIsolationLevel);

      if (con.getAutoCommit()) {
        txObject.setMustRestoreAutoCommit(true);
        if (logger.isDebugEnabled()) {
          logger.debug("Switching JDBC Connection [" + con + "] to manual
commit");
        }
        //开启事务，设置 autoCommit 为 false
        con.setAutoCommit(false);
      }

      prepareTransactionalConnection(con, definition);
```

```
            txObject.getConnectionHolder().setTransactionActive(true);

            int timeout = determineTimeout(definition);
            if (timeout != TransactionDefinition.TIMEOUT_DEFAULT) {
                txObject.getConnectionHolder().setTimeoutInSeconds(timeout);
            }

            //这里将当前的connection放入TransactionSynchronizationManager中持有
            //最终就是把Connection对象存入ThreadLocal中
            if (txObject.isNewConnectionHolder()) {
                TransactionSynchronizationManager.bindResource(obtainDataSource(),
txObject.getConnectionHolder());
            }
        }
        ......省略代码......
}
```

在 doGetTransaction() 方法中，如果同一个线程再次进入执行，就会获取到同一个 ConnectionHolder。

回到 TransactionAspectSupport 类的 invokeWithinTransaction() 方法，接下来将会调用 InvocationCallback 的 proceedWithInvocation() 方法，该方法的实现是调用了 ReflectiveMethodInvocation 类的 proceed() 方法，在 3.6 节的动态代理对象执行的部分已介绍过，此处不再赘述。

执行完代理对象的相关操作后，将会执行提交操作或者是回滚操作。

提交操作 commitTransactionAfterReturning() 方法如下：

```
protected void commitTransactionAfterReturning(@Nullable TransactionInfo txInfo) {
    if (txInfo != null && txInfo.getTransactionStatus() != null) {
        txInfo.getTransactionManager().commit(txInfo.getTransactionStatus());
    }
}
```

commitTransactionAfterReturning() 方法会调用 AbstractPlatformTransactionManager 类的 commit() 方法：

```
@Override
public final void commit(TransactionStatus status) throws TransactionException{
    if (status.isCompleted()) {
        throw new IllegalTransactionStateException(
            "Transaction is already completed - do not call commit or rollback more than once per transaction");
    }
```

```
    DefaultTransactionStatus defStatus = (DefaultTransactionStatus) status;
    if (defStatus.isLocalRollbackOnly()) {
        if (defStatus.isDebug()) {
            logger.debug("Transactional code has requested rollback");
        }
        processRollback(defStatus, false);
        return;
    }

    if (!shouldCommitOnGlobalRollbackOnly() && defStatus.isGlobalRollbackOnly()) {
        if (defStatus.isDebug()) {
            logger.debug("Global transaction is marked as rollback-only but transactional code requested commit");
        }
        processRollback(defStatus, true);
        return;
    }

    processCommit(defStatus);
}
```

commit()方法会调用 processCommit()方法，最终将调用 DataSourceTransactionManager()类的 doCommit()方法，这里将会提交事务：

```
@Override
protected void doCommit(DefaultTransactionStatus status) {
    DataSourceTransactionObject txObject = (DataSourceTransactionObject) status.getTransaction();
    Connection con = txObject.getConnectionHolder().getConnection();
    if (status.isDebug()) {
        logger.debug("Committing JDBC transaction on Connection [" + con + "]");
    }
    try {
        con.commit();
    }
    catch (SQLException ex) {
        throw new TransactionSystemException("Could not commit JDBC transaction", ex);
    }
}
```

如果代理对象执行操作过程中出现了异常，将会执行 completeTransactionAfterThrowing()方法执行回滚操作：

```
protected void completeTransactionAfterThrowing(@Nullable TransactionInfo txInfo, Throwable ex) {
```

```java
        if (txInfo != null && txInfo.getTransactionStatus() != null) {
            if (logger.isTraceEnabled()) {
                logger.trace("Completing transaction for [" +
txInfo.getJoinpointIdentification() + "] after exception: " + ex);
            }
            if (txInfo.transactionAttribute != null &&
txInfo.transactionAttribute.rollbackOn(ex)) {
                try {
                    txInfo.getTransactionManager().rollback(txInfo.
getTransactionStatus());
                }
                catch (TransactionSystemException ex2) {
                    logger.error("Application exception overridden by rollback
exception", ex);
                    ex2.initApplicationException(ex);
                    throw ex2;
                }
                catch (RuntimeException | Error ex2) {
                    logger.error("Application exception overridden by rollback
exception", ex);
                    throw ex2;
                }
            }
            else {
                // We don't roll back on this exception.
                // Will still roll back if TransactionStatus.isRollbackOnly() is true.
                try {
                    txInfo.getTransactionManager().commit(txInfo.
getTransactionStatus());
                }
                catch (TransactionSystemException ex2) {
                    logger.error("Application exception overridden by commit exception",
ex);
                    ex2.initApplicationException(ex);
                    throw ex2;
                }
                catch (RuntimeException | Error ex2) {
                    logger.error("Application exception overridden by commit exception",
ex);
                    throw ex2;
                }
            }
        }
    }
```

Spring 事务的执行过程如图 13-5 所示。

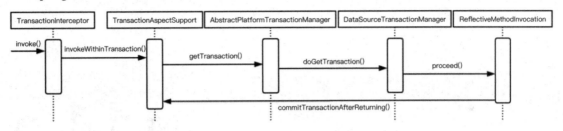

图 13-5　Spring 事务执行流程

13.8　小　　结

　　Spring 事务管理是企业开发中常用的技术，理解 Spring 事务管理的代码对于理解 Spring 事务管理有很大帮助。本章涉及的 Spring 事务隔离级别、Spring 事务传播行为以及 Spring 事务代码分析都是常见的面试题，希望读者务必掌握。

第 14 章

Spring 集成 Redis

Redis 是一个开源的使用 ANSI C 语言编写、支持网络、可基于内存也可以持久化的 Key-Value 数据库。

Redis 在企业开发中通常充当高速缓存的作用，用于保护接口或者数据库。在高并发场景、分布式场景下也可以充当分布式锁，避免多个 JVM 进程在同一时间对同一资源进行修改，从而造成数据不一致。

因为 Redis 是开发中最常用的缓存技术，本章将重点分析 Redis 常见操作命令和 Redis 常见架构以及 Spring 与 Redis 的集成开发。

14.1　Redis 单节点安装

Redis 下载地址为 https://redis.io/download，读者可以根据需要选择安装不同的 Redis 版本，本书使用的版本是 Redis 5.0.3。

Redis 企业应用中一般是在 Linux 服务器环境下部署安装。下面列出在 Linux 环境中下载和安装 Redis 需要用到的一些操作指令（Windows 操作系统可以通过 VMware 安装 Linux 虚拟机）：

```
//下载 Redis-5.0.3
wget http://download.redis.io/releases/redis-5.0.3.tar.gz
//解压 Redis-5.0.3
tar xzf Redis-5.0.3.tar.gz
//进入 Redis-5.0.3 解压后的目录
cd Redis-5.0.3
//编译 Redis-5.0.3
make
```

第 14 章 Spring 集成 Redis | 265

使用 make 命令编译 Redis 需要 C 语言环境，CentOS 自带 C 语言环境，若使用的 Linux 系统中没有 C 语言环境，则需要安装，如 yum 安装 yum install gcc-c++。

解压后的 Redis 目录下包含 Redis 核心配置文件 redis.conf。因为 Redis 默认并不是在后台运行的，要将 Redis 进程改为在后台运行，则要修改 redis.conf 中的配置项 daemonize，该配置项默认值为 no，要将其修改为 yes，如图 14-1 所示。

```
# By default Redis does not run as a daemon. Use 'yes' if you need it.
# Note that Redis will write a pid file in /var/run/redis.pid when daemonized.
daemonize yes
```

图 14-1 daemonize 配置项修改

再使用下面的命令启动 Redis 服务器端：

```
src/redis-server redis.conf
```

Redis 启动后如图 14-2 所示。

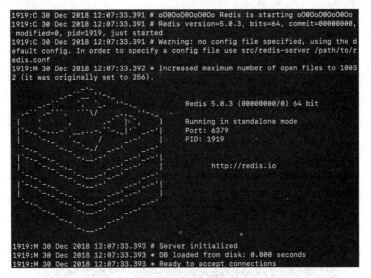

图 14-2 Redis 启动示意图

为了验证 Redis 服务启动正常，可以执行以下命令查看：

```
ps -ef | grep redis | grep -v grep
```

如果 Redis 服务正常启动，Redis 进程将会如图 14-3 所示。

```
501  1919   366   0 12:07下午 ttys000    0:00.41 src/redis-server *:6379
```

图 14-3 Redis 进程详情

Redis 安装文件中含有服务端启动程序，也有客户端程序 redis-cli，可以通过 Redis 客户端连接到 Redis 服务器端，来验证 Redis 服务是否正常启动。客户端启动命令如下：

```
src/redis-cli -h 127.0.0.1 -p 6379
```

Redis 客户端成功与 Redis 服务器端连接上，证明 Redis 服务启动正常。

14.2 Redis 支持的数据类型

相比于其他 NoSQL 数据库而言，Redis 能支持更多更丰富的数据类型，可达 5 种。

- String：字符串类型，一个 key 对应一个 vaule，是 Redis 最基本的数据类型。
- Hash：哈希类型，是一个键值对集合，适用于存储对象。
- List：列表类型，用于保存元素列表，可以在列表头部或尾部添加新元素。
- Set：集合类型，用于存放多个元素的无序集合。
- SortedSet：有序集合类型，每个元素都会关联一个 double 类型的分数。Redis 正是通过分数为集合中的成员进行从小到大排序的。

下面将详细讲解每种 Redis 数据类型的使用方式。

14.2.1 Redis String 类型的使用方式

Redis 字符串数据类型的相关命令是用于管理 Redis 字符串值的，基本语法如下：

```
COMMAND KEY_NAME
```

下面将介绍 Redis String 类型的常用操作。

（1）SET key value

创建 key 为 book，并设置 key 的值为 spring：

```
SET book spring
```

执行以上命令后，得到如图 14-4 所示的输出。

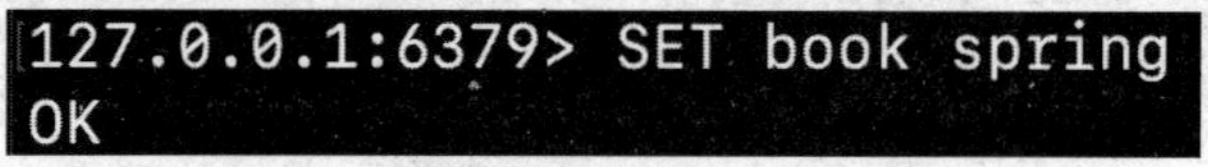

```
127.0.0.1:6379> SET book spring
OK
```

图 14-4　执行 SET book spring 后的结果

除此之外，还可以在 SET 命令后加入时间参数用于设置 key 的存活时间，语法如下：

```
SET key value [EX seconds] [PX milliseconds] [NX|XX]
```

其中各个参数含义如下。

- EX second：设置键的过期时间为 second。SET key value EX second 效果等同于 SETEX key second value。
- PX millisecond：设置键的过期时间为 millisecond。SET key value PX millisecond 效果等同于 PSETEX key millisecon dvalue。
- NX：键不存在时，对键进行设置操作。SET key value NX 等同于 SETNX key value。
- XX：只在键已经存在时，才对键进行设置操作。

第 14 章 Spring 集成 Redis

（2）GET key

查询 key 为 book 的 value 值：

```
GET book
```

执行以上命令后，得到如图 14-5 所示的输出。

```
127.0.0.1:6379> GET book
"spring"
```

图 14-5　执行 GET book 后的结果

（3）GETRANGE key start end

查询 key 对应的 value 字符串值的子字符：

```
GETRANGE book 0 3
```

执行以上命令后，得到如图 14-6 所示的输出。

```
127.0.0.1:6379> GETRANGE book 0 3
"spri"
```

图 14-6　执行 GETRANGE book 0 3 后的结果

（4）GETSET key value

将 key 的值设置为 value，并返回 key 的旧值：

```
GETSET book Spring 5 programming
```

执行以上命令后，得到如图 14-7 所示的输出。

```
127.0.0.1:6379> GETSET book "Spring5 programming"
"spring"
```

图 14-7　执行 GETSET book Spring 5 programming 后的结果

（5）MGET key1 key2 ...

批量获取多个 key 的值，下面创建一个 key 为 price，用于测试 MGET：

```
SET price 50
批量获取 book 和 price 两个 key 的值：
MGET book price
```

执行 MGET 指令后，得到如图 14-8 所示的输出。

```
127.0.0.1:6379> MGET book price
1) "Spring5 programming"
2) "50"
```

图 14-8　执行 MGET book price 后的结果

（6）SETEX key timeout value

创建 key-value 对，并设置 key 的过期时间为 timeout（单位为秒）。下面的命令行设置一个过期时间为 60s 的 key。

```
SETEX user 60 zhouguanya
```

执行这个命令后，得到如图 14-9 所示的输出。

```
127.0.0.1:6379> SETEX user 60 zhouguanya
OK
```

图 14-9　执行 SETEX user 60 zhouguanya 后的结果

（7）TTL key

以秒为单位，返回给定 key 的剩余生存时间（TTL，Time To Live）。通过以下命令查询 user 这个 key 的剩余生存时间。

```
TTL user
```

执行 TTL 命令后，得到如图 14-10 所示的输出。

```
127.0.0.1:6379> TTL user
(integer) 56
```

图 14-10　user 未失效时执行 TTL user 的结果

可以发现，此时 user 的剩余生存时间还剩下 56s，过一段时间再次执行这个命令，将会得到如图 14-11 所示的输出（-2 表示查询的 key 不存在）。

```
127.0.0.1:6379> TTL user
(integer) -2
```

图 14-11　user 失效后执行 TTL user 的结果

（8）SETNX key value

SETNX 的含义就是 Set if Not Exists。

考虑如下场景，需要先获取 key 的值，如果 key 不存在就设置 key 的值，否则不执行任何操作。这时查询 key 的值和设置 key 的值这两步操作不是原子性执行的。

SETNX 方法是原子性的，如果 key 不存在，则设置当前 key 成功。这个命令也是在 Redis 作为分布式锁时最常使用的。

首先尝试使用 EXISTS 查询 job 是否存在，如图 14-12 所示（EXISTS 返回 0 表示 key 不存在）。

```
127.0.0.1:6379> EXISTS job
(integer) 0
```

图 14-12　执行 EXISTS job 的结果

使用 SETNX 创建 job，执行如下命令：

```
SETNX job programmer
```

执行以上命令后得到如图 14-13 所示输出结果。

```
127.0.0.1:6379> SETNX job programmer
(integer) 1
```

图 14-13　执行 SETNX job programmer 的结果

执行 GET job 后得到如图 14-14 所示输出结果。

```
127.0.0.1:6379> GET job
"programmer"
```

图 14-14　执行 GET job 的结果

（9）SETRANGE key offtset value

用 value 参数覆盖给定 key 所储存的字符串值，从偏移量 offset 开始进行覆盖。下面使用 SETRANGE 覆盖 job 的值为 software engineer：

```
SETRANGE job 0 "software engineer"
```

执行以上命令后，得到如图 14-15 所示的输出。

```
127.0.0.1:6379> SETRANGE job 0 "software engineer"
(integer) 17
```

图 14-15　执行 SETRANGE job 0 "software engineer"的结果

执行 GET 命令查询 job 的值，得到如图 14-16 所示输出结果。

```
127.0.0.1:6379> GET job
"software engineer"
```

图 14-16　执行 get job 的结果

（10）STRLEN key

查询 key 对应的 value 的长度。使用以下执行命令查询 job 对应的 value 长度：

```
STRLEN job
```

执行以上命令得到如图 14-17 所示的结果。

```
127.0.0.1:6379> STRLEN job
(integer) 17
```

图 14-17　执行 STRLEN job 的结果

（11）MSET key1 value1 key2 value2...

同时设置一个或多个 key-value 对。使用如下命令同时设置 student 和 grade 两个 key 的 value 分别为 Tom 和 90：

```
MSET student Tom grade 90
```

执行以上命令，得到如图 14-18 所示输出结果。

```
127.0.0.1:6379> MSET student Tom grade 90
OK
```

图 14-18　执行 MSET student Tom grade 90 的结果

执行 MGET 命令验证执行结果，如图 14-19 所示。

```
127.0.0.1:6379> MGET student grade
1) "Tom"
2) "90"
```

图 14-19　执行 MGET student grade 的结果

（12）MSETEX key1 value1 key2 value2...

当且仅当所有给定 key 都不存在，同时设置一个或多个 key-value 对。通过以下命令同时设置 major 和 curriculum：

```
MSET major software curriculum java
```

执行以上命令，得到如图 14-20 所示输出结果。

```
127.0.0.1:6379> MSET major software curriculum java
OK
```

图 14-20　执行 MSET major software curriculum java 的结果

执行 MGET 命令得到如图 14-21 所示结果。

```
127.0.0.1:6379> MGET major curriculum
1) "software"
2) "java"
```

图 14-21　执行 MGET major curriculum 的结果

（13）INCR key

将 key 中存储的数字类型的 value 值增加 1。使用 SETNX 命令创建 salary，如图 14-22 所示。

```
127.0.0.1:6379> SETNX salary 1000
(integer) 1
```

图 14-22　执行 SETNX salary 1000 后的结果

执行以下命令对 salary 自增 1：

```
INCR salary
```

执行以上命令得到如图 14-23 所示的输出结果。

```
127.0.0.1:6379> INCR salary
(integer) 1001
```

图 14-23　执行 INCR salary 后的结果

（14）INCRBY key increment

将 key 所储存的 value 值加上给定的增量值（increment）。使用以下命令行对 salary 的值增加 1000：

```
INCRBY salary 1000
```

执行以上命令，得到如图 14-24 所示的结果。

```
127.0.0.1:6379> INCRBY salary 1000
(integer) 2001
```

图 14-24　执行 INCRBY salary 1000 后的结果

（15）DECR key

将 key 中储存的数字值减 1。执行以下命令将 salary 对应的 value 减 1：

```
DECR salary
```

执行以上命令得到如图 14-25 所示的结果。

```
127.0.0.1:6379> DECR salary
(integer) 2000
```

图 14-25　执行 INCRBY salary 1000 后的结果

（16）DECRBY key decrement

将 key 所储存的值减去给定的数值（decrement）。执行以下命令将 salary 对应的值减去 1000：

```
DECRBY salary 1000
```

执行以上命令得到如图 14-26 所示结果。

```
127.0.0.1:6379> DECRBY salary 1000
(integer) 1000
```

图 14-26　执行 DECRBY salary 1000 后的结果

（17）APPEND key value

如果 key 已经存在并且存储的是一个字符串值，APPEND 命令将指定的 value 追加到该 key 原始值的末尾。使用如下命令对 book 进行追加操作：

```
APPEND book "in Action"
```

执行以上命令得到如图 14-27 所示结果。

```
127.0.0.1:6379> APPEND book " in Action"
(integer) 29
```

图 14-27　执行 APPEND book "in Action"的结果

执行 GET 方法验证 APPEND 执行结果如图 14-28 所示。

```
127.0.0.1:6379> GET book
"Spring5 programming in Action"
```

图 14-28　执行 GET book 的结果

14.2.2　Redis Hash 类型的使用方式

Redis Hash 类型是一个 key-value 映射表，适合存储一个对象的一组属性。下面将介绍 Redis Hash 类型的常用操作。

(1) HSET key field value

设置 key 中 field 属性值为 value。下面命令将设置 book_spring5 的 name 属性为 Spring 5 programming：

```
HSET book_spring5 name "Spring 5 programming"
```

执行以上命令后，得到如图 14-29 所示输出结果。

```
127.0.0.1:6379> HSET book_spring5 name "Spring5 programming"
(integer) 1
```

图 14-29　执行 HSET book_spring5 name "Spring 5 programming" 的结果

(2) HGET key field

获取存储在哈希表中指定字段的值。通过以下命令查询 book_spring5 中 name 属性的值：

```
HGET book_spring5 name
```

执行以上命令得到如图 14-30 所示的输出结果。

```
127.0.0.1:6379> HGET book_spring5 name
"Spring5 programming"
```

图 14-30　执行 HGET book_spring5 name 的结果

(3) HEXISTS key field

查看哈希表中指定的字段是否存在。通过以下的命令查询 book_spring5 中 name 属性和 price 属性是否存在：

```
HEXISTS book_spring5 name
HEXISTS book_spring5 price
```

执行以上命令，得到如图 14-31 所示的结果。

```
127.0.0.1:6379> HEXISTS book_spring5 name
(integer) 1
127.0.0.1:6379> HEXISTS book_spring5 price
(integer) 0
```

图 14-31　执行 HEXISTS book_spring5 name 和 HEXISTS book_spring5 price 后的结果

(4) HINCRBY key field increment

为哈希表中指定字段的数值加上增量 increment。下面通过 HSET 为 book_spring5 新增价格属性，并设置价格为 50：

```
HSET book_spring5 price 50
```

通过以下命令将 book_spring5 的 price 属性加 10：

```
HINCRBY book_spring5 price 10
```

执行 HINCRBY 命令后得到如图 14-32 所示的结果：

```
127.0.0.1:6379> HINCRBY book_spring5 price 10
(integer) 60
```

图 14-32　执行 HINCRBY book_spring5 price 10 后的结果

（5）HGETALL key

获取在哈希表中指定 key 的所有字段和值。执行以下命令查询 book_spring5 中各个属性的值：

```
HGETALL book_spring5
```

执行以上命令得到如图 14-33 所示结果。

```
127.0.0.1:6379> HGETALL book_spring5
1) "name"
2) "Spring5 programming"
3) "price"
4) "60"
```

图 14-33　执行 HGETALL book_spring5 后的结果

（6）HKEYS key

获取所有哈希表中的字段。执行以下命令查询 book_spring5 中的所有属性：

```
HKEYS book_spring5
```

执行以上命令得到如图 14-34 所示结果。

```
127.0.0.1:6379> HKEYS book_spring5
1) "name"
2) "price"
```

图 14-34　执行 HKEYS book_spring5 后的结果

（7）HLEN key

获取哈希表中字段的数量。执行以下命令查询 book_spring5 中的属性数量：

```
HLEN book_spring5
```

执行以上命令得到如图 14-35 所示结果。

```
127.0.0.1:6379> HLEN book_spring5
(integer) 2
```

图 14-35　执行 HLEN book_spring5 的结果

（8）HMGET key field1 field2...

获取哈希表中所有给定字段的值。使用以下命令查询 book_spring5 的 name 和 price 属性的值：

```
HMGET book_spring5 name price
```

执行以上命令得到如图 14-36 所示结果。

```
127.0.0.1:6379> HMGET book_spring5 name price
1) "Spring5 programming"
2) "60"
```

图 14-36　执行 HMGET book_spring5 name price 的结果

（9）HMSET key field1 value1 field2 value2...

同时将多个 key-value 对设置到哈希表中。以下命令同时设置 book_spring5 的作者和出版社属性：

```
HMSET book_spring5 author michael press qinghua
```

执行以上命令后得到如图 14-37 所示的结果。

```
127.0.0.1:6379> HMSET book_spring5 author michael press qinghua
OK
```

图 14-37　执行 HMSET book_spring5 author michael press qinghua 的结果

（10）HVALS key

获取哈希表中所有值。以下命令用来查询 book_spring5 的所有 value：

```
HVALS book_spring5
```

执行以上命令后得到如图 14-38 所示的结果。

```
127.0.0.1:6379> HVALS book_spring5
1) "Spring5 programming"
2) "60"
3) "michael"
4) "qinghua"
```

图 14-38　执行 HVALS book_spring5 的结果

14.2.3　Redis List 类型的使用方式

Redis List 用于保存一组元素列表，下面介绍 Redis List 类型常用的操作。

（1）LPUSH key value1 [value2]

将一个或多个值插入到列表头部。以下命令将 Java 和 Spring 两个元素添加到 Redis List 类型的 technologyList 头部：

```
LPUSH technologyList Java Spring
```

执行以上命令后，得到如图 14-39 所示结果。

```
127.0.0.1:6379> LRANGE technologyList 0 1
1) "Spring"
2) "Java"
```

图 14-39　执行 LPUSH technologyList Java Spring 后的结果

（2）LRANGE key start stop

获取列表指定范围内的元素。以下命令从 technologyList 中获取 0～1 间的元素：

```
LRANGE technologyList 0 1
```

执行以上命令将会得到如图 14-40 所示结果。

```
127.0.0.1:6379> LRANGE technologyList 0 1
1) "Spring"
2) "Java"
```

图 14-40　执行 LPUSH technologyList Java Spring 后的结果

（3）LLEN key

查询列表长度。用下面命令查询 technologyList 列表的长度：

```
LLEN technologyList
```

执行以上命令得到如图 14-41 所示的结果。

```
127.0.0.1:6379> LLEN technologyList
(integer) 2
```

图 14-41　执行 LLEN technologyList 后的结果

（4）RPUSH key value1 [value2]

在列表末尾添加一个或多个值，与 LPUSH 不同的是，LPUSH 在列表头部添加元素，RPUSH 是在列表的末尾添加元素。以下命令将 technologyList 末尾添加 mybatis 和 redis 元素：

```
RPUSH technologyList mybatis redis
```

执行以上命令得到如图 14-42 所示的结果。

```
127.0.0.1:6379> RPUSH technologyList mybatis redis
(integer) 4
```

图 14-42　执行 RPUSH technologyList mybatis redis 后的结果

执行 LRANGE 验证结果，如图 14-43 所示。

```
127.0.0.1:6379> LRANGE technologyList 0 3
1) "Spring"
2) "Java"
3) "mybatis"
4) "redis"
```

图 14-43　执行 LRANGE technologyList 0 3 后的结果

（5）LINDEX key index

查询列表中的指定位置的元素。通过以下命令查询 technologyList 中位置为 2 的元素。

```
LINDEX technologyList 2
```

执行以上命令得到如图 14-44 所示的结果。

```
127.0.0.1:6379> LINDEX technologyList 2
"mybatis"
```

图 14-44　执行 LINDEX technologyList 2 后的结果

（6）BLPOP key1 [key2] timeout

移除并返回列表的第一个元素，如果列表没有元素会阻塞列表直到等待超时或发现有可移除元素为止。通过以下命令移除 technologyList 中的第一个元素：

```
BLPOP technologyList 10
```

执行结果如图 14-45 所示。

```
127.0.0.1:6379> BLPOP technologyList 10
1) "technologyList"
2) "Spring"
```

图 14-45　执行 BLPOP technologyList 10 后的结果

执行 LRANGE 命令验证 BLPOP 执行结果，如图 14-46 所示。

```
127.0.0.1:6379> LRANGE technologyList 0 3
1) "Java"
2) "mybatis"
3) "redis"
```

图 14-46　执行 LRANGE technologyList 0 3 后的结果

（7）BRPOP key1 [key2] timeout

移除并获取列表的最后一个元素，如果没有元素会阻塞列表直到超时或发现有可移除元素为止。通过以下命令移除 technologyList 中的最后一个元素：

```
BRPOP technologyList 10
```

执行结果如图 14-47 所示。

```
127.0.0.1:6379> BRPOP technologyList 10
1) "technologyList"
2) "redis"
```

图 14-47　执行 BRPOP technologyList 10 后的结果

执行 LRANGE 命令验证 BRPOP 执行结果，如图 14-48 所示。

```
127.0.0.1:6379> LRANGE technologyList 0 3
1) "Java"
2) "mybatis"
```

图 14-48　执行 LRANGE technologyList 0 3 后的结果

（8）LINSERT key BEFORE|AFTER pivot value

在列表元素的前或者后插入元素。执行以下命令在 Java 元素的后面加入 Kotlin。

```
LINSERT technologyList AFTER Java Kotlin
```

执行以上命令得到如图 14-49 所示的结果。

```
127.0.0.1:6379> LINSERT technologyList AFTER Java Kotlin
(integer) 3
```

图 14-49　执行 LINSERT technologyList AFTER Java Kotlin 后的结果

执行 LRANGE 验证 LINSERT 执行结果，如图 14-50 所示。

```
127.0.0.1:6379> LRANGE technologyList 0 3
1) "Java"
2) "Kotlin"
3) "mybatis"
```

图 14-50　执行 LRANGE technologyList 0 3 后的结果

（9）LTRIM key start stop

对一个列表进行截取，让列表只保留指定区间内的元素，不在指定区间之内的元素都将被删除。通过以下命令截取 technologyList 中 1~2 之间的元素。

```
LTRIM technologyList 1 2
```

执行以上命令得到如图 14-51 所示结果。

```
127.0.0.1:6379> LTRIM technologyList 1 2
OK
```

图 14-51　执行 LTRIM technologyList 1 2 后的结果

执行 LRANGE 验证 LTRIM 执行结果，如图 14-52 所示。

```
127.0.0.1:6379> LRANGE technologyList 0 2
1) "Kotlin"
2) "mybatis"
```

图 14-52　执行 LRANGE technologyList 0 2 后的结果

14.2.4　Redis Set 类型的使用方式

Redis Set 类型是无序集合，集合中的元素是唯一的不重复的。以下是 Redis Set 类型常用的操作。

（1）SADD key member1 member2...

向集合中添加一个或多个元素。使用下面命令向 company 中添加 boss、manager 和 staff 三个元素。

```
SADD company boss manager staff
```

执行结果如图 14-53 所示。

```
127.0.0.1:6379> SADD company boss manager staff
(integer) 3
```

图 14-53　执行 SADD company boss manager staff 后的结果

（2）SMEMBERS key

查询集合中的所有元素。通过以下命令查询 company 中的所有元素：

```
SMEMBERS company
```

执行以上命令得到如图 14-54 所示的结果。

```
127.0.0.1:6379> SMEMBERS company
1) "manager"
2) "boss"
3) "staff"
```

图 14-54 执行 SMEMBERS company 后的结果

（3）SCARD key

查询集合中元素的个数。通过以下命令查询 company 中元素的个数。

```
SCARD company
```

执行以上命令得到如图 14-55 所示的结果。

```
127.0.0.1:6379> SCARD company
(integer) 3
```

图 14-55　执行 SCARD company 后的结果

（4）SISMEMBER key member

判断 member 元素是否为集合中的成员。使用以下命令用于判断 boss 和 secretary 是否为 company 集合中的成员：

```
SISMEMBER company boss
SISMEMBER company secretary
```

执行以上命令得到如图 14-56 所示的结果。

```
127.0.0.1:6379> SISMEMBER company boss
(integer) 1
127.0.0.1:6379> SISMEMBER company secretary
(integer) 0
```

图 14-56　执行 SISMEMBER company boss 和 SISMEMBER company secretary 后的结果

（5）SDIFF key1 [key2]

查询给定所有集合的差集。

首先使用 SADD 创建一个集合 school：

```
SADD school president teacher staff
```

执行 SDIFF 命令查询 company 和 school 的差集：

```
SDIFF company school
```

执行 SDIFF 后得到如图 14-57 所示结果。

```
127.0.0.1:6379> SDIFF company school
1) "manager"
2) "boss"
```

图 14-57　执行 SDIFF company school 后的结果

（6）SINTER key1 [key2]

查询给定多个集合之间的交集。使用如下命令查询 company 和 staff 的交集：

```
SINTER company school
```

执行以上命令得到如图 14-58 所示结果。

```
127.0.0.1:6379> SINTER company school
1) "staff"
```

图 14-58　执行 SINTER company school 后的结果

（7）SRANDMEMBER key [count]

返回集合中的 count 个随机元素。执行以下命令获取 company 中的 1 个随机元素。

```
SRANDMEMBER company 1
```

执行以上命令得到如图 14-59 所示结果。

```
127.0.0.1:6379> SRANDMEMBER company 1
1) "boss"
```

图 14-59　执行 SRANDMEMBER company 1 后的结果

（8）SUNION key1 [key2]

查询多个集合的并集。执行以下命令查询 company 和 school 之间的并集。

```
SUNION company school
```

执行以上命令得到如图 14-60 所示结果。

```
127.0.0.1:6379> SUNION company school
1) "boss"
2) "staff"
3) "teacher"
4) "manager"
5) "president"
```

图 14-60　执行 SUNION company school 后的结果

（9）SPOP key

移除并返回集合中的一个随机元素。执行以下命令移除并返回 school 中的任意元素：

```
SPOP school
```

执行以上命令得到如图 14-61 所示结果。

```
127.0.0.1:6379> SPOP school
"teacher"
```

图 14-61　执行 SPOP school 后的结果

执行 SMEMBERS 验证 SPOP 执行结果如图 14-62 所示。

```
127.0.0.1:6379> SMEMBERS school
1) "president"
2) "staff"
```

图 14-62　执行 SPOP school 后的结果

（10）SREM key member1 [member2]

移除集合中一个或多个成员。执行以下命令移除 company 中的 manager 和 boss 元素：

```
SREM company manager boss
```

执行以上命令得到如图 14-63 所示结果。

```
127.0.0.1:6379> SREM company manager boss
(integer) 2
```

图 14-63　执行 SREM company manager boss 后的结果

执行 SMEMBERS company 验证 SREM 执行结果，得到如图 14-64 所示结果。

```
127.0.0.1:6379> SMEMBERS company
1) "staff"
```

图 14-64　执行 SREM company manager boss 后的结果

14.2.5　Redis SortedSet 类型的使用方式

Redis SortedSet 和 Redis Set 一样存储多个元素，且不存在重复元素。不同的是，Redis SortedSet 中的每个元素会关联一个分数，Redis 正是使用这个分数为 SortedSet 中的元素进行从小到大排序的。下面介绍一些 SortedSet 的常用操作。

（1）ZADD key score1 member1 [score2 member2]

向有序集合添加一个或多个元素，或者更新已存在成员的分数。通过以下命令添加和更新多个元素：

```
ZADD scores 100 Michael 80 Tom
ZADD scores 80 Jimmy 90 Tom
```

执行以上命令得到如图 14-65 所示的结果。

```
127.0.0.1:6379> ZADD scores 100 Michael 80 Tom
(integer) 2
127.0.0.1:6379> ZADD scores 80 Jimmy 90 Tom
(integer) 1
```

图 14-65　执行 ZADD scores 100 Michael 80 Tom 和 ZADD scores 80 Jimmy 90 Tom 后的结果

（2）ZCARD key

获取有序集合的所有元素个数。通过以下命令查询 scores 集合中所有元素的个数。

```
ZCARD scores
```

执行以上命令得到如图 14-66 所示的结果。

```
127.0.0.1:6379> ZCARD scores
(integer) 3
```

图 14-66　执行 ZCARD scores 后的结果

（3）ZRANGE key start stop [WITHSCORES]

通过索引区间返回有序集合中指定区间内的元素：

```
ZRANGE scores 0 2
```

执行以上命令得到如图 14-67 所示的结果。

```
127.0.0.1:6379> ZRANGE scores 0 2
1) "Jimmy"
2) "Tom"
3) "Michael"
```

图 14-67　执行 ZCARD scores 后的结果

（4）ZCOUNT key min max

计算在有序集合中指定分数区间范围内的元素个数。这里通过以下命令查询 scores 中分数在 80~100 之间的元素个数：

```
ZCOUNT scores 80 100
```

执行以上命令得到如图 14-68 所示结果。

```
127.0.0.1:6379> ZCOUNT scores 80 100
(integer) 3
```

图 14-68　执行 ZCARD scores 后的结果

（5）ZRANGE key start stop [WITHSCORES]

通过索引返回指定区间内有序集合内的元素。这里通过以下命令查询索引在 1~2 之间的元素：

```
ZRANGE scores 1 2
```

执行以上命令得到如图 14-69 所示结果。

```
127.0.0.1:6379> ZRANGE scores 1 2
1) "Tom"
2) "Michael"
```

图 14-69　执行 ZRANGE scores 1 2 后的结果

（6）ZRANGEBYSCORE key min max [WITHSCORES] [LIMIT]

通过分数查询有序集合指定区间内的元素。这里通过以下命令查询分数在 80~100 之间的元素：

```
ZRANGEBYSCORE scores 80 100
```

执行以上命令得到如图14-70所示结果。

```
127.0.0.1:6379> ZRANGEBYSCORE scores 80 100
1) "Jimmy"
2) "Tom"
3) "Michael"
```

图14-70　执行 ZRANGEBYSCORE scores 80 100 后的结果

（7）ZRANK key member

返回有序集合中指定成员的索引值。这里使用以下命令查询Tom在有序集合中的索引位置：

```
ZRANK scores Tom
```

执行以上命令得到如图14-71所示的结果。

```
127.0.0.1:6379> ZRANK scores Tom
(integer) 1
```

图14-71　执行 ZRANK scores Tom 后的结果

（8）ZREVRANGE key start stop [WITHSCORES]

通过索引区间，按照分数从高到底顺序查询有序集中指定区间内的成员。这里通过以下命令倒序查询 scores 中索引在 0~2 之间的元素：

```
ZREVRANGE scores 0 2
```

执行以上命令得到如图14-72所示的结果。

```
127.0.0.1:6379> ZREVRANGE scores 0 2
1) "Michael"
2) "Tom"
3) "Jimmy"
```

图14-72　执行 ZREVRANGE scores 0 2 后的结果

（9）ZREVRANGEBYSCORE key max min [WITHSCORES]

通过分数区间，按照从高到低的顺序查询有序集中指定区间内的成员。这里通过以下命令倒序查询 scores 中分数在 80~100 之间的元素：

```
ZREVRANGEBYSCORE scores 100 80
```

执行以上命令得到如图14-73所示结果。

```
127.0.0.1:6379> ZREVRANGEBYSCORE scores 100 80
1) "Michael"
2) "Tom"
3) "Jimmy"
```

图14-73　执行 ZREVRANGEBYSCORE scores 100 80 后的结果

（10）ZREVRANK key member

查询有序集合按分数值递减（从大到小）排序时，指定元素的排名。这里通过以下命令倒序查询 scores 中 Jimmy 元素的位置：

```
ZREVRANK scores Jimmy
```

执行以上命令得到如图 14-74 所示的结果。

```
127.0.0.1:6379> ZREVRANK scores Jimmy
(integer) 2
```

图 14-74　执行 ZREVRANK scores Jimmy 后的结果

（11）ZSCORE key member

查询有序集合中指定成员的分数值。这里通过以下命令查询 Michael 元素的分数值。

```
ZSCORE scores Michael
```

执行以上命令得到如图 14-75 所示的结果。

```
127.0.0.1:6379> ZSCORE scores Michael
"100"
```

图 14-75　执行 ZSCORE scores Michael 后的结果

（12）ZREM key member [member ...]

删除有序集合中的一个或多个成员。通过以下命令删除 scores 中 Michael 元素：

```
ZREM scores Michael
```

执行以上命令得到如图 14-76 所示的结果。

```
127.0.0.1:6379> ZREM scores Michael
(integer) 1
```

图 14-76　执行 ZREM scores Michael 后的结果

执行 ZRANGE 命令验证 ZREM 执行结果，如图 14-77 所示。

```
127.0.0.1:6379> ZRANGE scores 0 2
1) "Jimmy"
2) "Tom"
```

图 14-77　执行 ZRANGE scores 0 2 后的结果

（13）ZREMRANGEBYSCORE key min max

删除有序集合中给定的分数区间的所有元素。通过以下命令删除 scores 中 80~90 分之间的元素：

```
ZREMRANGEBYSCORE scores 80 90
```

执行以上命令得到如图 14-78 所示的结果。

```
127.0.0.1:6379> ZREMRANGEBYSCORE scores 80 90
(integer) 2
```

图 14-78　执行 ZREMRANGEBYSCORE scores 80 90 后的结果

执行 ZRANGE 命令验证 ZREMRANGEBYSCORE 执行结果，如图 14-79 所示。

```
127.0.0.1:6379> ZRANGE scores 0 2
(empty list or set)
```

图 14-79　执行 ZRANGE scores 0 2 后的结果

14.3　Redis 持久化策略

在运行情况下，Redis 将数据维持在内存中，为了让这些数据在 Redis 重启/死机之后仍然可用，Redis 分别提供了 RDB（Redis DataBase）和 AOF（Append Only File）两种持久化模式。

14.3.1　Redis RDB 持久化

在 Redis 运行时，RDB 程序将当前内存中的数据库快照保存到磁盘文件中，在 Redis 重新启动时，RDB 程序可以通过载入 RDB 文件来还原 Redis 中的数据。

RDB 的工作方式如下：在指定的时间间隔内，执行指定次数的写操作，将 Redis 内存中的数据写入到磁盘中保存起来，即生成一个 dump.rdb 文件。当 Redis 重新启动时，通过读取磁盘上的 dump.rdb 文件将磁盘中的数据恢复到内存中。

打开 Redis 目录下的 redis.conf 文件，找到 SNAPSHOTTING 相关默认配置项：

```
################################ SNAPSHOTTING  ################################
#
# Save the DB on disk:
#
#   save <seconds> <changes>
#
#   Will save the DB if both the given number of seconds and the given
#   number of write operations against the DB occurred.
#
#   In the example below the behaviour will be to save:
#   after 900 sec (15 min) if at least 1 key changed
#   after 300 sec (5 min) if at least 10 keys changed
#   after 60 sec if at least 10000 keys changed
#
#   Note: you can disable saving completely by commenting out all "save" lines.
#
#   It is also possible to remove all the previously configured save
#   points by adding a save directive with a single empty string argument
#   like in the following example:
#
#   save ""

save 900 1
save 300 10
save 60 10000
```

这是配置 RDB 持久化规则的，下面对配置项进行讲解：

```
save <指定时间间隔> <指定次数更新操作>
```

RDB 持久化表示在指定的时间间隔内发生指定次数的更新操作，那么将进行持久化操作。在 redis.conf 默认配置中，各配置项的含义如下：

```
# 900 秒内有 1 次更改即保存内存数据到磁盘
save 900 1
# 300 秒内有 10 次更改即保存内存数据到磁盘
save 300 10
# 60 秒内有 10000 次更改即保存内存数据到磁盘
save 60 10000
```

14.3.2 Redis AOF 持久化

AOF 默认是不开启的。持久化方式是以日志的形式记录每个写操作，并追加到文件中的。Redis 重启时，会根据日志文件的内容将保存的写操作执行一次，完成 Redis 内存数据的恢复。

打开 Redis 配置文件 redis.comf，找到 APPEND ONLY MODE 相关配置项：

```
############################## APPEND ONLY MODE ###############################

# By default Redis asynchronously dumps the dataset on disk. This mode is
# good enough in many applications, but an issue with the Redis process or
# a power outage may result into a few minutes of writes lost (depending on
# the configured save points).
#
# The Append Only File is an alternative persistence mode that provides
# much better durability. For instance using the default data fsync policy
# (see later in the config file) Redis can lose just one second of writes in a
# dramatic event like a server power outage, or a single write if something
# wrong with the Redis process itself happens, but the operating system is
# still running correctly.
#
# AOF and RDB persistence can be enabled at the same time without problems.
# If the AOF is enabled on startup Redis will load the AOF, that is the file
# with the better durability guarantees.
#
# Please check http://redis.io/topics/persistence for more information.

appendonly no

# The name of the append only file (default: "appendonly.aof")

appendfilename "appendonly.aof"

# The fsync() call tells the Operating System to actually write data on disk
```

```
            # instead of waiting for more data in the output buffer. Some OS will really flush
            # data on disk, some other OS will just try to do it ASAP.
            #
            # Redis supports three different modes:
            #
            # no: don't fsync, just let the OS flush the data when it wants. Faster.
            # always: fsync after every write to the append only log. Slow, Safest.
            # everysec: fsync only one time every second. Compromise.
            #
            # The default is "everysec", as that's usually the right compromise between
            # speed and data safety. It's up to you to understand if you can relax this to
            # "no" that will let the operating system flush the output buffer when
            # it wants, for better performances (but if you can live with the idea of
            # some data loss consider the default persistence mode that's snapshotting),
            # or on the contrary, use "always" that's very slow but a bit safer than
            # everysec.
            #
            # More details please check the following article:
            # http://antirez.com/post/redis-persistence-demystified.html
            #
            # If unsure, use "everysec".

            # appendfsync always
            appendfsync everysec
            # appendfsync no
```

可以从 redis.conf 中看到，默认情况下，Redis 没有开启 AOF 功能。如果想要开启 AOF 功能，可以将 appendonly 修改为 yes：

```
            appendonly yes
```

appendfilename 配置项控制 AOF 持久化文件的名称。appendfsync 配置项用于指定日志更新的条件：

```
            #每次发生数据变化会立刻写入到磁盘中
            # appendfsync always
            #默认配置，每秒异步记录一次
            appendfsync everysec
            #不同步
            # appendfsync no
```

14.4 Redis 主从复制模式

Redis 主从复制架构的特点是主节点负责接受写入数据的请求，从节点负责接受查询数据的请求，主节点定期把数据同步给从节点，以保证主从节点的一致性。

下面搭建 Redis 主从复制架构。

1. 创建 Redis 配置文件

进入 Redis 目录，复制原 redis.conf 为 redis6380.conf 文件，操作如下：

```
#进入Redis-5.0.3解压目录
cd redis-5.0.3
#复制一份新的配置文件
cp redis.confredis6380.conf
```

2. 修改配置文件

在 redis6380.conf 中修改启动端口和主从关系，代码如下：

```
#配置此Redis节点为127.0.0.1 6379节点的从节点
slaveof 127.0.0.1 6379
#配置启动端口为6380
port 6380
```

3. 启动 Redis 主从服务

分别启动 Redis 主节点 127.0.0.1 6379 和从节点 127.0.0.1 6380，验证主节点和从节点启动情况，如图 14-80 所示。

```
#启动Redis主节点127.0.0.1 6379
src/redis-server redis.conf
#启动Redis从节点127.0.0.1 6380
src/redis-server redis6380.conf
#查询Redis进程
ps -ef | grep redis | grep -v grep
```

```
MichaeldeMacBook-Pro:redis-5.0.3 michael$ ps -ef | grep redis | grep -v grep
  501  3864     1   0  9:44下午 ??         0:00.13 src/redis-server 127.0.0.1:6379
  501  3866     1   0  9:44下午 ??         0:00.06 src/redis-server 127.0.0.1:6380
```

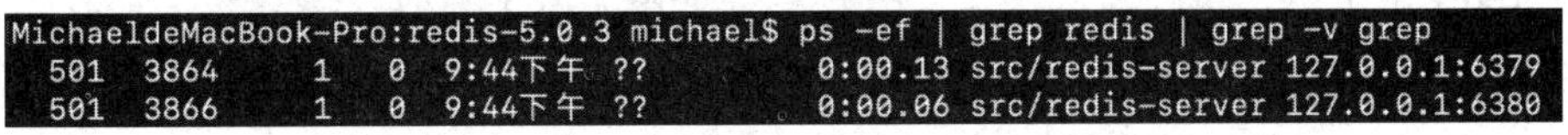

图 14-80　执行 ps -ef | grep redis | grep -v grep 后的结果

4. 查看主从状态

登录 Redis 主节点，执行 info replication 命令，如图 14-81 所示，从图 14-81 中可以看出当前节点是 master 节点：

```
#登录Redis主节点127.0.0.1 6379客户端
src/redis-cli -h 127.0.0.1 -p 6379
```

```
#Redis 主节点执行 info replication
info replication
```

```
127.0.0.1:6379> info replication
# Replication
role:master
connected_slaves:1
slave0:ip=127.0.0.1,port=6380,state=online,offset=1834,lag=1
master_replid:45bc28c101e915acc87b444009be0c6b3eaa9ab1
master_replid2:0000000000000000000000000000000000000000
master_repl_offset:1834
second_repl_offset:-1
repl_backlog_active:1
repl_backlog_size:1048576
repl_backlog_first_byte_offset:1
repl_backlog_histlen:1834
```

图 14-81 Redis 主节点执行 info replication 后的结果

登录 Redis 从节点，执行 info replication 命令，如图 14-82 所示，从图中可以看出当前节点是 slave 节点。

```
127.0.0.1:6380> info replication
# Replication
role:slave
master_host:127.0.0.1
master_port:6379
master_link_status:up
master_last_io_seconds_ago:2
master_sync_in_progress:0
slave_repl_offset:2282
slave_priority:100
slave_read_only:1
connected_slaves:0
master_replid:45bc28c101e915acc87b444009be0c6b3eaa9ab1
master_replid2:0000000000000000000000000000000000000000
master_repl_offset:2282
second_repl_offset:-1
repl_backlog_active:1
repl_backlog_size:1048576
repl_backlog_first_byte_offset:1
repl_backlog_histlen:2282
```

图 14-82 Redis 从节点执行 info replication 后的结果

5. Redis 主节点写入操作

登录 Redis 主节点，执行写入操作，执行结果如图 14-83 所示。

```
#登录 redis 主节点客户端
src/redis-cli -h 127.0.0.1 -p 6379
#在 redis 主节点写入
```

#写入 Hash master_slave 中 key-value 对 master ："127.0.0.1 6379"

```
HMSET master_slave master "127.0.0.1 6379"
```

#写入 Hash master_slave 中 key-value 对 slave ："127.0.0.1 6380"

```
HMSET master_slave slave "127.0.0.1 6380"
```

```
127.0.0.1:6379> HMSET master_slave master "127.0.0.1 6379"
OK
127.0.0.1:6379> HMSET master_slave slave "127.0.0.1 6380"
OK
127.0.0.1:6379> HGETALL master_slave
1) "master"
2) "127.0.0.1 6379"
3) "slave"
4) "127.0.0.1 6380"
```

图 14-83　主节点写入和查询 Hash

6. 查询从节点查询同步状态

登录从节点客户端，查询从节点同步状态，如图 14-84 所示：

```
127.0.0.1:6380> HGETALL master_slave
1) "master"
2) "127.0.0.1 6379"
3) "slave"
4) "127.0.0.1 6380"
```

图 14-84　主节点写入和查询 Hash 后的结果

从以上步骤看出，Redis 从节点 127.0.0.1 6380 虽然没有发生写入操作，但执行查询可以发现，Redis 主节点 127.0.0.1 6379 发生写入的操作已经同步到 Redis 从节点 127.0.0.1 6380。

Redis 主从架构有多种不同的拓补结构，以下是一些常见的主从拓扑结构。

14.4.1　Redis 一主一从拓扑结构

Redis 一主一从拓扑结构主要用于主节点故障转移到从节点。当主节点的写入操作并发高且需要持久化时，可以只在从节点开启 AOF（主节点不需要），这样即保证了数据的安全性，也避免了持久化对主节点性能的影响。Redis 一主一从拓扑结构如图 14-85 所示。

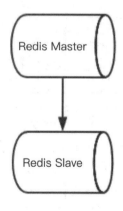

图 14-85　Redis 一主一从拓扑结构

14.4.2　Redis 一主多从拓扑结构

针对读取操作并发较高的场景，读取操作由多个从节点来分担；但节点越多，主节点同步到多节点的次数也越多，影响带宽，也对主节点的稳定性造成负担。

Redis 一主多从拓扑结构如图 14-86 所示。

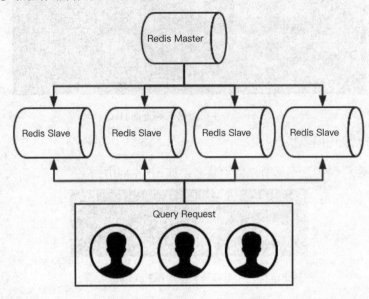

图 14-86　Redis 一主多从拓扑结构

14.4.3　Redis 树形拓扑结构

一主多从拓扑结构的缺点是主节点推送次数多、压力大，可用树形拓扑结构解决，主节点只负责推送数据到从节点 A，再由从节点 A 推送到 B、C 和 D，可以减轻主节点推送的压力。

Redis 树形拓扑结构如图 14-87 所示。

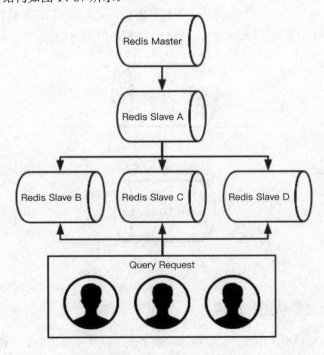

图 14-87　Redis 树形拓扑结构

14.4.4 Redis 主从架构的缺点

主从复制架构虽然可以提交读并发，但这种方式也有缺点。

（1）主从复制架构中，如果主节点出现问题，则不能提供服务，需人工修改重新设置主节点。
（2）主从复制架构中，主节点单机写能力有限。

14.5 Redis 哨兵模式

从 Redis 主从架构的缺点可以看出，当主节点 Master 出现故障后，Redis 新的主节点必须由开发人员手动修改，这显然不满足高可用的特性。因此在 Redis 主从架构的基础上，演变出了 Redis 哨兵机制。

哨兵机制（sentinel）的高可用原理是：当主节点出现故障时，由 Redis 哨兵自动完成故障发现和转移，并通知 Redis 客户端，实现高可用性。

14.5.1 Redis 哨兵模式简介

Redis 哨兵进程是用于监控 Redis 集群中 Master 主服务器工作状态的。在主节点 Master 发生故障的时候，可以实现 Master 和 Slave 服务器的自动切换，保证系统的高可用性。

Redis 哨兵是一个分布式系统，可以在一个架构中运行多个 Redis 哨兵进程，这些进程使用流言协议（gossipprotocols）来接收关于 Master 主服务器是否下线的信息，并使用投票协议（Agreement Protocols）来决定是否执行自动故障迁移，以及选择某个 Slave 节点作为新的 Master 节点。

每个 Redis 哨兵进程会向其他 Redis 哨兵、Master 主节点、Slave 从节点定时发送消息，以确认被监控的节点是否"存活着"。如果发现对方在指定配置时间（可配置的）内未得到回应，那么暂时认为被监控节点已死机，即所谓的"客观下线（Subjective Down，简称 SDOWN）"。

与"主观下线"对应的是"客观下线（Objectively Down，简称 ODOWN）"。当"哨兵群"中的多数 Redis 哨兵进程在对 Master 主节点做出 SDOWN 的判断，并且通过 SENTINEL is-master-down-by-addr 命令互相交流之后，得出 Master Server 的下线判断，此时认为主节点 Master 发生"客观下线"。通过一定的选举算法，从剩下的存活的从节点中选出一台晋升为 Master 主节点，然后自动修改相关配置，并开启故障转移（failover）。

Redis 哨兵虽然由一个单独的可执行文件 redis-sentinel 控制启动，但实际上 Redis 哨兵只是一个运行在特殊模式下的 Redis 服务器，可以在启动一个普通 Redis 服务器时通过指定 sentinel 选项来启动 Redis 哨兵，Redis 哨兵的一些设计思路和 Zookeeper 非常类似。

Redis 哨兵集群之间会互相通信，交流 Redis 节点的状态，做出相应的判断并进行处理。这里的"主观下线"和"客观下线"是比较重要的状态，这两个状态决定了是否进行故障转移，可以通过订阅指定的频道信息，当服务器出现故障的时候通知管理员。客户端可以将 Redis 哨兵看作是一个只提供了订阅功能的 Redis 服务器，客户端不可以使用 PUBLISH 命令向这个服务器发送信息，但是客户端可以用 SUBSCRIBE/PSUBSCRIBE 命令，通过订阅指定的频道来获取相应的事件提醒。

Redis 哨兵架构可以用图 14-88 表示。

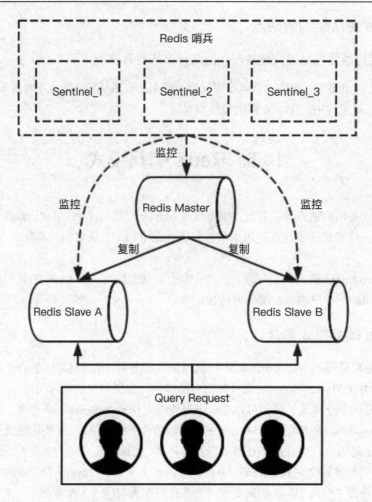

图 14-88　Redis 哨兵拓扑结构

14.5.2　Redis 哨兵定时监控任务

1. 任务 1

每个 Redis 哨兵节点每 10s 会向主节点和从节点发送 info 命令获取拓扑结构图，Redis 哨兵配置时只要配置对主节点的监控即可，可以通过向主节点发送 info 命令获取从节点的信息，并当有新的从节点加入时可以立刻感知到，如图 14-89 所示。

2. 任务 2

每个 Redis 哨兵节点每隔 2s 会向 Redis 数据节点的指定频道上发送该 Redis 哨兵节点对于主节点的状态判断以及当前 Redis 哨兵节点自身的信息，同时每个哨兵节点也会订阅该频道用来获取其他 Redis 哨兵节点的信息及对主节点的状态判断，如图 14-90 所示。

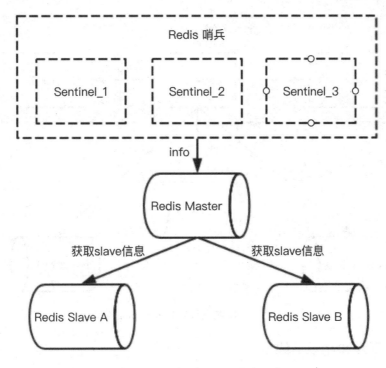

图 14-89　Redis 哨兵每隔 10s 执行一次 info

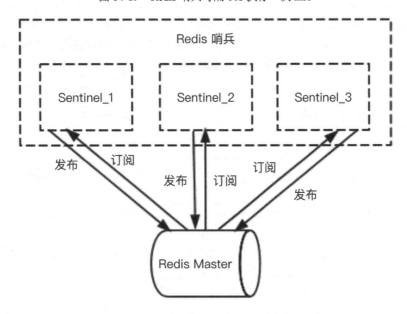

图 14-90　Redis 哨兵每隔 2s 执行一次发布和订阅

3. 任务 3

每隔 1s 每个 Redis 哨兵会向主节点、从节点及其余 Redis 哨兵节点发送一次 ping 命令做一次"心跳"检测，这也是 Redis 哨兵用来判断节点是否正常的重要依据，如图 14-91 所示。

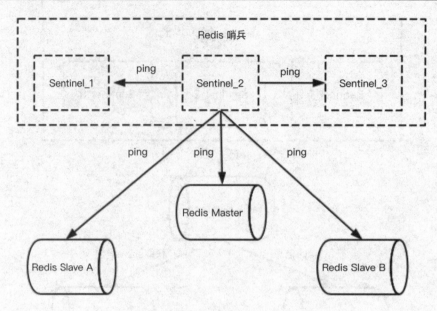

图 14-91　Redis 哨兵每隔 1s 执行 ping 命令

14.5.3　主观下线和客观下线

当主观下线的节点是主节点时，此时探测到主节点主观下线的 Redis 哨兵节点会通过指令 sentinel is-masterdown-by-addr 寻求其他 Redis 哨兵节点对主节点状态做出的判断，当超过 quorum（选举）个数，此时 Redis 哨兵节点则认为该主节点确实有问题，这样就客观下线了，大部分哨兵节点都同意下线操作，即发生客观下线，如图 14-92 所示。

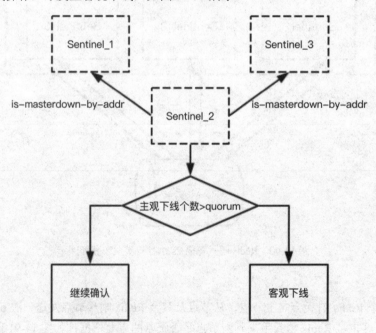

图 14-92　主观下线和客观下线

14.5.4 Redis 哨兵选举领导者

Redis 哨兵选举领导者的步骤如下：

（1）每个在线的哨兵节点都可以成为领导者，当此 Redis 哨兵（如图 14-92 所示 Sentinel 2）确认主节点主观下线时，会向其他哨兵发送 is-master-down-by-addr 命令，征求判断并要求将自己设置为 Redis 哨兵集群的领导者，由领导者处理故障转移。

（2）当其他 Redis 哨兵收到 is-master-down-by-addr 命令时，可以同意或者拒绝此 Redis 哨兵成为领导者。

（3）当此 Redis 哨兵得到的票数 >= max(quorum,num(sentinels)/2+1)时，Redis 哨兵将成为 Redis 哨兵集群。如果没有超过，则继续选举。

Redis 哨兵选举领导者的过程如图 14-93 所示。

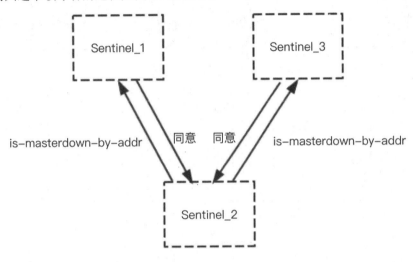

图 14-93 Redis 哨兵选举领导者

14.5.5 故障转移

故障转移的步骤如下：

（1）将 Slave A 脱离原从节点，升级主节点。
（2）将从节点 Slave B 指向新的主节点。
（3）通知客户端主节点已更换。
（4）如果主节点故障恢复，则设置成为新的主节点的从节点。

故障转移过程（假设图 14-92 中 Sentinel 2 成为领导者）如图 14-94 所示。
经过故障转移后，Redis 哨兵架构的拓扑结果将发生变化，如图 14-95 所示。

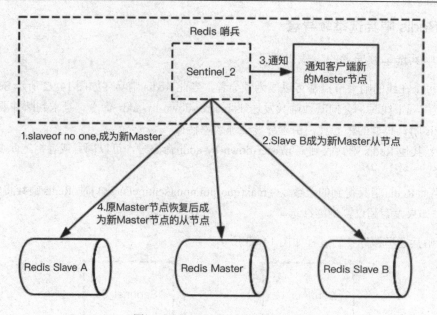

图 14-94　Redis 哨兵机制故障转移

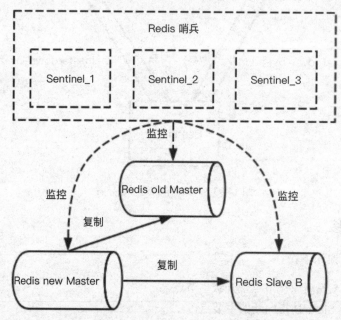

图 14-95　Redis 哨兵机制故障转移后的拓扑图

14.5.6　Redis 哨兵模式安装部署

本节按照图 14-88 所示的拓扑结构安装部署 Redis 哨兵模式。

1. 创建 Redis 主从节点配置文件

进入 Redis 目录，将 redis.conf 文件复制 3 份，分别命名为 redis6379.conf、redis6380.conf 和 redis6381.conf。

```
#创建 Redis 127.0.0.1 6379 配置文件
cp redis.conf redis6379.conf
#创建 Redis 127.0.0.1 6380 配置文件
cp redis.conf redis6380.conf
#创建 Redis 127.0.0.1 6381 配置文件
cp redis.conf redis6381.conf
```

2. 修改各个 Redis 配置文件

修改 redis6379.conf 配置文件，配置启动端口为 6379：

```
port 6379
```

修改 redis6380.conf 配置文件，配置启动端口为 6380，并配置此节点为 127.0.0.1 6379 的从节点。

```
port 6380
slaveof 127.0.0.1 6379
```

修改 redis6380.conf 配置文件，配置启动端口为 6381，并配置此节点为 127.0.0.1 6379 的从节点。

```
port 6381
slaveof 127.0.0.1 6379
```

3. 分别启动 Redis 主从节点

分别启动 Redis Master 节点 127.0.0.1 6379，Redis Slave 节点 127.0.0.1 6380 和节点 127.0.0.1 6381。启动主节点 127.0.0.1 6379，如图 14-96 所示。

```
src/redis-server redis6379.conf
```

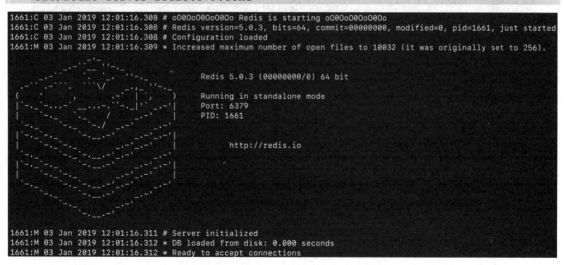

图 14-96　启动节点 127.0.0.1 6379

启动主节点 127.0.0.1 6380，如图 14-97 所示。

```
src/redis-server redis6380.conf
```

```
1666:C 03 Jan 2019 12:04:37.291 # o0O0o00Oo0O0o Redis is starting o0O0o00Oo0O0o
1666:C 03 Jan 2019 12:04:37.291 # Redis version=5.0.3, bits=64, commit=00000000, modified=0, pid=1666, just started
1666:C 03 Jan 2019 12:04:37.291 # Configuration loaded
1666:S 03 Jan 2019 12:04:37.292 * Increased maximum number of open files to 10032 (it was originally set to 256).
```

```
           Redis 5.0.3 (00000000/0) 64 bit
           Running in standalone mode
           Port: 6380
           PID: 1666

           http://redis.io
```

```
1666:S 03 Jan 2019 12:04:37.293 # Server initialized
1666:S 03 Jan 2019 12:04:37.293 * DB loaded from disk: 0.000 seconds
```

图 14-97　启动节点 127.0.0.1 6380

启动主节点 127.0.0.1 6381，如图 14-98 所示。

```
src/redis-server redis6381.conf
```

```
1679:C 03 Jan 2019 12:08:01.219 # o0O0o00Oo0O0o Redis is starting o0O0o00Oo0O0o
1679:C 03 Jan 2019 12:08:01.219 # Redis version=5.0.3, bits=64, commit=00000000, modified=0, pid=1679, just started
1679:C 03 Jan 2019 12:08:01.219 # Configuration loaded
1679:S 03 Jan 2019 12:08:01.220 * Increased maximum number of open files to 10032 (it was originally set to 256).
```

```
           Redis 5.0.3 (00000000/0) 64 bit
           Running in standalone mode
           Port: 6381
           PID: 1679

           http://redis.io
```

```
1679:S 03 Jan 2019 12:08:01.222 # Server initialized
1679:S 03 Jan 2019 12:08:01.222 * DB loaded from disk: 0.000 seconds
```

图 14-98　启动 127.0.0.1 6381

验证 Redis Master 节点和 Slave 节点启动状况，如图 14-99 所示。

```
ps -ef | grep redis | grep -v grep
```

```
501  1661  796  0 12:01下午 ttys000    0:00.49 src/redis-server 127.0.0.1:6379
501  1666  858  0 12:04下午 ttys001    0:00.34 src/redis-server 127.0.0.1:6380
501  1679  875  0 12:08下午 ttys003    0:00.19 src/redis-server 127.0.0.1:6381
```

图 14-99　查看启动的 Redis 进程

下面验证主从节点之间的状态。使用客户端连接 Master 节点 127.0.0.1 6379 执行 info 命令，如图 14-100 所示。

```
src/redis-cli -h 127.0.0.1 -p 6379
```

使用客户端连接 Slave 节点 127.0.0.1 6380 执行 info 命令，如图 14-101 所示。

```
src/redis-cli -h 127.0.0.1 -p 6380
```

```
127.0.0.1:6379> info replication
# Replication
role:master
connected_slaves:2
slave0:ip=127.0.0.1,port=6380,state=online,offset=532,lag=1
slave1:ip=127.0.0.1,port=6381,state=online,offset=532,lag=1
master_replid:9d499822105c21c008589d51c5b64355678dfc9f
master_replid2:0000000000000000000000000000000000000000
master_repl_offset:532
second_repl_offset:-1
repl_backlog_active:1
repl_backlog_size:1048576
repl_backlog_first_byte_offset:1
repl_backlog_histlen:532
```

图 14-100 连接 Master 节点 127.0.0.1 6379 执行 info 命令

```
127.0.0.1:6380> info replication
# Replication
role:slave
master_host:127.0.0.1
master_port:6379
master_link_status:up
master_last_io_seconds_ago:5
master_sync_in_progress:0
slave_repl_offset:560
slave_priority:100
slave_read_only:1
connected_slaves:0
master_replid:9d499822105c21c008589d51c5b64355678dfc9f
master_replid2:0000000000000000000000000000000000000000
master_repl_offset:560
second_repl_offset:-1
repl_backlog_active:1
repl_backlog_size:1048576
repl_backlog_first_byte_offset:57
repl_backlog_histlen:504
```

图 14-101 连接 Slave 节点 127.0.0.1 6380 执行 info 命令

使用客户端连接 Slave 节点 127.0.0.1 6381 执行 info 命令，如图 14-102 所示。

```
src/redis-cli -h 127.0.0.1 -p 6381
```

```
127.0.0.1:6381> info replication
# Replication
role:slave
master_host:127.0.0.1
master_port:6379
master_link_status:up
master_last_io_seconds_ago:3
master_sync_in_progress:0
slave_repl_offset:574
slave_priority:100
slave_read_only:1
connected_slaves:0
master_replid:9d499822105c21c008589d51c5b64355678dfc9f
master_replid2:0000000000000000000000000000000000000000
master_repl_offset:574
second_repl_offset:-1
repl_backlog_active:1
repl_backlog_size:1048576
repl_backlog_first_byte_offset:57
repl_backlog_histlen:518
```

图 14-102 连接 Slave 节点 127.0.0.1 6381 执行 info 命令

4. 创建 Redis 哨兵配置文件

进入 Redis 目录，将 sentinel.conf 配置文件复制 3 份，分别命名为 sentinel26379.conf、sentinel26380.conf 和 sentinel26381.conf。

```
#创建Redis 127.0.0.1 6379 配置文件
cp sentinel.conf sentinel26379.conf
#创建Redis 127.0.0.1 6380 配置文件
cp sentinel.conf sentinel26380.conf
#创建Redis 127.0.0.1 6381 配置文件
cp sentinel.conf sentinel26381.conf
```

5. 修改 Redis 哨兵配置文件

修改 sentinel26379.conf 配置文件，配置启动端口为 26379，并配置监听 127.0.0.1 6379 主节点。

```
#配置哨兵端口号 26379
port 26379
#配置监听 master 节点 127.0.0.1 6379
#最后一个参数 2 表示当集群中有两个 Redis 哨兵认为 master 下线，才能真正认为该 master 已经不可用
sentinel monitor mymaster 127.0.0.1 6379 2
```

修改 sentinel26380.conf 配置文件，配置启动端口为 26380，并配置监听 127.0.0.1 6379 主节点。

```
#配置哨兵端口号 26380
port 26380
#配置监听 master 节点 127.0.0.1 6379
#最后一个参数 2 表示当集群中有两个 Redis 哨兵认为 master 下线，才能真正认为该 master 已经不可用
sentinel monitor mymaster 127.0.0.1 6379 2
```

修改 sentinel26381.conf 配置文件，配置启动端口为 26381，并配置监听 127.0.0.1 6379 主节点。

```
#配置哨兵端口号 26381
port 26381
#配置监听 master 节点 127.0.0.1 6379
#最后一个参数 2 表示当集群中有两个 Redis 哨兵认为 master 下线，才能真正认为该 master 已经不可用
sentinel monitor mymaster 127.0.0.1 6379 2
```

6. 分别启动 Redis 哨兵

```
src/redis-sentinel sentinel26379.conf
src/redis-sentinel sentinel26380.conf
src/redis-sentinel sentinel26381.conf
```

验证 Redis 哨兵启动，如图 14-103 所示。

```
ps -ef | grep sentinel | grep -v grep
501  1743  1  0  1:15下午 ??    0:02.91 src/redis-sentinel *:26379 [sentinel]
501  1767  1  0  1:22下午 ??    0:00.82 src/redis-sentinel *:26380 [sentinel]
501  1771  1  0  1:22下午 ??    0:00.79 src/redis-sentinel *:26381 [sentinel]
```

图 14-103　Redis 哨兵进程

7. 验证故障转移

停止 Redis Master 主节点 127.0.0.1 6379，用于模拟 Redis 主节点下线。从图 14-99 可知，Redis 主节点进程号是 1661，使用以下命令关闭 Redis 主节点所在的进程：

```
kill -9 1661
```

3 个 Redis 哨兵节点的日志输出如图 14-104 所示。

```
1833:X 03 Jan 2019 13:54:57.202 # +sdown master mymaster 127.0.0.1 6379
1833:X 03 Jan 2019 13:54:57.283 # +odown master mymaster 127.0.0.1 6379 #quorum 2/2
1833:X 03 Jan 2019 13:54:57.283 # +new-epoch 1
1833:X 03 Jan 2019 13:54:57.283 # +try-failover master mymaster 127.0.0.1 6379
1833:X 03 Jan 2019 13:54:57.285 # +vote-for-leader 1d376b8a71e3c171ca0b1243d7a3275bfb6ce820 1
1833:X 03 Jan 2019 13:54:57.286 # c5334cbb5f902a4615d32fc54ea8cc08320a05f7 voted for c5334cbb5f902a4615d32fc54ea8cc08320a05f7 1
1833:X 03 Jan 2019 13:54:57.287 # a32ecd87d79db1c7bd6a0bc31e7875d517c32f2c voted for 1d376b8a71e3c171ca0b1243d7a3275bfb6ce820 1
1833:X 03 Jan 2019 13:54:57.360 # +elected-leader master mymaster 127.0.0.1 6379
1833:X 03 Jan 2019 13:54:57.360 # +failover-state-select-slave master mymaster 127.0.0.1 6379
1833:X 03 Jan 2019 13:54:57.463 # +selected-slave slave 127.0.0.1:6381 127.0.0.1 6381 @ mymaster 127.0.0.1 6379
1833:X 03 Jan 2019 13:54:57.463 * +failover-state-send-slaveof-noone slave 127.0.0.1:6381 127.0.0.1 6381 @ mymaster 127.0.0.1 6379
1833:X 03 Jan 2019 13:54:57.558 * +failover-state-wait-promotion slave 127.0.0.1:6381 127.0.0.1 6381 @ mymaster 127.0.0.1 6379
1833:X 03 Jan 2019 13:54:57.860 # +promoted-slave slave 127.0.0.1:6381 127.0.0.1 6381 @ mymaster 127.0.0.1 6379
1833:X 03 Jan 2019 13:54:57.860 # +failover-state-reconf-slaves master mymaster 127.0.0.1 6379
1833:X 03 Jan 2019 13:54:57.912 * +slave-reconf-sent slave 127.0.0.1:6380 127.0.0.1 6380 @ mymaster 127.0.0.1 6379
1833:X 03 Jan 2019 13:54:58.420 # -odown master mymaster 127.0.0.1 6379
1833:X 03 Jan 2019 13:54:58.872 * +slave-reconf-inprog slave 127.0.0.1:6380 127.0.0.1 6380 @ mymaster 127.0.0.1 6379
1833:X 03 Jan 2019 13:54:58.873 * +slave-reconf-done slave 127.0.0.1:6380 127.0.0.1 6380 @ mymaster 127.0.0.1 6379
1833:X 03 Jan 2019 13:54:58.927 # +failover-end master mymaster 127.0.0.1 6379
1833:X 03 Jan 2019 13:54:58.927 # +switch-master mymaster 127.0.0.1 6379 127.0.0.1 6381
1833:X 03 Jan 2019 13:54:58.927 * +slave slave 127.0.0.1:6380 127.0.0.1 6380 @ mymaster 127.0.0.1 6381
1833:X 03 Jan 2019 13:54:58.928 * +slave slave 127.0.0.1:6379 127.0.0.1 6379 @ mymaster 127.0.0.1 6381
1833:X 03 Jan 2019 13:55:28.968 # +sdown slave 127.0.0.1:6379 127.0.0.1 6379 @ mymaster 127.0.0.1 6381
```

图 14-104 Redis 故障转移

从 Redis 哨兵节点的日志输出可以看出，哨兵监控到了 Redis Master 主节点 127.0.0.1 6379 从 SDOWN 状态变成了 ODOWN 状态，并成功执行故障转换，新的 Master 主节点是 127.0.0.1 6381。

8. 重启旧的 Redis 主节点

执行以下命令重启 Redis 节点 127.0.0.1 6379，该节点会作为从节点加入其中：

```
src/redis-server redis6379.conf
```

9. 验证故障转以后的 Redis 拓扑结构

分别使用 Redis 客户端连接节点 127.0.0.1 6379、节点 127.0.0.1 6380 和节点 127.0.0.1 6381。

```
src/redis-cli -h 127.0.0.1 -p 6379
src/redis-cli -h 127.0.0.1 -p 6380
src/redis-cli -h 127.0.0.1 -p 6381
```

用 Redis 客户端连接新的 Redis Master 主节点 127.0.0.1 6381 执行 info 命令，如图 14-105 所示。

```
info replication
```

```
127.0.0.1:6381> info replication
# Replication
role:master
connected_slaves:2
slave0:ip=127.0.0.1,port=6380,state=online,offset=273174,lag=1
slave1:ip=127.0.0.1,port=6379,state=online,offset=273174,lag=1
master_replid:5e62a8ebd5ae5bfbba94e0f8631d786ffc498980
master_replid2:99cdb2f139437ee48b0fdcbd2a0a92729b9adb99
master_repl_offset:273174
second_repl_offset:16005
repl_backlog_active:1
repl_backlog_size:1048576
repl_backlog_first_byte_offset:1
repl_backlog_histlen:273174
```

图 14-105 新的 Redis 主节点执行 info replication 命令

从 info replication 命令的输出可以看出，当前节点 127.0.0.1 6381 是 Master 主节点，且此节点含有两个从节点，分别是 127.0.0.1 6379 和 127.0.0.1 6380。

用 Redis 客户端连接新的 Redis Slave 从节点 127.0.0.1 6379 执行 info 命令，如图 14-106 所示。

从图 14-106 可知，127.0.0.1 6379 节点是从节点，与之对应的是主节点 127.0.0.1 6381。

```
127.0.0.1:6379> info replication
# Replication
role:slave
master_host:127.0.0.1
master_port:6381
master_link_status:up
master_last_io_seconds_ago:0
master_sync_in_progress:0
slave_repl_offset:909026
slave_priority:100
slave_read_only:1
connected_slaves:0
master_replid:5e62a8ebd5ae5bfbba94e0f8631d786ffc498980
master_replid2:0000000000000000000000000000000000000000
master_repl_offset:909026
second_repl_offset:-1
repl_backlog_active:1
repl_backlog_size:1048576
repl_backlog_first_byte_offset:179940
repl_backlog_histlen:729087
```

图 14-106　Redis 从节点 127.0.0.1 6379 执行 info replication 命令

用 Redis 客户端连接新的 Redis Slave 从节点 127.0.0.1 6380 执行 info 命令，如图 14-107 所示。

```
127.0.0.1:6380> info replication
# Replication
role:slave
master_host:127.0.0.1
master_port:6381
master_link_status:up
master_last_io_seconds_ago:0
master_sync_in_progress:0
slave_repl_offset:968218
slave_priority:100
slave_read_only:1
connected_slaves:0
master_replid:5e62a8ebd5ae5bfbba94e0f8631d786ffc498980
master_replid2:99cdb2f139437ee48b0fdcbd2a0a92729b9adb99
master_repl_offset:968218
second_repl_offset:16005
repl_backlog_active:1
repl_backlog_size:1048576
repl_backlog_first_byte_offset:1
repl_backlog_histlen:968218
```

图 14-107　Redis 从节点 127.0.0.1 6380 执行 info replication 命令

通过以上步骤可知，Redis 哨兵模式搭建成功，并且此哨兵模式可以实现自动故障转移。通过 Redis 哨兵实现的故障转移降低了开发人员对 Redis 的维护成本，同时也增强了 Redis 的高可用性。

14.6　Redis 集群模式

Redis 集群是可以在多个 Redis 节点之间进行数据共享的架构。Redis 集群通过分区容错（partition）来提高可用性（availability），即使集群中有一部分节点失效或者无法进行通信，集群也可以继续处理命令请求。

14.6.1 Redis 集群模式数据共享

Redis 集群有如下特点。

（1）将数据切分到多个 Redis 节点。
（2）当集群中部分节点失效或者无法通讯时，整个集群仍可以处理请求。

Redis 将数据进行分片，每个 Redis 集群包含 16384 个哈希槽（hash slot），Redis 中存储的每个 key 都属于这些哈希槽中的一个。可通过计算得知每个 key 应该存放的具体哈希槽：

```
key 存放的哈希槽 = CRC16(key) % 16384
```

其中 CRC16(key)是计算 key 的 CRC16 校验和的。

Redis 集群中的每个 Redis 节点负责处理一部分哈希槽。假设 1 个 Redis 集群包含 3 个 Redis 节点，每个节点可能处理的哈希槽如下。

（1）Redis 节点 A 负责处理 0~5500 号哈希槽。
（2）Redis 节点 B 负责处理 5501~11000 号哈希槽。
（3）Redis 节点 C 负责处理 11001~ 16384 号哈希槽。

通过这种将哈希槽分布到不同 Redis 节点的做法使得用户可以很容易地向集群中添加或者删除 Redis 节点。如向 Redis 集群中加入节点 D，只需将节点 A、B 和 C 中的部分哈希槽移动到节点 D 即可。

14.6.2 Redis 集群中的主从复制

为了使 Redis 集群在出现问题时仍然可以正常运行，Redis 集群对节点使用了主从复制功能。即集群中的每个节点有 1 个 Master 主节点和若干个从节点。

在 14.6.1 的案例中，一个 Redis 集群有 A、B 和 C 3 个节点，当节点 B 下线时，整个集群将无法正常工作。如果在创建 Redis 集群时，为节点 B 创建了从节点 Slave_B，那么当主节点 B 下线时，集群就可以将 Slave_B 作为新的主节点，并让其替代主节点 B，这样整个集群就不会因为主节点 B 下线而无法正常工作，即 Redis 集群拥有分区容错性。

但是如果 Redis 集群中的主节点 B 和其从节点 Slave_B 都下线，还是会导致 Redis 集群无法正常工作。

14.6.3 Redis 集群中的一致性问题

在分析 Redis 集群一致性问题前，先要了解 CAP 原则。CAP 原则又称 CAP 定理，指的是在一个分布式系统中，一致性（Consistency）、可用性（Availability）、分区容错性（Partition Tolerance），三者不可兼得。Redis 集群模式也是一个分布式系统，因此也存在相应的问题。

之前对 Redis 集群的分析中可知，Redis 集群对可用性和分区容错性有较好的支持。因此 Redis 集群模式下数据一致性存在一定的问题，Redis 集群不保证强一致性。

因在 Redis 集群中，主从节点之间的复制是异步执行的，即主节点对命令的复制工作发生在返回命令回复给客户端之后，如果每次处理命令请求都需要等待复制操作完成，那么主节点处理命令请求的速度将极大地降低——必须在性能和一致性之间做出权衡。这种情况下会存在数据一致性问题，即集群中部分节点短时间内获取不到最新的主节点新增的数据。

另一种存在数据一致性的情况是 Redis 集群出现了网络分区。假设有这样一个 Redis 集群，集群中含有 A、A1、B、B1、C 和 C1 共 6 个节点，其中节点 A、B 和 C 是主节点，A1、B1 和 C1 是从节点，另有一个客户端 X。如在某一时刻 Redis 集群发生了网络分区，整个集群分为两方，多数的一方（Majority）包含节点 A、A1、B、B1 和 C1，少数的一方（Minority）包含主节点 C 和客户端 X。在网络分区期间，主节点 C 仍然能接受客户端 C 的请求，此时就会出现 Minority 和 Majority 数据一致性问题。

如果网络分区持续时间较短，集群仍可正常运行。如果网络分区时间足够长，Minority 分区中的节点标记节点 C 为下线状态，并使用从节点 C1 替换原主节点 C，将导致客户端 X 发送给原主节点 C 的写入数据丢失。

对于 Majority 一方，如果一个主节点未能在节点超时所设定的时限内重新联系上集群，那么集群会将这个主节点视为下线，并使用从节点来代替这个主节点继续工作。

对于 Minority 一方，如果一个主节点未能在节点超时所设定的时限内重新联系上集群，那么它将停止处理写命令，并向客户端报告错误。

14.6.4　Redis 集群架构

Redis 集群中所有的节点彼此之间互相通信，使用二进制协议优化传输速度和带宽。集群中过半数的节点检测到某个节点失效时，集群会将这个节点标记为失败（fail）。因为 Redis 客户端与 Redis 集群中的节点直连，所以 Redis 客户端只要连接到集群中的任一个节点即可。Redis 集群的架构如图 14-108 所示。

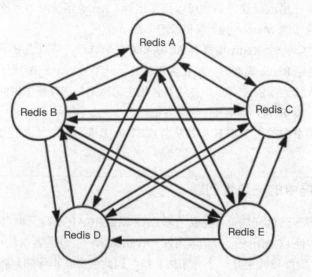

图 14-108　Redis 集群架构

14.6.5　Redis 集群容错

判断当前节点是否下线需要集群中所有的 Redis Master 节点参与。如果集群中半数以上的

Master 节点与当前节点通信超时，则认为当前节点下线。如图 14-109 所示，当虚线部分通信超时节点个数大于集群中节点半数时，就认为 Redis A 节点下线。

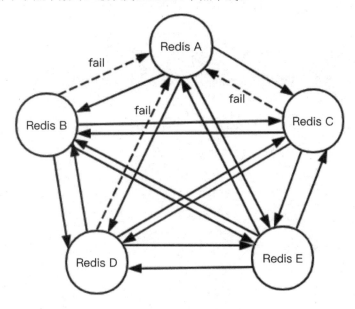

图 14-109　Redis 集群架构

如下两种情况任何一种发生时，整个集群不可用。

（1）某个 Master 节点下线，并且这个 Master 节点没有可用的 Slave 节点。

（2）集群中过半数以上的 Master 节点下线，无论 Master 节点是否有 Slave 节点。

14.6.6　Redis 集群环境搭建

本节将介绍 Redis 集群的搭建。本案例中启动 6 个 Redis 节点，这 6 个节点会分为两种，其中 3 个是 Master 主节点，另外 3 个是 Master 主节点对应的从节点。具体环境搭建如下。

（1）进入 Redis 目录，创建一个新的目录 cluster，在这个目录中创建 Redis 集群所需的配置文件：

```
mkdir cluster
```

（2）进入 cluster 目录后，将 redis.conf 文件复制 6 份，每份配置文件对应一个 Redis 节点：

```
cd cluster
cp ../redis.conf ./redis6001.conf
cp ../redis.conf ./redis6002.conf
cp ../redis.conf ./redis6003.conf
cp ../redis.conf ./redis6004.conf
cp ../redis.conf ./redis6005.conf
cp ../redis.conf ./redis6006.conf
```

（3）分别修改 redis6001.conf~redis6006.conf 文件。每个配置文件具体修改如下：

```
#修改redis6001.conf
vim redis6001.conf
#redis6001.conf中做如下修改
#配置启动端口
port 6001
#开启集群配置
cluster-enabled yes
#集群的配置，配置文件首次启动自动生成
cluster-config-file nodes-6001.conf

#################################################

#修改redis6002.conf
vim redis6002.conf
#redis6002.conf中做如下修改
#配置启动端口
port 6002
#开启集群配置
cluster-enabled yes
#集群的配置，配置文件首次启动自动生成
cluster-config-file nodes-6002.conf

#################################################

#修改redis6003.conf
vim redis6003.conf
#redis6003.conf中做如下修改
#配置启动端口
port 6003
#开启集群配置
cluster-enabled yes
#集群的配置，配置文件首次启动自动生成
cluster-config-file nodes-6003.conf

#################################################

#修改redis6004.conf
vim redis6004.conf
#redis6004.conf中做如下修改
#配置启动端口
port 6004
#开启集群配置
cluster-enabled yes
#集群的配置，配置文件首次启动自动生成
cluster-config-file nodes-6004.conf
```

```
##########################################

#修改 redis6005.conf
vim redis6005.conf
#redis6005.conf 中做如下修改
#配置启动端口
port 6005
#开启集群配置
cluster-enabled yes
#集群的配置，配置文件首次启动自动生成
cluster-config-file nodes-6005.conf

##########################################

#修改 redis6006.conf
vim redis6006.conf
#redis6006.conf 中做如下修改
#配置启动端口
port 6006
#开启集群配置
cluster-enabled yes
#集群的配置，配置文件首次启动自动生成
cluster-config-file nodes-6006.conf
```

（4）由于需要启动的节点较多，可以使用 Shell 脚本管理。下面的命令将会创建两个 Shell 脚本，start-all.sh 是用来启动 6 个 Redis 节点的，stop-all.sh 是用来停止 6 个 Redis 节点的：

```
#当前所在目录是 cluster
#启动 6 个 Redis 节点
vim start-all.sh
../src/redis-server redis6001.conf
../src/redis-server redis6002.conf
../src/redis-server redis6003.conf
../src/redis-server redis6004.conf
../src/redis-server redis6005.conf
../src/redis-server redis6006.conf

##########################################

#停止 6 个 Redis 节点
vim stop-all.sh
../src/redis-cli -p 6001 shutdown
../src/redis-cli -p 6002 shutdown
../src/redis-cli -p 6003 shutdown
../src/redis-cli -p 6004 shutdown
```

```
../src/redis-cli -p 6005 shutdown
../src/redis-cli -p 6006 shutdown
```

（5）执行 start-all.sh 脚本启动 6 个节点：

```
start-all.sh
```

（6）查看 6 个 Redis 节点的运行进程。如图 14-110 所示。

```
ps -ef | grep redis | grep -v grep
```

```
501  5901  1  0  8:03下午 ??    0:06.67 ../src/redis-server 127.0.0.1:6001 [cluster]
501  5932  1  0  8:10下午 ??    0:05.43 ../src/redis-server 127.0.0.1:6002 [cluster]
501  5934  1  0  8:10下午 ??    0:05.43 ../src/redis-server 127.0.0.1:6003 [cluster]
501  5936  1  0  8:10下午 ??    0:05.42 ../src/redis-server 127.0.0.1:6004 [cluster]
501  5938  1  0  8:10下午 ??    0:05.40 ../src/redis-server 127.0.0.1:6005 [cluster]
501  5940  1  0  8:10下午 ??    0:05.39 ../src/redis-server 127.0.0.1:6006 [cluster]
```

图 14-110　查看启动的 6 个 Redis 节点的进程

（7）使用启动的 6 个 Redis 节点创建 Redis 集群。其中 "--cluster-replicas 1" 表示自动为每个 Master 节点分配一个 Slave 节点。在这个案例中有 6 个节点，因此这个 Redis 集群会生成 3 个 Master 节点和 3 个 Slave 节点。

```
../src/redis-cli --cluster create --cluster-replicas 1 127.0.0.1:6001 127.0.0.1:6002 127.0.0.1:6003 127.0.0.1:6004 127.0.0.1:6005 127.0.0.1:6006
```

执行以上命令，得到如下的输出：

```
>>> Performing hash slots allocation on 6 nodes...
Master[0] -> Slots 0 - 5460
Master[1] -> Slots 5461 - 10922
Master[2] -> Slots 10923 - 16383
Adding replica 127.0.0.1:6004 to 127.0.0.1:6001
Adding replica 127.0.0.1:6005 to 127.0.0.1:6002
Adding replica 127.0.0.1:6006 to 127.0.0.1:6003
>>> Trying to optimize slaves allocation for anti-affinity
[WARNING] Some slaves are in the same host as their master
M: 411cc025c8e6109e9cb600b68a533576b9cf6188 127.0.0.1:6001
   slots:[0-5460] (5461 slots) master
M: 4aa18df4f17af30e81364af5e762af15b28b0338 127.0.0.1:6002
   slots:[5461-10922] (5462 slots) master
M: 71c3e50c020a74fe6bb43523b84f9d879465d97e 127.0.0.1:6003
   slots:[10923-16383] (5461 slots) master
S: e063c5d4ace5b62816a108f28023a760b82ba494 127.0.0.1:6004
   replicates 411cc025c8e6109e9cb600b68a533576b9cf6188
S: d96b2dbf3abc2078e9dedad6be49c97672473696 127.0.0.1:6005
   replicates 4aa18df4f17af30e81364af5e762af15b28b0338
S: cf55560081b842bc7d2843ed17798082d2c875f4 127.0.0.1:6006
   replicates 71c3e50c020a74fe6bb43523b84f9d879465d97e
Can I set the above configuration? (type 'yes' to accept): yes
>>> Nodes configuration updated
```

```
>>> Assign a different config epoch to each node
>>> Sending CLUSTER MEET messages to join the cluster
Waiting for the cluster to join
.....
>>> Performing Cluster Check (using node 127.0.0.1:6001)
M: 411cc025c8e6109e9cb600b68a533576b9cf6188 127.0.0.1:6001
   slots:[0-5460] (5461 slots) master
   1 additional replica(s)
S: cf55560081b842bc7d2843ed17798082d2c875f4 127.0.0.1:6006
   slots: (0 slots) slave
   replicates 71c3e50c020a74fe6bb43523b84f9d879465d97e
S: e063c5d4ace5b62816a108f28023a760b82ba494 127.0.0.1:6004
   slots: (0 slots) slave
   replicates 411cc025c8e6109e9cb600b68a533576b9cf6188
S: d96b2dbf3abc2078e9dedad6be49c97672473696 127.0.0.1:6005
   slots: (0 slots) slave
   replicates 4aa18df4f17af30e81364af5e762af15b28b0338
M: 4aa18df4f17af30e81364af5e762af15b28b0338 127.0.0.1:6002
   slots:[5461-10922] (5462 slots) master
   1 additional replica(s)
M: 71c3e50c020a74fe6bb43523b84f9d879465d97e 127.0.0.1:6003
   slots:[10923-16383] (5461 slots) master
   1 additional replica(s)
[OK] All nodes agree about slots configuration.
>>> Check for open slots...
>>> Check slots coverage...
[OK] All 16384 slots covered.
```

在以上输出中，M 代表 Master，S 代表 Slave，可以看出，主节点 127.0.0.1 6001 覆盖了 0~5460 的哈希槽，主节点 127.0.0.1 6002 覆盖了 5461~10922 的哈希槽，主节点 127.0.0.1 6003 覆盖了 10923~16383 的哈希槽。

集群启动后，127.0.0.1 6004、127.0.0.1 6005 和 127.0.0.1 6006 这 3 个节点成了集群中的从节点：

```
Can I set the above configuration? (type 'yes' to accept): yes
```

其中这一行是与用户交互的，输入 yes 的含义是在 nodes.conf 配置文件中保存更新的配置。集群启动后，生成 6 个 nodes-6001.conf~nodes-6006.conf 配置文件。

（8）执行以下命令查看当前集群的状态。如图 14-111 所示。

```
../src/redis-cli --cluster info 127.0.0.1:6001
```

```
127.0.0.1:6001 (411cc025...) -> 0 keys | 5461 slots | 1 slaves.
127.0.0.1:6002 (4aa18df4...) -> 0 keys | 5462 slots | 1 slaves.
127.0.0.1:6003 (71c3e50c...) -> 0 keys | 5461 slots | 1 slaves.
[OK] 0 keys in 3 masters.
0.00 keys per slot on average.
```

图 14-111　检查集群状态

（9）重复步骤（2）和步骤（3），创建新的配置文件 redis6007.conf 和 redis6008.conf，并启动两个新的 Redis 节点，节点 127.0.0.1 6007 和节点 127.0.0.1 6008，此时 Redis 进程如图 12-112 所示。

```
501 12847    1    0 11:15下午    ??    0:04.34 ../src/redis-server 127.0.0.1:6001 [cluster]
501 12849    1    0 11:15下午    ??    0:04.38 ../src/redis-server 127.0.0.1:6002 [cluster]
501 12851    1    0 11:15下午    ??    0:04.33 ../src/redis-server 127.0.0.1:6003 [cluster]
501 12853    1    0 11:15下午    ??    0:04.33 ../src/redis-server 127.0.0.1:6004 [cluster]
501 12855    1    0 11:15下午    ??    0:04.33 ../src/redis-server 127.0.0.1:6005 [cluster]
501 12857    1    0 11:15下午    ??    0:04.32 ../src/redis-server 127.0.0.1:6006 [cluster]
501 12864    1    0 11:18下午    ??    0:03.79 ../src/redis-server 127.0.0.1:6007 [cluster]
501 12866    1    0 11:18下午    ??    0:03.55 ../src/redis-server 127.0.0.1:6008 [cluster]
```

图 14-112　启动新的 Redis 节点

（10）执行以下命令将 127.0.0.1 6007 节点加入到 Redis 集群中：

```
../src/redis-cli --cluster add-node 127.0.0.1:6007 127.0.0.1:6001
```

执行以上命令得到如下输出：

```
>>> Adding node 127.0.0.1:6007 to cluster 127.0.0.1:6001
>>> Performing Cluster Check (using node 127.0.0.1:6001)
M: ccddaf23f65e1a12059fc65bfaff7b8e2e7cd675 127.0.0.1:6001
   slots:[0-5460] (5461 slots) master
   1 additional replica(s)
M: f0f2e40e6c45eb3feebc38339626a4aa94cbce5f 127.0.0.1:6002
   slots:[5461-10922] (5462 slots) master
   1 additional replica(s)
S: 9711e49639d8d9ef8149dc1bce1f9aabb20d4d1a 127.0.0.1:6006
   slots: (0 slots) slave
   replicates f0f2e40e6c45eb3feebc38339626a4aa94cbce5f
S: a59721da681cc9dbce00dc1a013361998951738b 127.0.0.1:6005
   slots: (0 slots) slave
   replicates ccddaf23f65e1a12059fc65bfaff7b8e2e7cd675
S: 7fd22130bfc628da8d81f35972d2bb9466e3fbf4 127.0.0.1:6004
   slots: (0 slots) slave
   replicates 139b8552ca7569eb71119c7709b4252511e00f2d
M: 139b8552ca7569eb71119c7709b4252511e00f2d 127.0.0.1:6003
   slots:[10923-16383] (5461 slots) master
   1 additional replica(s)
[OK] All nodes agree about slots configuration.
>>> Check for open slots...
>>> Check slots coverage...
[OK] All 16384 slots covered.
>>> Send CLUSTER MEET to node 127.0.0.1:6007 to make it join the cluster.
[OK] New node added correctly.
```

从输出结果中可知，节点 127.0.0.1 6007 已成功加入集群。此时节点 127.0.0.1 6007 成为主节点，且没有从节点。查询生成的 nodes-6007.conf 配置文件包含主节点 127.0.0.1 6007，其在集群中的 id 为 3a3387a7b0864fe60019283d417dceabed8cda4c。

下面命令为其在集群中创建从节点,其中"--cluster-master-id"参数指定当前节点所属的主节点即节点 127.0.0.1 6007,命令如下:

```
../src/redis-cli --cluster add-node --cluster-slave --cluster-master-id
3a3387a7b0864fe60019283d417dceabed8cda4c 127.0.0.1:6008 127.0.0.1:6001
```

执行以上命令得到如下输出:

```
>>> Adding node 127.0.0.1:6008 to cluster 127.0.0.1:6001
>>> Performing Cluster Check (using node 127.0.0.1:6001)
M: ccddaf23f65e1a12059fc65bfaff7b8e2e7cd675 127.0.0.1:6001
   slots:[0-5460] (5461 slots) master
   1 additional replica(s)
M: f0f2e40e6c45eb3feebc38339626a4aa94cbce5f 127.0.0.1:6002
   slots:[5461-10922] (5462 slots) master
   1 additional replica(s)
S: 9711e49639d8d9ef8149dc1bce1f9aabb20d4d1a 127.0.0.1:6006
   slots: (0 slots) slave
   replicates f0f2e40e6c45eb3feebc38339626a4aa94cbce5f
S: a59721da681cc9dbce00dc1a013361998951738b 127.0.0.1:6005
   slots: (0 slots) slave
   replicates ccddaf23f65e1a12059fc65bfaff7b8e2e7cd675
M: 3a3387a7b0864fe60019283d417dceabed8cda4c 127.0.0.1:6007
   slots: (0 slots) master
S: 7fd22130bfc628da8d81f35972d2bb9466e3fbf4 127.0.0.1:6004
   slots: (0 slots) slave
   replicates 139b8552ca7569eb71119c7709b4252511e00f2d
M: 139b8552ca7569eb71119c7709b4252511e00f2d 127.0.0.1:6003
   slots:[10923-16383] (5461 slots) master
   1 additional replica(s)
[OK] All nodes agree about slots configuration.
>>> Check for open slots...
>>> Check slots coverage...
[OK] All 16384 slots covered.
>>> Send CLUSTER MEET to node 127.0.0.1:6008 to make it join the cluster.
Waiting for the cluster to join
>>> Configure node as replica of 127.0.0.1:6007.
[OK] New node added correctly.
```

再次检查 Redis 集群的状态,可得如图 14-113 所示结果。

```
127.0.0.1:6001 (ccddaf23...) -> 0 keys | 5461 slots | 1 slaves.
127.0.0.1:6002 (f0f2e40e...) -> 0 keys | 5462 slots | 1 slaves.
127.0.0.1:6007 (3a3387a7...) -> 0 keys | 0 slots | 1 slaves.
127.0.0.1:6003 (139b8552...) -> 0 keys | 5461 slots | 1 slaves.
[OK] 0 keys in 4 masters.
```

图 14-113 重新检查集群状态

此时节点 127.0.0.1 6007 已成功加入集群，并且有一个从节点。但节点 127.0.0.1 6007 并没有覆盖任何哈希槽。下面讲解为节点 127.0.0.1 6007 重新分配哈希槽的过程。

（11）执行以下命令对 Redis 集群中的主节点 127.0.0.1 6007 分配哈希槽：

```
../src/redis-cli --cluster reshard 127.0.0.1 6007
```

执行以上命令得到如下输出：

```
>>> Performing Cluster Check (using node 127.0.0.1:6007)
M: 3a3387a7b0864fe60019283d417dceabed8cda4c 127.0.0.1:6007
   slots: (0 slots) master
   1 additional replica(s)
M: f0f2e40e6c45eb3feebc38339626a4aa94cbce5f 127.0.0.1:6002
   slots:[5461-10922] (5462 slots) master
   1 additional replica(s)
S: 7fd22130bfc628da8d81f35972d2bb9466e3fbf4 127.0.0.1:6004
   slots: (0 slots) slave
   replicates 139b8552ca7569eb71119c7709b4252511e00f2d
M: 139b8552ca7569eb71119c7709b4252511e00f2d 127.0.0.1:6003
   slots:[10923-16383] (5461 slots) master
   1 additional replica(s)
M: ccddaf23f65e1a12059fc65bfaff7b8e2e7cd675 127.0.0.1:6001
   slots:[0-5460] (5461 slots) master
   1 additional replica(s)
S: 9711e49639d8d9ef8149dc1bce1f9aabb20d4d1a 127.0.0.1:6006
   slots: (0 slots) slave
   replicates f0f2e40e6c45eb3feebc38339626a4aa94cbce5f
S: 80651ab13a571a59dfec787394095dffa21afd1f 127.0.0.1:6008
   slots: (0 slots) slave
   replicates 3a3387a7b0864fe60019283d417dceabed8cda4c
S: a59721da681cc9dbce00dc1a013361998951738b 127.0.0.1:6005
   slots: (0 slots) slave
   replicates ccddaf23f65e1a12059fc65bfaff7b8e2e7cd675
[OK] All nodes agree about slots configuration.
>>> Check for open slots...
>>> Check slots coverage...
[OK] All 16384 slots covered.
How many slots do you want to move (from 1 to 16384)? 4096
```

最后数字是要用户指定 Redis 节点 127.0.0.1 6007 覆盖的哈希槽数量。因为整个集群中有 4 个 Master 节点，且哈希槽的数量是 16384 个，因此给 127.0.0.1 6007 分配 4096 个哈希槽（取平均值）。输入 4096 后，得到如下输出：

```
What is the receiving node ID? 3a3387a7b0864fe60019283d417dceabed8cda4c
```

这里需要指定节点 127.0.0.1 6007 在集群中的 id（同步骤 10 中的 cluster-master-id）。输入节点 id 后，得到如下输出：

```
Please enter all the source node IDs.
  Type 'all' to use all the nodes as source nodes for the hash slots.
  Type 'done' once you entered all the source nodes IDs.
Source node #1: all
```

这里需要指定从哪些主节点将哈希槽转移给当前主节点。这里输入 all 表示从全部的其余主节点中转移哈希槽给当前主节点。以下是部分输出：

```
Ready to move 4096 slots.
  Source nodes:
    M: f0f2e40e6c45eb3feebc38339626a4aa94cbce5f 127.0.0.1:6002
      slots:[5461-10922] (5462 slots) master
      1 additional replica(s)
    M: 139b8552ca7569eb71119c7709b4252511e00f2d 127.0.0.1:6003
      slots:[10923-16383] (5461 slots) master
      1 additional replica(s)
    M: ccddaf23f65e1a12059fc65bfaff7b8e2e7cd675 127.0.0.1:6001
      slots:[0-5460] (5461 slots) master
      1 additional replica(s)
  Destination node:
    M: 3a3387a7b0864fe60019283d417dceabed8cda4c 127.0.0.1:6007
      slots: (0 slots) master
      1 additional replica(s)
  Resharding plan:
    Moving slot 5461 from f0f2e40e6c45eb3feebc38339626a4aa94cbce5f
    Moving slot 5462 from f0f2e40e6c45eb3feebc38339626a4aa94cbce5f
    Moving slot 5463 from f0f2e40e6c45eb3feebc38339626a4aa94cbce5f
    Moving slot 5464 from f0f2e40e6c45eb3feebc38339626a4aa94cbce5f
```

再次检查集群状态，可得如图 14-114 所示结果。

```
127.0.0.1:6001 (ccddaf23...) -> 0 keys | 4096 slots | 1 slaves.
127.0.0.1:6002 (f0f2e40e...) -> 0 keys | 4096 slots | 1 slaves.
127.0.0.1:6007 (3a3387a7...) -> 0 keys | 4096 slots | 1 slaves.
127.0.0.1:6003 (139b8552...) -> 0 keys | 4096 slots | 1 slaves.
[OK] 0 keys in 4 masters.
```

图 14-114　重新检查集群状态

从图 14-114 可知，新加入集群的节点 127.0.0.1 6007 获取到了 4096 个哈希槽。支持 Redis 集群环境搭建完成，删除节点与新增节点过程类似，此处不再赘述。

14.7　Spring、MyBatis 和 Redis 集成快速体验

本节介绍使用 Spring 集成 MyBatis 和 Redis 进行开发的过程，该过程将 Redis 作为缓存，可减少对数据库查询的次数，降低数据库负载。

1. 准备环境

在 maven 项目的 pom.xml 文件中加入以下依赖,本例中使用 Jedis 实现对 Redis 集群的连接:

```xml
<dependency>
    <groupId>org.springframework.data</groupId>
    <artifactId>spring-data-redis</artifactId>
    <version>2.1.3.RELEASE</version>
</dependency>
<dependency>
    <groupId>redis.clients</groupId>
    <artifactId>jedis</artifactId>
    <version>3.0.1</version>
</dependency>
<dependency>
    <groupId>com.alibaba</groupId>
    <artifactId>fastjson</artifactId>
    <version>1.2.33</version>
</dependency>
<dependency>
  <groupId>org.mybatis</groupId>
  <artifactId>mybatis-spring</artifactId>
  <version>1.3.2</version>
</dependency>
<dependency>
  <groupId>mysql</groupId>
  <artifactId>mysql-connector-java</artifactId>
  <version>8.0.12</version>
</dependency>
<dependency>
  <groupId>org.mybatis</groupId>
  <artifactId>mybatis</artifactId>
  <version>3.4.6</version>
</dependency>
```

2. 创建 book 表

创建 book 表,代码如下:

```sql
CREATE TABLE `book` (
  `id` int(11) unsigned NOT NULL AUTO_INCREMENT COMMENT '主键',
  `name` varchar(50) NOT NULL COMMENT '书名',
  `price` int(11) DEFAULT NULL COMMENT '价格',
  `adddate` timestamp NOT NULL DEFAULT CURRENT_TIMESTAMP COMMENT '添加时间',
  `updatedate` timestamp NOT NULL DEFAULT CURRENT_TIMESTAMP ON UPDATE CURRENT_TIMESTAMP COMMENT '修改时间',
  PRIMARY KEY (`id`),
```

```
  KEY (`id`)
) ENGINE=InnoDB DEFAULT CHARSET=utf8
```

3. 创建 Book 实体类

创建实体类 Book 与 book 表一一对应，代码如下：

```java
/**
 * @Author: zhouguanya
 * @Date: 2019/01/05
 * @Description: 实体类
 */
public class Book {
    private int id;
    private String name;
    private int price;
    private Date addDate;
    private Date updateDate;
    public int getId() {
        return id;
    }

    public void setId(int id) {
        this.id = id;
    }

    public String getName() {
        return name;
    }

    public void setName(String name) {
        this.name = name;
    }

    public int getPrice() {
        return price;
    }

    public void setPrice(int price) {
        this.price = price;
    }

    public Date getAddDate() {
        return addDate;
    }

    public void setAddDate(Date addDate) {
        this.addDate = addDate;
    }
```

```
    public Date getUpdateDate() {
        return updateDate;
    }
    public void setUpdateDate(Date updateDate) {
        this.updateDate = updateDate;
    }
}
```

4. 创建 DAO 接口

创建 BookDao，有两种方法，保存 Book 的 save()方法和查询 Book 的 query()方法，代码如下：

```
/**
 * @Author: zhouguanya
 * @Date: 2019/01/05
 * @Description: dao 接口
 */
public interface BookDao {
    int save(Book book);
    Book query(int id);
}
```

5. 创建 Mapper 文件

创建 mybatis-book-mapper.xml 文件，其中包含对 Book 对象的存储和查询操作：

```
<?xml version="1.0" encoding="UTF-8"?>
<!DOCTYPE mapper PUBLIC "-//mybatis.org//DTD Mapper 3.0//EN"
"http://mybatis.org/dtd/mybatis-3-mapper.dtd">
<mapper namespace="com.test.redis.demo.dao.BookDao">
    <resultMap id="BaseResultMap" type="com.test.redis.demo.model.Book">
        <id column="id" jdbcType="INTEGER" property="id" />
        <result column="name" jdbcType="VARCHAR" property="name" />
        <result column="price" jdbcType="INTEGER" property="price" />
        <result column="adddate" jdbcType="TIMESTAMP" property="addDate" />
        <result column="updatedate" jdbcType="TIMESTAMP" property="updateDate" />
    </resultMap>
    <select id="query" parameterType="java.lang.Integer" resultMap="BaseResultMap">
        select
        *
        from book
        where id = #{id,jdbcType=BIGINT}
    </select>
```

```xml
    <insert id="save" parameterType="com.test.redis.demo.model.Book" keyProperty="id">
        insert into book (name, price)
        values (#{name,jdbcType=VARCHAR}, #{price,jdbcType=INTEGER})
    </insert>
</mapper>
```

6. 配置 MySQL 和 Redis

创建 MyBatis 配置文件 mybatis_jdbc.properties,代码如下:

```
mysql.driver=com.mysql.jdbc.Driver
#jdbc 连接
mysql.url=jdbc:mysql://127.0.0.1:3306/test
#MySQL 用户名
mysql.username=root
#MySQL 密码
mysql.password=123456
```

创建 Redis 配置文件 redis.properties,代码如下:

```
#属性文件
redis.host1=127.0.0.1
redis.port1=6001
redis.host2=127.0.0.1
redis.port2=6002
redis.host3=127.0.0.1
redis.port3=6003
redis.host4=127.0.0.1
redis.port4=6004
redis.host5=127.0.0.1
redis.port5=6005
redis.host6=127.0.0.1
redis.port6=6006
redis.host7=127.0.0.1
redis.port7=6007
redis.host8=127.0.0.1
redis.port8=6008
redis.maxIdle=50
redis.maxActive=100
redis.maxWait=5000
redis.testOnBorrow=true
```

7. Spring 集成 MyBatis

在 Spring 中集成 MyBatis,代码如下:

```xml
    <context:component-scan base-package="com.test"/>
    <context:property-placeholder location="classpath:mybatis_jdbc.properties" ignore-unresolvable="true"/>
    <bean id="dataSource" class="org.springframework.jdbc.datasource.DriverManagerDataSource">
        <property name="driverClassName" value="${mysql.driver}" />
        <property name="url" value="${mysql.url}" />
        <property name="username" value="${mysql.username}" />
        <property name="password" value="${mysql.password}" />
    </bean>
    <!-- spring 和 MyBatis 整合-->
    <bean id="sqlSessionFactory" class="org.mybatis.spring.SqlSessionFactoryBean">
        <property name="dataSource" ref="dataSource" />
        <!-- 自动扫描 mapping.xml 文件，**表示迭代查找 -->
        <property name="mapperLocations">
            <array>
                <value>classpath:mybatis-book-mapper.xml</value>
            </array>
        </property>
    </bean>

    <!-- DAO 接口所在包名，Spring 会自动查找其下的类，包下的类需要使用@MapperScan注解,否则容器注入会失败 -->
    <bean class="org.mybatis.spring.mapper.MapperScannerConfigurer">
        <property name="basePackage" value="com.test.redis.demo.dao" />
        <property name="sqlSessionFactoryBeanName" value="sqlSessionFactory" />
    </bean>

    <!--<tx:annotation-driven transaction-manager="transactionManager"/>-->
    <!-- 事务管理 -->
    <bean id="transactionManager" class="org.springframework.jdbc.datasource.DataSourceTransactionManager">
        <property name="dataSource" ref="dataSource" />
    </bean>
    <bean id="transactionInterceptor" class="org.springframework.transaction.interceptor.TransactionInterceptor">
        <property name="transactionManager" ref="transactionManager"/>
        <property name="transactionAttributes">
            <props>
                <prop key="delete*">PROPAGATION_REQUIRED</prop>
                <prop key="add*">PROPAGATION_REQUIRED</prop>
                <prop key="update*">PROPAGATION_REQUIRED</prop>
                <prop key="save*">PROPAGATION_REQUIRED</prop>
                <prop key="find*">PROPAGATION_REQUIRED,readOnly</prop>
```

```
        </props>
    </property>
</bean>
```

8. Spring 集成 Redis

分别将多个 Redis 节点配置到 Redis 集群中，并配置 Jedis 连接池，代码如下：

```
<context:component-scan base-package="com.test"/>
<context:property-placeholder location="classpath:redis.properties" ignore-unresolvable="true"/>
<bean id="jedisPoolConfig" class="redis.clients.jedis.JedisPoolConfig">
    <property name="maxIdle" value="${redis.maxIdle}"/>
    <property name="maxTotal" value="${redis.maxActive}" />
    <property name="maxWaitMillis" value="${redis.maxWait}" />
    <property name="testOnBorrow" value="${redis.testOnBorrow}"/>
</bean>
<bean id="redisHost1" class="redis.clients.jedis.HostAndPort">
    <constructor-arg name="host" value="${redis.host1}" />
    <constructor-arg name="port" value="${redis.port1}" />
</bean>
<bean id="redisHost2" class="redis.clients.jedis.HostAndPort">
    <constructor-arg name="host" value="${redis.host2}" />
    <constructor-arg name="port" value="${redis.port2}" />
</bean>
<bean id="redisHost3" class="redis.clients.jedis.HostAndPort">
    <constructor-arg name="host" value="${redis.host3}" />
    <constructor-arg name="port" value="${redis.port3}" />
</bean>
<bean id="redisHost4" class="redis.clients.jedis.HostAndPort">
    <constructor-arg name="host" value="${redis.host4}" />
    <constructor-arg name="port" value="${redis.port4}" />
</bean>
<bean id="redisHost5" class="redis.clients.jedis.HostAndPort">
    <constructor-arg name="host" value="${redis.host5}" />
    <constructor-arg name="port" value="${redis.port5}" />
</bean>
<bean id="redisHost6" class="redis.clients.jedis.HostAndPort">
    <constructor-arg name="host" value="${redis.host6}" />
    <constructor-arg name="port" value="${redis.port6}" />
</bean>
<bean id="redisHost7" class="redis.clients.jedis.HostAndPort">
    <constructor-arg name="host" value="${redis.host7}" />
    <constructor-arg name="port" value="${redis.port7}" />
</bean>
<bean id="redisHost8" class="redis.clients.jedis.HostAndPort">
```

```xml
        <constructor-arg name="host" value="${redis.host8}" />
        <constructor-arg name="port" value="${redis.port8}" />
</bean>
<bean id="redisCluster" class="redis.clients.jedis.JedisCluster">
    <constructor-arg name="jedisClusterNode">
        <set>
            <ref bean="redisHost1"/>
            <ref bean="redisHost2"/>
            <ref bean="redisHost3"/>
            <ref bean="redisHost4"/>
            <ref bean="redisHost5"/>
            <ref bean="redisHost6"/>
        </set>
    </constructor-arg>
    <constructor-arg name="connectionTimeout" value="6000" />
    <constructor-arg name="soTimeout" value="2000" />
    <constructor-arg name="maxAttempts" value="3" />
    <constructor-arg name="poolConfig">
        <ref bean="jedisPoolConfig"/>
    </constructor-arg>
</bean>
```

9. 创建 Service 接口和实现

创建 BookService 接口，其中含有 save()方法和 query()方法，代码如下：

```java
/**
 * @Author: zhouguanya
 * @Date: 2019/01/05
 * @Description: BookService 接口
 */
public interface BookService {
    int save(Book book);
    Book query(int id);
}
```

创建 BookServiceImpl 实现类，实现 BookService 接口。其中 save()方法直接将 Book 对象持久化到数据库中；在 query()方法中，首先从 Redis 集群中查询对应 bookId 的 Book 对象，如果 Redis 缓存未命中，将从数据库中查询对应 bookId 的 Book 对象。

```java
/**
 * @Author: zhouguanya
 * @Date: 2019/01/05
 * @Description: BookService 接口实现类
 */
@Service
public class BookServiceImpl implements BookService {
```

```java
    @Autowired
    private JedisCluster jedisCluster;
    @Autowired
    private BookDao bookDao;
    @Override
    public int save(Book book) {
        return bookDao.save(book);
    }

    @Override
    public Book query(int bookId) {
        //从缓存中查询bookId对应的Book信息
        String cachedBook = jedisCluster.get("book_" + bookId);
        //缓存未命中
        if (StringUtils.isEmpty(cachedBook)) {
            System.out.println("未命中缓存bookId = " + bookId);
            Book book = bookDao.query(bookId);
            //写入缓存，设置过期时间60s
            jedisCluster.setex("book_" + bookId, 60, JSON.toJSONString(book));
            return book;
        } else {
            System.out.println("命中缓存bookId = " + bookId);
            return JSON.parseObject(cachedBook, Book.class);
        }
    }
}
```

10. 测试代码

编写单元测试，首先在book表中插入数据，然后从book表查询记录，代码如下：

```java
/**
 * @Author: zhouguanya
 * @Date: 2019/01/05
 * @Description: 测试
 */
@RunWith(SpringJUnit4ClassRunner.class)
@ContextConfiguration(locations = {"classpath:spring-redis.xml",
"classpath:spring-book.xml"})
public class BookServiceTest {
    @Autowired
    private BookService bookService;
    @Test
    public void testSave() {
        Book book = new Book();
        book.setName("Spring 5 Programming");
```

```
        book.setPrice(50);
        bookService.save(book);
    }
    @Test
    public void testQuery() {
        Book book = bookService.query(1);
        System.out.println("bookId = 1 的详情 = " + JSON.toJSONString(book));
    }
}
```

首先执行 testSave() 方法，将 Book 对象持久化到数据库中，执行结果如图 14-115 所示。

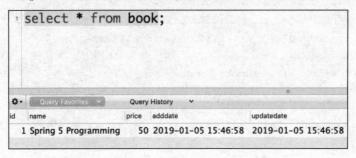

图 14-115　执行 testSave() 方法后的结果

下面执行 testQuery() 方法，查询 id 为 1 的 Book 对象。当第一次执行 testQuery() 方法时，得到如下输出：

```
未命中缓存 bookId = 1
bookId = 1 的详情 = {"addDate":1546724818000,"id":1,"name":"Spring 5 
Programming",
"price":50,"updateDate":1546724818000}
```

此次查询未命中缓存，因此 testQuery() 方法会从数据查询 id 为 1 的 Book 对象，并将此对象序列化后保存到 Redis 中。

此时通过以下命令查询 Redis 集群状态，发现集群中有一个 Redis 节点包含了 1 个 key，如图 14-116 所示。

```
../src/redis-cli --cluster info 127.0.0.1:6001
127.0.0.1:6001 (ccddaf23...) -> 0 keys | 4096 slots | 1 slaves.
127.0.0.1:6002 (f0f2e40e...) -> 0 keys | 4096 slots | 1 slaves.
127.0.0.1:6007 (3a3387a7...) -> 0 keys | 4096 slots | 1 slaves.
127.0.0.1:6003 (139b8552...) -> 1 keys | 4096 slots | 1 slaves.
```

图 14-116　查询集群状态

使用以下命令查询节点 127.0.0.1 6003 的 book_1 剩余存活时间，如图 14-117 所示。

```
127.0.0.1:6003> ttl book_1
(integer) 52
```

图 14-117　查询 book_1 剩余存活时间

可见 book_1 已经成功保存到 Redis 集群中，且还剩下 52s 存活时间。

再次执行 testQuery()方法（book_1 失效时间以内再次执行），得到如下输出：

```
命中缓存 bookId = 1
bookId = 1 的详情 = {"addDate":1546724818000,"id":1,"name":"Spring 5
Programming",
 "price":50,"updateDate":1546724818000}
```

可见第二次执行 testQuery()方法返回的数据是从 Redis 缓存中获取到的。

14.8　Redis 缓存穿透和雪崩

14.8.1　Redis 缓存穿透

正常的缓存使用场景是，所有的查询请求先经过缓存，当缓存命中后，直接返回缓存中的数据；在缓存未命中的情况下，去数据库查询数据，并写入缓存。缓存的目的是为了尽可能将请求在缓存层处理，避免大量的请求进入存储层，以达到保护存储层的效果，如图 14-118 所示。

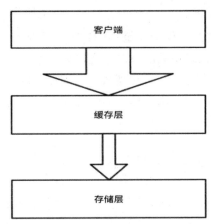

图 14-118　缓存架构模型

缓存穿透的含义是频繁查询根本不存在的数据，会导致缓存层和存储层都不会命中，因为这部分数据频繁查询，缓存不能有效命中，导致存储层负载加大，缓存架构模型如图 14-119 所示。

通常可以在应用程序中分别统计总调用数、缓存层命中数和存储层命中数。如果发现大量存储层空命中，有可能就是出现了缓存穿透问题。造成缓存穿透的原因有以下两点：

- 应用程序自身的问题，如缓存设计或者数据存储问题。
- 黑客恶意攻击，已感染网络爬虫等。

下面分析常见的缓存穿透问题的解决办法。

1. 缓存空对象

在图 14-119 中，由于存储层大量的请求不能命中，无法填充缓存层，造成了恶性循环。缓存空对象的方式,就是在存储层未命中的情况下,仍然将空对象存储到缓存层中,之后再次访问到这条数据将会在缓存层命中,有效保护了存储层。缓存空对象的架构设计如图 14-120 所示。

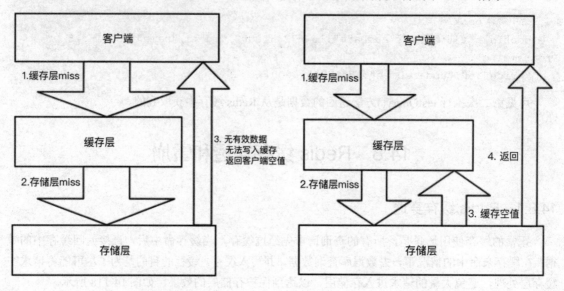

图 14-119　缓存架构模型　　　　　　　图 14-120　缓存空对象

缓存空对象的解决方案有两个问题:第一个问题是缓存为空会存储更多的空对象,因此缓存层需要更多的存储空间,比较有效的办法是为这类数据设计合理的过期时间,节约缓存层的空间;第二个问题是缓存层和存储层会出现数据一致性问题,即在缓存有效期内,存储层这个数据可能已经被更新,此时可以使用消息队列或者其他当时刷新缓存层的对象。下面是缓存空对象这种解决方案的伪代码:

```
/**
 * 根据 key 查询对象,缓存空对象
 */
public Object get(String key) {
    // 获取缓存中的数据
    Object cacheValue = cache.get(key);
    // 缓存为空
    if (cacheValue == null) {
        // 获取存储层数据
        Object storeValue = db.get(key);
        // 存储层未命中,设置空对象
        if (storeValue == null) {
            storeValue = new Blank();
        }
        // 存储层数据写入缓存
        cache.set(key, storeValue);
        // 如果 storeValue 为空对象 Blank 类型,设置超时时间为 600s
        if (storeValue instanceof Blank) {
```

```
                cache.expire(key, 60 * 10);
            }
            return storeValue;
        }
        // 缓存非空直接返回
        return cacheValue;
    }
```

2. 布隆过滤器拦截

布隆过滤器（Bloom Filter）是由布隆（Burton Howard Bloom）于 1970 年提出的。布隆过滤器由一个很长的二进制向量和一系列随机映射函数组成，布隆过滤器可以检索一个元素是否在一个集合中。

布隆过滤器算法的核心思想是使用 M 个 Hash 函数，通过每个 Hash 函数对每个 key 生成 1 个整数值。在初始状态下，需要一个长度为 N 的比特数组，比特数组每一位都是 0。当某个 key 加入布隆过滤器时，使用 M 个 Hash 函数计算出 M 个 Hash 值，并且根据 K 个 Hash 值将比特数组中对应位置的比特位设置为 1。当查询某个 key 是否在布隆过滤器中时，可通过 M 个 Hash 函数计算出 M 个 Hash 值，并根据生成的 M 个 Hash 值查找比特数组中对应的比特位，只有当所有的 Hash 值对应的比特位都为 1 时，可认为此 key 在布隆过滤器中，否则认为 key 不在布隆过滤器中。

初始状态下布隆过滤器如图 14-121 所示，此时使用的比特位都为 0。

图 14-121　布隆过滤器初始状态

当 K1 加入布隆过滤器后，布隆过滤器的状态变成如图 14-122 所示的结果。

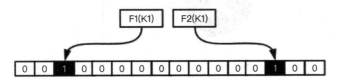

图 14-122　布隆过滤器加入 K1 时的结果

当 K2 加入布隆过滤器后，布隆过滤器的状态变成如图 14-123 所示的结果。

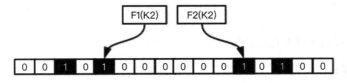

图 14-123　布隆过滤器加入 K2 时的结果

按照 K1 和 K2 的添加步骤，依次将所有存储层已经存在的 key 以及存储层新增的 key 都加入到布隆过滤器中。

当用户请求携带 Kn 经过布隆过滤器时，如图 14-124 所示。

由于 F2（Kn）对应的比特位为 0，此时认为 Kn 不在布隆过滤器中。因此可以在布隆过滤器这一层将请求拦截住，在一定程度上保护了存储层。

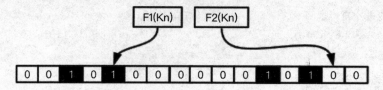

图 14-124　查询请求 Kn 经过布隆过滤器

可以在应用层面使用 Google Guava 框架实现布隆过滤器，也可以利用 Redis 的 Bitmaps 实现布隆过滤器，GitHub 上已经开源了类似的方案，读者可以进行参考，地址为 https://github.com/erikdubbelboer/Redis-Lua-scaling-bloom-filter。

14.8.2　Redis 缓存雪崩

缓存层承载着大量的请求，有效保护了存储层。但是如果由于缓存大量失效或者缓存整体不能提供服务，导致大量的请求到达存储层，会使存储层负载增加。这就是缓存雪崩的场景，如图 14-125 所示。

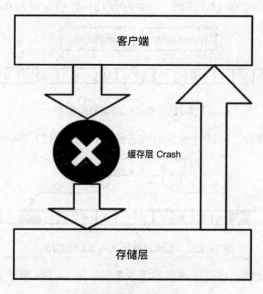

图 14-125　缓存雪崩

解决缓存雪崩可以从以下几点着手。

1. 保持缓存层的高可用性

使用 Redis 哨兵模式或者 Redis 集群部署方式，即便个别 Redis 节点下线，整个缓存层依然可以使用。除此之外还可以在多个机房部署 Redis，这样即便是机房死机，依然可以实现缓存层的高可用。

2. 限流降级组件

无论是缓存层还是存储层都会有出错的概率，可以将它们视为资源。作为并发量较大的分布式系统，假如有一个资源不可用，可能会造成所有线程在获取这个资源时异常，造成整个系统不可用。降级在高并发系统中是非常正常的，比如推荐服务中，如果个性化推荐服务不可用，可以降级补充热点数据，不至于造成整个推荐服务不可用。此处推荐一个常见的限流降级的组件——Hystrix，有关 Hystrix 的资料请参考 https://github.com/Netflix/Hystrix。

3. 缓存不过期

Redis 中保存的 key 永不失效，这样就不会出现大量缓存同时失效的问题，但是随之而来的就是 Redis 需要更多的存储空间。

4. 优化缓存过期时间

设计缓存时，为每一个 key 选择合适的过期时间，避免大量的 key 在同一时刻同时失效，造成缓存雪崩。

5. 使用互斥锁重建缓存

在高并发场景下，为了避免大量的同时请求到达存储层查询数据、重建缓存，可以使用互斥锁控制。如根据 key 去缓存层查询数据，当缓存层未命中时，对 key 加锁，然后从存储层查询数据，将数据写入缓存层，最后释放锁。若其他线程发现获取锁失败，则让线程休眠一段时间后重试。对于锁的类型，如果是在单机环境下可以使用 Java 并发包下的 Lock，如果是在分布式环境下，可以使用分布式锁（Redis 中的 SETNX 方法）。

分布式环境下使用 Redis 分布式锁实现缓存重建如以下伪代码所示：

```java
/**
 * 使用互斥锁重建缓存伪代码
 */
public String get(String key) {
    // redis 中查询 key 对应的 value
    String value = redis.get(key);
    // 缓存未命中
    if (value == null) {
        // 互斥锁
        String key_mutex = "mutex_lock" + key;
        // 互斥锁加锁成功
        if (redis.setnx(key_mutex, "1")) {
            try {
                // 设置互斥锁超时时间
                redis.expire(key_mutex, 3 * 60);
                // 从数据库查询数据
                value = db.get(key);
                // 数据写入缓存
                redis.set(key, value);
```

```
        } finally {
            //释放锁
            redis.delete(key_mutex);
        }


    } else {
        //加锁失败,线程休息 50ms 后重试
        Thread.sleep(50);
        return get(key);
    }
  }
}
```

这种方式重建缓存的优点是设计思路简单,对数据一致性有保障;缺点是代码复杂度增加,有可能会造成用户等待。假设在高并发下,缓存重建期间 key 是锁着的,如果当前并发 1000 个请求,其中 999 个都在阻塞,会导致 999 个用户请求阻塞而等待。

6. 异步重建缓存

在这种方案下构建缓存采取异步策略,会从线程池中获取线程来异步构建缓存,从而不会让所有的请求直接到达存储层。该方案中每个 Redis key 维护逻辑超时时间,当逻辑超时时间小于当前时间时,则说明当前缓存已经失效,应当进行缓存更新,否则说明当前缓存未失效,直接返回缓存中的 value 值。如在 Redis 中将 key 的过期时间设置为 60 min,在对应的 value 中设置逻辑过期时间为 30 min。这样当 key 到了 30 min 的逻辑过期时间,就可以异步更新这个 key 的缓存,但是在更新缓存的这段时间内,旧的缓存依然可用。这种异步重建缓存的方式可以有效避免大量的 key 同时失效。

```
/**
 * 异步重建缓存伪代码
 */
public String get(String key) {
    // 从缓存中查询 key 对应的 ValueObject 对象
    ValueObject valueObject = redis.get(key);
    // 缓存中对应的 value
    String value = valueObject.getValue();
    // 逻辑过期时间
    long logicTimeout = valueObject.getTimeout();
    // 当前 key 在逻辑上失效
    if (logicTimeout <= System.currentTimeMillis()) {
        // 异步更新缓存
        threadPool.execute(new Runnable() {
            public void run() {
                String mutex_lock = "mutex_lock" + key;
                        // 加分布式锁成功
```

```
                if (redis.setnx(mutex_lock, "1")) {
                    try {
                        // 设置分布式锁的超时时间
                        redis.expire(mutex_lock, 3 * 60);
                        // 从存储层查询数据
                        String dbValue = db.get(key);
                        // 设置缓存
                        redis.set(key, dbValue);
                    } finally {
                        redis.delete(mutex_lock);
                    }

                } else {
                    // TODO：等待锁或者什么都不做
                }
            }
        });
    }
    return value;
}
```

14.9 小　　结

本章讲解了 Redis 常见的 API、Redis 多种部署方式、Redis 与 Spring、MyBatis 集成开发和 Redis 在高并发场景下的缓存穿透、雪崩等问题。这些都是企业开发过程中常见的场景，也是面试时经常被问到的考点。

第 15 章

Spring 集成 ZooKeeper

ZooKeeper 是开放代码的分布式协调服务框架,是一个为分布式应用提供一致性服务的组件。

在分布式环境中协调和管理服务是一个非常复杂的过程,ZooKeeper 通过其简单的架构和 API 解决了这个问题。ZooKeeper 允许开发人员专注于核心应用程序逻辑,而不必担心应用程序的分布式特性。

15.1 ZooKeeper 集群安装

本书以 ZooKeeper 在 Linux 环境下的安装为例,如果使用的是 Windows 操作系统,可以在 Windows 上安装 Linux 虚拟机完成 ZooKeeper 集群的安装。

在下面的安装步骤中,ZooKeeper 集群中共有 5 个 ZooKeeper Server 节点,其中一个 Server 节点充当 ZooKeeper 集群的 Leader,其他两个 Server 节点充当 ZooKeeper 集群的 Follower,剩下的两个 Server 节点充当 Zookeeper 集群的 Observer。集群中每种角色的功能请参考本书 15.2 节。

ZooKeeper 安装步骤如下。

1. 下载 ZooKeeper

ZooKeeper 官网下载地址为 https://www.apache.org/dyn/closer.cgi/zookeeper/,读者可以根据需要选择合适的 ZooKeeper 版本进行下载。本书使用的版本是 Zookeeper-3.4.13。

2. 解压 ZooKeeper

使用 tar 命令解压 ZooKeeper 到当前目录下。

```
tar -zxvf zookeeper-3.4.13.tar.gz
```

3. 创建配置文件

进入 ZooKeeper 解压目录，进入 ZooKeeper 目录下的 conf 目录，创建 zoo1.cfg~zoo5.cfg 5 个配置文件：

```
cd zookeeper-3.4.13
cd conf
```

创建 zoo1.cfg 配置文件，文件中保存以下内容：

```
#zoo1.cfg 配置文件
#Client 与 Server 通信心跳时间
tickTime=2000
#Leader 与 Follower 初始连接时能容忍的最多心跳数
initLimit=10
#Leader 与 Follower 请求和应答之间能容忍的最多心跳数
syncLimit=5
#数据文件目录
dataDir=../data/zk1
#客户端连接端口
clientPort=2182
#日志地址
dataLogDir=../logs/zk1/logs
#服务器名称与地址，集群信息（服务器编号，服务器地址，通信端口，选举端口，运行角色）
server.1=127.0.0.1:2111:3111
server.2=127.0.0.1:2222:3222
server.3=127.0.0.1:2333:3333
server.4=127.0.0.1:2444:3444:observer
server.5=127.0.0.1:2555:3555:observer
```

配置文件中每个配置项的含义如下：

- tickTime：Client 与 Server 的通信心跳时间。Zookeeper 服务器之间或客户端与服务器之间维持心跳的时间间隔，即每隔 tickTime 时间就会发送一个心跳。tickTime 以毫秒为单位。
- initLimit：Leader 与 Follower 初始通信时限。集群中的 Follower 服务器与 Leader 服务器之间初始连接时能容忍的最多心跳数。
- syncLimit：Leader 与 Follower 同步通信时限。集群中的 Follower 服务器与 Leader 服务器之间请求和应答之间能容忍的最多心跳数。
- dataDir：数据文件目录。Zookeeper 保存数据的目录。
- clientPort：客户端连接端口。客户端连接 Zookeeper 服务器的端口，Zookeeper 会监听这个端口，接受客户端的访问请求。
- dataLogDir：日志存放地址。
- 服务器名称与地址：集群信息。这个配置项的书写格式比较特殊，各个配置项含义如下：

服务器编号=服务器地址：Leader 与 Follower 通信端口：选举端口：角色

复制 4 份 zoo1.cfg 文件，分别命名为 zoo2.cfg、zoo3.cfg、zoo4.cfg 和 zoo5.cfg，一次修改每个配置文件。

```
######################zoo2.cfg 配置文件######################
#Client 与 Server 通信心跳时间
tickTime=2000
#Leader 与 Follower 初始连接时能容忍的最多心跳数
initLimit=10
#Leader 与 Follower 请求和应答之间能容忍的最多心跳数
syncLimit=5
#数据文件目录
dataDir=../data/zk2
#客户端连接端口
clientPort=2183
#日志地址
dataLogDir=../logs/zk2/logs
#服务器名称与地址，集群信息（服务器编号，服务器地址，通信端口，选举端口）
server.1=127.0.0.1:2111:3111
server.2=127.0.0.1:2222:3222
server.3=127.0.0.1:2333:3333
server.4=127.0.0.1:2444:3444:observer
server.5=127.0.0.1:2555:3555:observer
######################zoo3.cfg 配置文件######################
#Client 与 Server 通信心跳时间
tickTime=2000
#Leader 与 Follower 初始连接时能容忍的最多心跳数
initLimit=10
#Leader 与 Follower 请求和应答之间能容忍的最多心跳数
syncLimit=5
#数据文件目录
dataDir=../data/zk3
#客户端连接端口
clientPort=2184
#日志地址
dataLogDir=../logs/zk3/logs
#服务器名称与地址，集群信息（服务器编号，服务器地址，通信端口，选举端口）
server.1=127.0.0.1:2111:3111
server.2=127.0.0.1:2222:3222
server.3=127.0.0.1:2333:3333
server.4=127.0.0.1:2444:3444:observer
server.5=127.0.0.1:2555:3555:observer
######################zoo4.cfg 配置文件######################
#Client 与 Server 通信心跳时间
tickTime=2000
#Leader 与 Follower 初始连接时能容忍的最多心跳数
```

```
initLimit=10
#Leader 与 Follower 请求和应答之间能容忍的最多心跳数
syncLimit=5
#数据文件目录
dataDir=../data/zk4
#客户端连接端口
clientPort=2185
#日志地址
dataLogDir=../logs/zk4/logs
#Observer 模式启动
peerType=observer
#服务器名称与地址，集群信息（服务器编号，服务器地址，通信端口，选举端口）
server.1=127.0.0.1:2111:3111
server.2=127.0.0.1:2222:3222
server.3=127.0.0.1:2333:3333
server.4=127.0.0.1:2444:3444:observer
server.5=127.0.0.1:2555:3555:observer
####################zoo5.cfg 配置文件####################
#Client 与 Server 通信心跳时间
tickTime=2000
#Leader 与 Follower 初始连接时能容忍的最多心跳数
initLimit=10
#Leader 与 Follower 请求和应答之间能容忍的最多心跳数
syncLimit=5
#数据文件目录
dataDir=../data/zk5
#客户端连接端口
clientPort=2186
#日志地址
dataLogDir=../logs/zk5/logs
#Observer 模式启动
peerType=observer
#服务器名称与地址，集群信息（服务器编号，服务器地址，通信端口，选举端口）
server.1=127.0.0.1:2111:3111
server.2=127.0.0.1:2222:3222
server.3=127.0.0.1:2333:3333
server.4=127.0.0.1:2444:3444:observer
server.5=127.0.0.1:2555:3555:observer
```

zoo4.cfg 和 zoo5.cfg 中配置的 peerType 表示此 ZooKeeper 节点以 Observer 的模式加入集群中。Observer 详细介绍可参考 15.2 节。

4. 创建 myid 文件

在每个配置文件中都有一个 dataDir 配置项，这个配置项是用来配置数据文件的目录。在 data 目录下分别创建目录 zk1、zk2、zk3、zk4 和 zk5，再在每个目录下创建一个 myid 文件。

```
cd ../data
mkdir zk1
mkdir zk2
mkdir zk3
mkdir zk4
mkdir zk5
echo '1' > zk1/myid
echo '2' > zk2/myid
echo '3' > zk3/myid
echo '4' > zk4/myid
echo '5' > zk5/myid
```

5. 创建启动脚本

进入 Zookeeper 解压目录下的 bin 目录，创建 start-all.sh 脚本用于启动 5 个 ZooKeeper 进程：

```
./zkServer.sh start ../conf/zoo1.cfg
./zkServer.sh start ../conf/zoo2.cfg
./zkServer.sh start ../conf/zoo3.cfg
./zkServer.sh start ../conf/zoo4.cfg
./zkServer.sh start ../conf/zoo5.cfg
```

创建 stop-all.sh 脚本用于停止 5 个 ZooKeeper 进程：

```
./zkServer.sh stop ../conf/zoo1.cfg
./zkServer.sh stop ../conf/zoo2.cfg
./zkServer.sh stop ../conf/zoo3.cfg
./zkServer.sh stop ../conf/zoo4.cfg
./zkServer.sh stop ../conf/zoo5.cfg
```

创建 status-all.sh 脚本用于检查当前集群的状态：

```
./zkServer.sh status ../conf/zoo1.cfg
./zkServer.sh status ../conf/zoo2.cfg
./zkServer.sh status ../conf/zoo3.cfg
./zkServer.sh status ../conf/zoo4.cfg
./zkServer.sh status ../conf/zoo5.cfg
```

为 start-all.sh、stop-all.sh 和 status-all.sh 添加执行权限：

```
chmod +x start-all.sh stop-all.sh
```

6. 启动 ZooKeeper 集群

执行 start-all.sh 脚本启动 ZooKeeper 集群：

```
./start-all.sh
```

执行结果如图 15-1 所示。

```
ZooKeeper JMX enabled by default
Using config: ../conf/zoo1.cfg
Starting zookeeper ... STARTED
ZooKeeper JMX enabled by default
Using config: ../conf/zoo2.cfg
Starting zookeeper ... STARTED
ZooKeeper JMX enabled by default
Using config: ../conf/zoo3.cfg
Starting zookeeper ... STARTED
ZooKeeper JMX enabled by default
Using config: ../conf/zoo4.cfg
Starting zookeeper ... STARTED
ZooKeeper JMX enabled by default
Using config: ../conf/zoo5.cfg
Starting zookeeper ... STARTED
```

图 15-1　启动 ZooKeeper 集群

7. 验证启动

执行以下命令查询启动的 ZooKeeper 进程：

```
ps -ef | grep zookeeper | grep -v grep
```

通过以上命令可知，当前启动了 5 个 ZooKeeper 进程。

8. 查询集群状态

执行以下命令查看 ZooKeeper 集群状态：

```
./status-all.sh
```

ZooKeeper 集群状态如图 15-2 所示。

```
Using config: ../conf/zoo1.cfg
Mode: follower
ZooKeeper JMX enabled by default
Using config: ../conf/zoo2.cfg
Mode: leader
ZooKeeper JMX enabled by default
Using config: ../conf/zoo3.cfg
Mode: follower
ZooKeeper JMX enabled by default
Using config: ../conf/zoo4.cfg
Mode: observer
ZooKeeper JMX enabled by default
Using config: ../conf/zoo5.cfg
Mode: observer
```

图 15-2　ZooKeeper 集群状态

从集群状态可以看出，当前 ZooKeeper 集群中有一个 Leader 节点、两个 Follower 节点和两个 Observer 节点。下一节将介绍 Leader 节点产生的过程以及每种类型的 ZooKeeper 节点的特性。

15.2 ZooKeeper 总体架构

ZooKeeper 集群中的节点有以下几种角色，如表 15-1 所示。

表 15-1 Zookeeper 集群中的角色

角色	描述
领导者(Leader)	领导者负责投票的发起和决议，更新系统状态
跟随者(Follower)	接受客户端请求并返回结果，在选举阶段参与投票
观察者(Observer)	接受客户端连接，将写请求转发给 Leader，不参与选举阶段投票
客户端(Client)	请求的发起方

ZooKeeper 集群由一组 Server 节点组成，这一组 Server 节点中存在一个角色为 Leader 的节点，其他节点为 Follower 或 Observer。ZooKeeper 总体架构如图 15-3 所示。

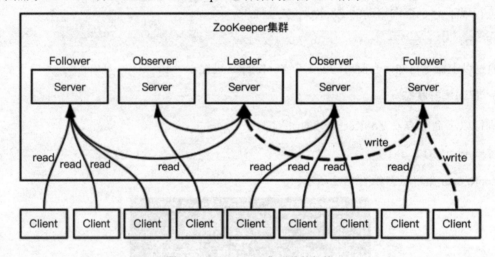

图 15-3 ZooKeeper 集群总体架构

15.2.1 ZooKeeper 选举机制

下面分析 15.1 中安装的 ZooKeeper 集群的 Leader 的选举过程。

当启动 Zookeeper 集群时，首先需要做的一件事就是 Leader 选举。Zookeeper 中 Leader 默认的选举算法是 FastLeaderElection，可通过 electionAlg 配置项选择不同的 Zookeeper 选举算法。

当集群中不存在 Leader 服务器时集群会进行 Leader 服务器的选举，通常存在两种情况，一是集群刚启动时，二是集群运行时，Leader 服务器因故退出。集群中的服务器会向其他所有的 Follower 服务器发送消息，这个消息可以形象化的称之为投票。

投票主要由两个信息组成，推举的 Leader 服务器的 ID（即配置在 myid 文件中的数字），以及该服务器的事务 ID，事务表示对服务器状态变更的操作，一个服务器的事务 ID 越大，则其数据越新。

ZooKeeper 集群启动时的 Leader 选举过程如下：

（1）在集群初始化阶段，当有一台服务器 Server1 启动时，其单独无法进行和完成 Leader 选举；当第二台服务器 Server2 启动时，此时两台机器可以相互通信，每台机器都试图找到 Leader，于是进入 Leader 选举过程。

（2）每个 Server 发出一个投票。由于是初始情况，Server1 和 Server2 都会将票投给本服务器，以便本服务器成为 Leader 服务器。每个投票会包含所推举的服务器的 SID（服务器的唯一标识）和 ZXID，使用（SID, ZXID）来表示，此时 Server1 的投票为（1, 0），Server2 的投票为（2, 0），然后各自将这个投票发给集群中其他机器。

（3）每个 Server 接受来自各个服务器的投票。集群的每个服务器收到投票后，首先判断该投票的有效性，如检查是否为本轮投票、是否来自 LOOKING 状态的服务器。

（4）处理投票。针对每一个投票，服务器都需要将其他服务器的投票和本服务器投票进行对比。对比过程中涉及如下一些术语。

- vote_sid：接收到的投票中所推举 Leader 服务器的 SID。
- vote_zxid：接收到的投票中所推举 Leader 服务器的 ZXID。
- self_sid：当前服务器的 SID。
- self_zxid：当前服务器的 ZXID。

每次对收到的投票处理，都是对（vote_sid, vote_zxid）和（self_sid, self_zxid）对比的过程：

- 如果 vote_zxid 大于 self_zxid，那么认可当前收到的投票，并再次将该投票发送出去。
- 如果 vote_zxid 小于 self_zxid，那么坚持自己的投票，不做任何变更。
- 如果 vote_zxid 等于 self_zxid，那么就对比两者的 SID，如果 vote_sid 大于 self_sid，那么就认可当前收到的投票，并再次将该投票发送出去。
- 如果 vote_zxid 等于 self_zxid，并且 vote_sid 小于 self_sid，那么保持原投票，不做任何变更。

（5）统计投票。每次投票后，服务器都会统计投票信息，判断是否已经有过半机器接受到相同的投票信息，Server1 和 Server2 统计出集群中已经有两台机器接受了（2, 0）的投票信息，此时便认为已经选出了 Leader。

（6）改变服务器状态。一旦确定了 Leader，每个服务器就会更新自己的状态。如果服务器是 Follower，那么就变更为 FOLLOWING，如果服务器是 Leader，就变更为 LEADING。

ZooKeeper 集群启动时的 Leader 选举过程如图 15-4 所示。

处理 ZooKeeper 集群启动时需要进行 Leader 选举，在集群运行过程中，如果集群中的 Leader 下线也会触发 Leader 选举。

下面分析在 ZooKeeper 集群运行时的 Leader 选举过程。

假设一个 ZooKeeper 集群有 5 台 Server，在 ZooKeeper 集群运行时，Server2 个是 Leader，其余 4 台 Server 是 Follower。在某一时刻，Server1 和 Server2 因故障下线。此时 Server3、Server4 和 Server5 对应的投票状态（SID，ZXID）为（3, 9）、（4, 8）和（5, 8）。ZooKeeper 运行时期 Leader 的选举过程如图 15-5 所示。

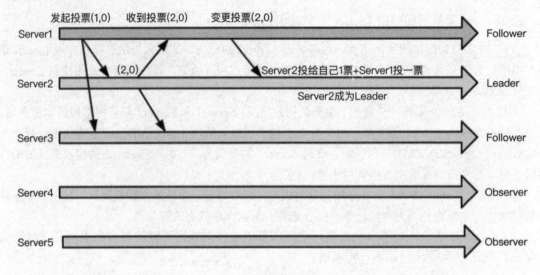

图 15-4　ZooKeeper 集群启动时期的 Leader 选举过程

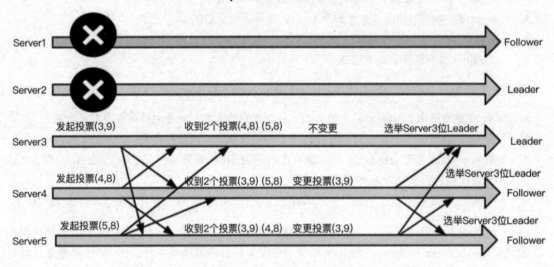

图 15-5　ZooKeeper 运行时期 Leader 选举过程

15.2.2　ZooKeeper 数据模型

ZooKeeper 拥有一个层次的命名空间，这和标准的文件系统非常相似。ZooKeeper 中的每个节点被称为 Znode，每个节点可以拥有子节点。ZooKeeper 数据模型架构如图 15-6 所示。

ZooKeeper 命名空间中的 Znode 兼具文件和目录两种特点。既可以像文件一样维护数据、元信息、ACL、时间戳等数据结构，又可以像目录一样作为路径标识的一部分。每个 Znode 由 3 部分组成。

- stat: 存储状态信息，用于描述该 Znode 的版本、权限等信息。
- data: 存储与该 Znode 关联的数据。
- children: 存储该 Znode 下的子节点。

第 15 章 Spring 集成 ZooKeeper | 339

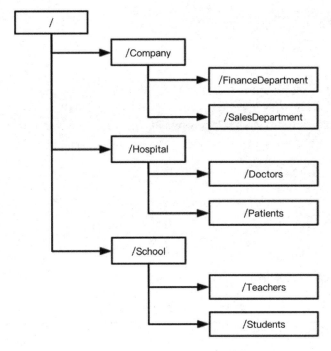

图 15-6 ZooKeeper 数据模型

ZooKeeper 虽然可以关联一些数据，但并没有被设计为常规的关系型数据库或者大数据存储，相反的是，Znode 用来管理调度数据，比如分布式应用中的配置文件信息、状态信息、汇集位置等。ZooKeeper 规定节点的数据大小不能超过 1MB，但在实际使用中 Znode 的数据量应该尽可能小，因为数据过大会导致 ZooKeeper 性能明显下降。

Znode 有以下 4 种类型。

- PERSISTENT：持久节点。ZooKeeper 客户端与 ZooKeeper 服务器端断开连接后，该节点依旧存在。
- PERSISTENT_SEQUENTIAL：持久顺序节点。ZooKeeper 客户端与 ZooKeeper 服务器端断开连接后，该节点依旧存在，并且 Zookeeper 给该节点名称进行顺序编号。
- EPHEMERAL：临时节点。和持久节点不同的是，临时节点的生命周期和客户端会话绑定。如果客户端会话失效，那么这个节点会被自动清除。在临时节点下面不能创建子节点。
- EPHEMERAL_SEQUENTIAL：临时顺序节点。临时顺序节点的生命周期和客户端会话绑定。如果客户端会话失效，那么这个节点就会被自动清除。创建的节点会自动加上编号。

使用 ZooKeeper 自带的客户端连接到 ZooKeeper 集群，查看当前集群中节点状态：

```
./zkCli.sh -server 127.0.0.1:2181
```

执行以下命令查询当前根节点下的 Znode 节点：

```
ls /
```

可以发现当前节点 ZooKeeper 集群当前只有 "/zookeeper" 一个 Znode 节点。通过以下命令查询当前节点的状态：

```
stat /zookeeper
```

执行 stat 命令后得到如下信息:

```
cZxid = 0x0
ctime = Thu Jan 01 08:00:00 CST 1970
mZxid = 0x0
mtime = Thu Jan 01 08:00:00 CST 1970
pZxid = 0x0
cversion = -1
dataVersion = 0
aclVersion = 0
ephemeralOwner = 0x0
dataLength = 0
numChildren = 1
```

各字段的含义如表 15-2 所示。

表 15-2　stat 命令字段的含义

字　段	描　述
czxid	创建该节点的事物 ID
ctime	创建该节点的时间
mZxid	更新该节点的事物 ID
mtime	更新该节点的时间
pZxid	添加和移除子结点更改的事务 ID
cversion	当前节点的子节点版本号
dataVersion	当前节点的数据版本号
aclVersion	当前节点的 ACL 权限版本号
ephemeralowner	如果是临时节点，该属性是临时节点的事物 ID。如果不是临时节点，这个值为 0
dataLength	当前节点的数据长度
numchildren	当前节点的子节点个数

下面使用 ZooKeeper 的客户端创建 4 种类型的 Znode。

1. 创建持久节点

使用以下命令在根目录下创建"/School"节点，存放数据为 QingHua：

```
create /School QingHua
```

使用 get 命令查询"/School"节点的内容，"/School"节点内容如下：

```
[zk: 127.0.0.1:2182(CONNECTED) 30] get /School
QingHua
cZxid = 0xa00000008
ctime = Tue Jan 08 15:25:04 CST 2019
mZxid = 0xa00000008
```

```
mtime = Tue Jan 08 15:25:04 CST 2019
pZxid = 0xa00000008
cversion = 0
dataVersion = 0
aclVersion = 0
ephemeralOwner = 0x0
dataLength = 7
numChildren = 0
```

执行 close 命令关闭会话。再次使用 ZooKeeper 客户端连接到 ZooKeeper 服务器端，查看 "/School" 节点的信息，发现 "/School" 节点仍然存在。

2. 创建持久顺序节点

在 "/School" 节点下创建 "/Student" 子节点，在 "/Student" 子节点下创建多个持久顺序节点。使用 "-s" 参数表示创建顺序节点：

```
create /School/Student AllTheStudents
create -s /School/Student/s_ Michael
create -s /School/Student/s_ Jack
create -s /School/Student/s_ Tom
```

验证以上创建的多个持久顺序节点：

```
ls /School/Student
```

可以看到 "/School/Student" 节点下有以下几个节点：

```
[s_0000000000, s_0000000001, s_0000000002]
```

结果中的 0000000000~0000000002 都是自动添加的序列号。

执行 close 命令关闭会话。再次使用 ZooKeeper 客户端连接到 ZooKeeper 服务器端，查看 "/School/Student/s_0000000000~/School/Student/s_0000000002" 节点的信息，发现节点依然存在。

3. 创建临时节点

在 "/School" 节点下创建临时节点 "/Teacher"。参数 "-e" 表示创建临时节点：

```
create -e /School/Teacher EphemeralTeachers
```

查看 "/School/Teacher" 临时节点的信息，如下所示：

```
[zk: 127.0.0.1:2182(CONNECTED) 1] get /School/Teacher
EphemeralTeachers
cZxid = 0xa0000002d
ctime = Tue Jan 08 16:07:36 CST 2019
mZxid = 0xa0000002d
mtime = Tue Jan 08 16:07:36 CST 2019
pZxid = 0xa0000002d
cversion = 0
dataVersion = 0
```

```
aclVersion = 0
ephemeralOwner = 0x100012781570003
dataLength = 17
numChildren = 0
```

执行 close 命令关闭会话：

```
[zk: 127.0.0.1:2182(CONNECTED) 2] close
2019-01-08 16:10:01,141 [myid:] - INFO  [main:ZooKeeper@693] - Session: 0x100012781570003 closed
[zk: 127.0.0.1:2182(CLOSED) 3] 2019-01-08 16:10:01,143 [myid:] - INFO  [main-EventThread:ClientCnxn$EventThread@522] - EventThread shut down for session: 0x100012781570003
```

重新连接到 ZooKeeper 服务器端，查看临时节点"/School/Teacher"信息。发现临时节点"/School/Teacher"已经不存在。

```
[zk: 127.0.0.1:2182(CONNECTED) 0] get /School/Teacher
Node does not exist: /School/Teacher
```

4. 创建临时顺序节点

在"/School"节点下创建永久节点"/Teacher"。参数"-e"表示创建临时节点：

```
create /School/Teacher AllTheTeachers
```

在"/School/Teacher"节点下创建若干个临时顺序节点：

```
create -e -s /School/Teacher/t_ TeacherHuang
create -e -s /School/Teacher/t_ TeacherZhou
create -e -s /School/Teacher/t_ TeacherZhang
```

关闭会话后，再次连接 ZooKeeper 服务器，发现临时顺序节点消失。

15.3　Spring 集成 ZooKeeper 快速体验

在本章 15.1 节中创建了 ZooKeeper 集群。要想在 Spring 中使用 ZooKeeper 进行开发，需要使用 ZooKeeper 客户端连接到 ZooKeeper 集群中。

ZooKeeper 的常用客户端有 3 种，分别是 ZooKeeper 原生的客户端、Apache Curator 客户端和开源 zkclient 客户端。

（1）ZooKeeper 原生客户端。ZooKeeper 自带的客户端是官方提供的，是比较简单的功能，编程烦琐，不够直接。

（2）Apache Curator。Apache Curator 是 Apache 的开源项目，封装了 ZooKeeper 自带的客户端，使用相对简便，易于使用。

（3）zkclient。zkclient 是另一个开源的 ZooKeeper 客户端，其地址为 https://github.com/adyliu/zkclient，生产环境不推荐使用。

本书使用 Apache Curator 客户端集成 Spring 进行开发。

1. 添加 Curator 相关的依赖

在 pom.xml 中添加以下依赖关系：

```xml
<dependency>
    <groupId>org.apache.curator</groupId>
    <artifactId>curator-framework</artifactId>
    <version>2.11.1</version>
</dependency>
<dependency>
    <groupId>org.apache.curator</groupId>
    <artifactId>curator-recipes</artifactId>
    <version>2.11.1</version>
</dependency>
```

2. 创建自定义客户端

创建自定义 ZooKeeper 客户端，封装 Curator 中的 ZooKeeper 客户端，并提供一些对 ZooKeeper 的操作：

```java
/**
 * @Author: zhouguanya
 * @Date: 2019/01/08
 * @Description: 自定义 Zookeeper 客户端
 */
public class ZookeeperClient {
    /**
     * Zookeeper 客户端
     */
    private CuratorFramework curatorFramework = null;

    public CuratorFramework getCuratorFramework() {
        return curatorFramework;
    }

    /**
     * 构造函数输入
     */
    public ZookeeperClient(CuratorFramework curatorFramework) {
        this.curatorFramework = curatorFramework;
    }

    /**
     * 创建节点
     */
```

```java
        public void save(String path, String data, CreateMode createMode) {
            try {
                curatorFramework.create().creatingParentContainersIfNeeded()
.withMode(createMode).forPath(path, data.getBytes());
            } catch (Exception e) {
                e.printStackTrace();
            }
        }

        /**
         * 查询节点信息
         */
        public String query(String path) {
            try {
                byte[] data = curatorFramework.getData().forPath(path);
                if (data != null && data.length > 0) {
                    return new String(data);
                }
            } catch (Exception e) {
                e.printStackTrace();
            }
            return null;
        }
    }
```

3. 在 Spring 中集成 Curator

创建配置文件 spring-zookeeper.xml，集成 Curator：

```xml
    <!--重试策略-->
    <bean id="retryPolicy" class="org.apache.curator.retry.RetryNTimes">
        <!--重试次数-->
        <constructor-arg index="0" value="10"/>
        <!--每次间隔ms-->
        <constructor-arg index="1" value="5000"/>
    </bean>

    <!--Curator ZooKeeper 客户端-->
    <bean id="client"
class="org.apache.curator.framework.CuratorFrameworkFactory"
    factory-method="newClient" init-method="start">
        <!--ZK 服务地址，集群使用逗号分隔-->
        <constructor-arg index="0" value="127.0.0.1:2182,127.0.0.1:2183,
127.0.0.1:2184,127.0.0.1:2185,127.0.0.1:2186"/>
        <!--session timeout 会话超时时间-->
        <constructor-arg index="1" value="10000"/>
```

```xml
<!--ConnectionTimeout 创建连接超时时间-->
<constructor-arg index="2" value="5000"/>
<!--重试策略-->
<constructor-arg index="3" ref="retryPolicy"/>
</bean>

<!--自定义 ZooKeeper 客户端-->
<bean id="zookeeperClient" class="com.test.zk.demo.ZookeeperClient">
    <constructor-arg index="0" ref="client"/>
</bean>
```

4. 单元测试

创建单元测试，在 ZooKeeper 服务器端创建"/spring5/test"节点：

```java
/**
 * @Author: zhouguanya
 * @Date: 2019/01/08
 * @Description: 测试 ZooKeeper 客户端
 */
@RunWith(SpringJUnit4ClassRunner.class)
@ContextConfiguration("classpath:spring-zookeeper.xml")
public class ZookeeperClientTest {
    @Autowired
    private ZookeeperClient zookeeperClient;

    @Test
    public void test() {
        String path = "/spring5/test";
        //保存
        zookeeperClient.save(path, "Spring 5 Zookeeper Test", CreateMode.PERSISTENT);
        //查询
        String data = zookeeperClient.query(path);
        System.out.println("data = " + data);
    }
}
```

执行单元测试，得到如下所示的输出结果：

```
data = Spring 5 Zookeeper Test
```

登录 ZooKeeper 客户端查看"/spring5/test"节点信息，如下所示：

```
[zk: 127.0.0.1:2182(CONNECTED) 17] get /spring5/test
Spring 5 Zookeeper Test
cZxid = 0xb00000037
ctime = Tue Jan 08 20:16:01 CST 2019
```

```
mZxid = 0xb00000037
mtime = Tue Jan 08 20:16:01 CST 2019
pZxid = 0xb00000037
cversion = 0
dataVersion = 0
aclVersion = 0
ephemeralOwner = 0x0
dataLength = 22
numChildren = 0
```

通过测试结果可知，Spring 集成 Curator 成功，ZooKeeper 客户端与 ZooKeeper 服务端连接成功，创建节点和查询节点信息成功。

15.4　ZooKeeper 发布订阅

Curator 可以监听变动的节点路径、节点值等。Curator 的 API 提供了 3 个接口，分别如下。

（1）NodeCache：NodeCache 对一个节点进行监听，监听事件包括指定路径的增、删、改操作。

（2）PathChildrenCache：PathChildrenCache 可以对指定路径节点的一级子目录进行监听，对其子目录的增、删、改操作进行监听，不对该节点的操作进行监听。

（3）TreeCache：TreeCache 综合了 NodeCache 和 PathChildrenCahce 的特性，对整个目录进行监听，可以设置监听深度。

15.4.1　NodeCache

使用 NodeCache 监听一个节点的变更情况。下面通过案例说明如何通过 NodeCache 监听"/NodeCache/PubSub"节点变化。

1. 创建发布者

创建发布者，在 ZooKeeper 服务端进行写入操作：

```java
/**
 * @Author: zhouguanya
 * @Date: 2019/01/08
 * @Description: 发布者
 */
public class Publisher {
    private ZookeeperClient zookeeperClient;

    public Publisher (ZookeeperClient zookeeperClient) {
        this.zookeeperClient = zookeeperClient;
    }
```

```java
/**
 * 发布信息
 */
public void publish(String path, String data) {
    try {
        Stat status = zookeeperClient.getCuratorFramework().checkExists().forPath(path);
        if (status == null) {
            zookeeperClient.getCuratorFramework().create().creatingParentContainersIfNeeded().forPath(path, data.getBytes());
        } else {
            zookeeperClient.getCuratorFramework().setData().forPath(path, data.getBytes());
        }

    } catch (Exception e) {
        e.printStackTrace();
    }
}
```

2. 创建订阅者

创建订阅者，使用 NodeCache 订阅节点变化：

```java
/**
 * @Author: zhouguanya
 * @Date: 2019/01/08
 * @Description: NodeCache 订阅
 */
public class NodeCacheSubscriber {

    private ZookeeperClient zookeeperClient;

    private String name;
    public NodeCacheSubscriber(String name, ZookeeperClient zookeeperClient){
        this.name = name;
        this.zookeeperClient = zookeeperClient;
    }
    /**
     * 订阅
     */
    public void subscribe(String path) {
        NodeCache nodeCache = new NodeCache(zookeeperClient.getCuratorFramework(), path);
        nodeCache.getListenable().addListener(() ->
```

```
                System.out.printf("%s 监听到节点信息发生变化,当前数据=%s%n",
name,new String(nodeCache.getCurrentData().getData())));
        try {
            nodeCache.start();
        } catch (Exception e) {
            e.printStackTrace();
        }
    }
}
```

3. 单元测试

创建单元测试,其中一个发布者负责对节点进行写入和更新操作,两个订阅者监听订阅节点的变化。

```
/**
 * @Author: zhouguanya
 * @Date: 2019/01/08
 * @Description: 发布订阅测试
 */
@RunWith(SpringJUnit4ClassRunner.class)
@ContextConfiguration("classpath:spring-zookeeper.xml")
public class NodeCachePubSubTest {
    @Autowired
    private ZookeeperClient zookeeperClient;
    @Test
    public void test() throws InterruptedException {
        String path = "/NodeCache/PubSub";
        Publisher publisher = new Publisher(zookeeperClient);
        //写入数据 100
        publisher.publish(path, String.valueOf(100));
        NodeCacheSubscriber subscriber1 = new NodeCacheSubscriber("订阅者 1",
zookeeperClient);
        subscriber1.subscribe(path);
        NodeCacheSubscriber subscriber2 = new NodeCacheSubscriber("订阅者 2",
zookeeperClient);
        subscriber2.subscribe(path);
        Thread.sleep(100);
        System.out.println("----------------分割线----------------");
        //更新数据 200
        publisher.publish(path, String.valueOf(200));
    }
}
```

执行以上单元测试代码,查看输出结果如下:

订阅者 1 监听到节点信息发生变化，当前数据=100
订阅者 2 监听到节点信息发生变化，当前数据=100
------------------分割线------------------
订阅者 2 监听到节点信息发生变化，当前数据=200
订阅者 1 监听到节点信息发生变化，当前数据=200

可以发现写入数据 100 时，订阅者 1 和订阅者 2 都监听到了数据写入。当数据节点的数据为 200 时，订阅者 1 和订阅者 2 都接到了数据更新。

使用 ZooKeeper 客户端连接到 ZooKeeper 服务器，查看当前节点的状态，如下所示：

```
[zk: 127.0.0.1:2182(CONNECTED) 27] get /NodeCache/PubSub
200
cZxid = 0xb000000a4
ctime = Wed Jan 09 11:12:21 CST 2019
mZxid = 0xb000000b4
mtime = Wed Jan 09 11:27:55 CST 2019
pZxid = 0xb000000a4
cversion = 0
dataVersion = 8
aclVersion = 0
ephemeralOwner = 0x0
dataLength = 3
numChildren = 0
```

15.4.2　PathChildrenCache

本节使用 15.4.1 小节中的发布者，使用 PathChildrenCache 创建监听者进行验证。

1. 创建订阅者

使用 PathChildrenCache 订阅子节点变化。

```
/**
 * @Author: zhouguanya
 * @Date: 2019/01/08
 * @Description: PathChildrenCache 订阅
 */
public class PathChildrenCacheSubscriber {

    private ZookeeperClient zookeeperClient;

    private String name;
    public PathChildrenCacheSubscriber(String name, ZookeeperClient zookeeperClient) {
        this.name = name;
        this.zookeeperClient = zookeeperClient;
    }
```

```java
    /**
     * 订阅
     */
    public void subscribe(String path) {
        PathChildrenCache pathChildrenCache = new PathChildrenCache
(zookeeperClient.getCuratorFramework(), path, true);
        pathChildrenCache.getListenable().addListener((client, event) -> {
            // 当前节点的所有子节点
            List<ChildData> childDataList = pathChildrenCache.getCurrentData();
            for (ChildData childData : childDataList) {
                System.out.printf("%s 监听子节点更新,当前子节点path=%s,子节点数据=%s%n", name, childData.getPath(), new String(childData.getData()));
            }
        });
        try {
            pathChildrenCache.start();
        } catch (Exception e) {
            e.printStackTrace();
        }
    }
}
```

2. 单元测试

创建单元测试,其中含有一个发布者,两个订阅者。发布者依次修改子节点信息,观察订阅者的监听情况:

```java
/**
 * @Author: zhouguanya
 * @Date: 2019/01/08
 * @Description: 发布订阅测试
 */
@RunWith(SpringJUnit4ClassRunner.class)
@ContextConfiguration("classpath:spring-zookeeper.xml")
public class PathChildrenCachePubSubTest {

    @Autowired
    private ZookeeperClient zookeeperClient;

    @Test
    public void test() throws InterruptedException {
        String basePath = "/PathChildrenCache/PubSub";
        String firstPath = basePath + "/first";
        String secondPath = basePath + "/second";
        Publisher publisher = new Publisher(zookeeperClient);
```

```
            //写入数据100
            publisher.publish(basePath, String.valueOf(100));
            PathChildrenCacheSubscriber subscriber1 = new
PathChildrenCacheSubscriber("订阅者1", zookeeperClient);
            subscriber1.subscribe(basePath);
            PathChildrenCacheSubscriber subscriber2 = new
PathChildrenCacheSubscriber("订阅者2", zookeeperClient);
            subscriber2.subscribe(basePath);
            //创建节点:/PathChildrenCache/PubSub/first
            publisher.publish(firstPath, String.valueOf(200));
            Thread.sleep(100);
            System.out.println("-----------------------分割线
-----------------------");
            //创建节点:/PathChildrenCache/PubSub/second
            publisher.publish(secondPath, String.valueOf(300));
     }

}
```

执行单元测试，得到如下输出结果：

```
    订阅者2监听子节点更新，当前子节点path=/PathChildrenCache/PubSub/first,子节点数据
=200
    订阅者1监听子节点更新，当前子节点path=/PathChildrenCache/PubSub/first,子节点数据
=200
    -----------------------分割线-----------------------
    订阅者2监听子节点更新，当前子节点path=/PathChildrenCache/PubSub/first,子节点数据
=200
    订阅者1监听子节点更新，当前子节点path=/PathChildrenCache/PubSub/first,子节点数据
=200
    订阅者2监听子节点更新，当前子节点path=/PathChildrenCache/PubSub/second,子节点数
据=300
    订阅者1监听子节点更新，当前子节点path=/PathChildrenCache/PubSub/second,子节点数
据=300
```

可以看到，每次对父节点"/PathChildrenCache/PubSub"添加子节点，订阅者1和订阅者2都可以监听到子节点变化。监听者订阅到父节点所有的子节点信息。

使用ZooKeeper客户端连接到ZooKeeper，查询子节点列表，发现父节点下多出两个子节点。分别查询子节点的信息，如下所示：

```
[zk: 127.0.0.1:2182(CONNECTED) 57] get /PathChildrenCache/PubSub
100
cZxid = 0xb0000012f
ctime = Wed Jan 09 17:26:05 CST 2019
mZxid = 0xb0000012f
mtime = Wed Jan 09 17:26:05 CST 2019
```

```
pZxid = 0xb00000131
cversion = 2
dataVersion = 0
aclVersion = 0
ephemeralOwner = 0x0
dataLength = 3
numChildren = 2
[zk: 127.0.0.1:2182(CONNECTED) 54] ls /PathChildrenCache/PubSub
[first, second]
[zk: 127.0.0.1:2182(CONNECTED) 55] get /PathChildrenCache/PubSub/first
200
cZxid = 0xb00000130
ctime = Wed Jan 09 17:26:05 CST 2019
mZxid = 0xb00000130
mtime = Wed Jan 09 17:26:05 CST 2019
pZxid = 0xb00000130
cversion = 0
dataVersion = 0
aclVersion = 0
ephemeralOwner = 0x0
dataLength = 3
numChildren = 0
[zk: 127.0.0.1:2182(CONNECTED) 56] get /PathChildrenCache/PubSub/second
300
cZxid = 0xb00000131
ctime = Wed Jan 09 17:26:05 CST 2019
mZxid = 0xb00000131
mtime = Wed Jan 09 17:26:05 CST 2019
pZxid = 0xb00000131
cversion = 0
dataVersion = 0
aclVersion = 0
ephemeralOwner = 0x0
dataLength = 3
numChildren = 0
```

15.4.3 TreeCache

本节中使用15.4.1一节中的发布者。使用TreeCache创建监听者进行验证。

1. 创建订阅者

使用TreeCache订阅子节点变化，并设置监听的目录深度为2：

```
/**
 * @Author: zhouguanya
 * @Date: 2019/01/08
```

```java
 * @Description: TreeCache 订阅
 */
public class TreeCacheSubscriber {

    private ZookeeperClient zookeeperClient;

    private String name;
    public TreeCacheSubscriber(String name, ZookeeperClient zookeeperClient){
        this.name = name;
        this.zookeeperClient = zookeeperClient;
    }
    /**
     * 订阅
     */
    public void subscribe(String path) {
        // 创建 TreeCache 并前天最大深度为 2
        TreeCache treeCache = TreeCache.newBuilder(zookeeperClient.getCuratorFramework(), path)
                .setCacheData(true).setMaxDepth(2).build();
        treeCache.getListenable().addListener((client, event) -> {
            ChildData data = event.getData();
            System.out.printf("%s 监听到节点变更,节点路径=%s,节点值=%s%n", name,data.getPath(),new String(data.getData()));

        });
        try {
            treeCache.start();
        } catch (Exception e) {
            e.printStackTrace();
        }
    }
}
```

2. 单元测试

创建单元测试,其中含有一个发布者,两个订阅者,监听"/TreeCache/PubSub"节点下的 2 级目录,并验证监听"/TreeCache/PubSub"节点下的 3 级目录是否正常:

```java
/**
 * @Author: zhouguanya
 * @Date: 2019/01/08
 * @Description: TreeCacheSubscriber 发布订阅测试
 */
@RunWith(SpringJUnit4ClassRunner.class)
@ContextConfiguration("classpath:spring-zookeeper.xml")
```

```java
public class TreeCachePubSubTest {

    @Autowired
    private ZookeeperClient zookeeperClient;

    @Test
    public void test() throws InterruptedException {
        String basePath = "/TreeCache/PubSub";
        String firstPath = basePath + "/first";
        String secondPath = firstPath + "/second";
        String thirdPath = secondPath + "/third";
        Publisher publisher = new Publisher(zookeeperClient);
        //写入数据100
        publisher.publish(basePath, String.valueOf(100));
        TreeCacheSubscriber subscriber1 = new TreeCacheSubscriber("订阅者1", zookeeperClient);
        subscriber1.subscribe(basePath);
        TreeCacheSubscriber subscriber2 = new TreeCacheSubscriber("订阅者2", zookeeperClient);
        subscriber2.subscribe(basePath);
        //创建节点:/PathChildrenCache/PubSub/first
        publisher.publish(firstPath, String.valueOf(200));
        Thread.sleep(100);
        System.out.println("------------------------分割线----------------------");
        //创建节点:/PathChildrenCache/PubSub/first/second
        publisher.publish(secondPath, String.valueOf(300));
        Thread.sleep(100);
        System.out.println("------------------------分割线----------------------");
        //创建节点:/PathChildrenCache/PubSub/first/second/third
        publisher.publish(thirdPath, String.valueOf(400));
    }

}
```

执行单元测试，得到如下所示结果：

```
订阅者1监听到节点变更，节点路径=/TreeCache/PubSub，节点值=100
订阅者2监听到节点变更，节点路径=/TreeCache/PubSub，节点值=100
订阅者1监听到节点变更，节点路径=/TreeCache/PubSub/first，节点值=200
订阅者2监听到节点变更，节点路径=/TreeCache/PubSub/first，节点值=200
------------------------分割线----------------------
订阅者1监听到节点变更，节点路径=/TreeCache/PubSub/first/second，节点值=300
订阅者2监听到节点变更，节点路径=/TreeCache/PubSub/first/second，节点值=300
------------------------分割线----------------------
```

从以上单元测试的结果可以看到，"/TreeCache/PubSub" 节点及其 1 级和 2 级目录的修改都会被订阅者监听到。

使用 ZooKeeper 客户端连接到 ZooKeeper，查询子节点列表，发现 "/TreeCache/PubSub" 节点及其 1 级、2 级和 3 级子节点都创建成功。分别查询各个子节点的信息，如下所示：

```
[zk: 127.0.0.1:2182(CONNECTED) 85] get /TreeCache/PubSub
100
cZxid = 0xb000001b5
ctime = Wed Jan 09 20:12:23 CST 2019
mZxid = 0xb000001b5
mtime = Wed Jan 09 20:12:23 CST 2019
pZxid = 0xb000001b6
cversion = 1
dataVersion = 0
aclVersion = 0
ephemeralOwner = 0x0
dataLength = 3
numChildren = 1
[zk: 127.0.0.1:2182(CONNECTED) 86] get /TreeCache/PubSub/first
200
cZxid = 0xb000001b6
ctime = Wed Jan 09 20:12:23 CST 2019
mZxid = 0xb000001b6
mtime = Wed Jan 09 20:12:23 CST 2019
pZxid = 0xb000001b7
cversion = 1
dataVersion = 0
aclVersion = 0
ephemeralOwner = 0x0
dataLength = 3
numChildren = 1
[zk: 127.0.0.1:2182(CONNECTED) 87] get /TreeCache/PubSub/first/second
300
cZxid = 0xb000001b7
ctime = Wed Jan 09 20:12:23 CST 2019
mZxid = 0xb000001b7
mtime = Wed Jan 09 20:12:23 CST 2019
pZxid = 0xb000001b8
cversion = 1
dataVersion = 0
aclVersion = 0
ephemeralOwner = 0x0
dataLength = 3
numChildren = 1
[zk: 127.0.0.1:2182(CONNECTED) 88] get /TreeCache/PubSub/first/second/third
```

```
400
cZxid = 0xb000001b8
ctime = Wed Jan 09 20:12:23 CST 2019
mZxid = 0xb000001b8
mtime = Wed Jan 09 20:12:23 CST 2019
pZxid = 0xb000001b8
cversion = 0
dataVersion = 0
aclVersion = 0
ephemeralOwner = 0x0
dataLength = 3
numChildren = 0
```

从"/TreeCache/PubSub"节点及其各个子节点的信息可知，"/TreeCache/PubSub"节点的 3 级子节点"/TreeCache/PubSub/first/second/third"创建成功，但是 3 级子节点的变更并未被订阅者监听到，这是因为 TreeCache 中 maxDepth=2 控制的。

15.5　ZooKeeper 分布式锁

在单个 JVM 进程内，可以通过 Java 提供的 Lock 实现在多线程场景下对共享资源的加锁，以确保线程安全。但是到了分布式场景下，部署多个 JVM 进程（Java 应用）时，Java 提供的 Lock 将无法实现分布式场景下的共享资源的线程安全。使用 ZooKeeper 可以实现分布式锁。

下面描述 Zookeeper 实现分布式锁的算法流程。

（1）假设锁空间的根节点为"/lock"。多个客户端连接 Zookeeper，并在"/lock"节点下创建临时顺序的子节点。第一个客户端创建的子节点为"/lock/lock-0000000000"，第二个客户端创建的子节点为"/lock/lock-0000000001"，以此类推。

（2）客户端执行业务逻辑之前需要获取分布式锁。客户端获取"/lock"下的子节点列表，判断自己创建的子节点是否为当前子节点列表中序号最小的子节点（序号越小说明创建的时间越早，获取锁的时间也越早）。如果是则认为客户端获得锁，否则监听序号刚好在创建的节点之前一位（如 /lock/lock-0000000001 监听 /lock/lock-0000000000）的子节点删除消息。

（3）客户端执行业务逻辑代码。

（4）完成业务流程后，删除对应的子节点释放锁。

在步骤（1）中创建的临时节点的目的是为了能够保证在故障的情况下锁也能被释放，考虑如下场景：

假如客户端 A 当前创建的子节点为序号最小的节点，若获得锁之后客户端所在机器死机，客户端没有主动删除子节点。如果创建的是永久的节点，那么这个锁永远不会释放，将导致死锁；由于创建的是临时节点，客户端死机后，过一定时间若 Zookeeper 没有收到客户端的心跳包则判断会话失效，会将临时节点删除从而释放锁。

在步骤（2）中获取子节点列表与设置监听这两步操作的原子性问题。考虑如下场景：

客户端 A 对应子节点为"/lock/lock-0000000000",客户端 B 对应子节点为"/lock/lock-0000000001",客户端 B 获取子节点列表时发现对应的子节点不是序号最小的,于是尝试对客户端 A 对应的子节点监听。但是在客户端 B 设置监听器前客户端 A 完成业务流程删除了子节点"/lock/lock-0000000000",客户端 B 设置的监听器岂不是丢失了这个事件从而导致永远等待了?这个问题并不存在。因为 Zookeeper 提供的 API 中设置监听器的操作和读取子节点列表的操作是原子执行的,即在读子节点列表的同时设置监听器,保证不会丢失事件。

ZooKeeper 实现分布式锁的流程如图 15-7 所示。

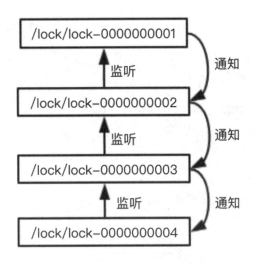

图 15-7 ZooKeeper 数据模型

下面通过案例说明 ZooKeeper 分布式锁的使用。

1. 创建共享资源

创建共享资源,模拟多个线程同时操作共享资源:

```
/**
 * @Author: zhouguanya
 * @Date: 2019/01/09
 * @Description: 模拟一个共享资源,只能单线程访问
 */
public class SharedResource {
    private final AtomicBoolean shareResourceInUse = new AtomicBoolean(false);

    public void use() throws InterruptedException {
        // 在真实环境中会在这里访问/维护一个共享的资源
        if (!shareResourceInUse.compareAndSet(false, true)) {
            throw new IllegalStateException("Needs to be used by one client at a time");
        }
        try {
```

```
            System.out.println("共享资源操作中");
            Thread.sleep(1);
        } finally {
            shareResourceInUse.set(false);
        }
    }
}
```

2. 创建客户端

实现一个分布式客户端,其中对分布式锁的加锁、业务逻辑处理和分布式锁的解锁操作。

```java
/**
 * @Author: zhouguanya
 * @Date: 2019/01/09
 * @Description: 请求锁,使用资源,释放锁
 */
public class DistributeClient {
    private final InterProcessMutex lock;
    private final SharedResource resource;
    private final String clientName;

    public DistributeClient(CuratorFramework client, String lockPath, SharedResource resource, String clientName) {
        this.resource = resource;
        this.clientName = clientName;
        lock = new InterProcessMutex(client, lockPath);
    }

    /**
     * 客户端执行方法
     */
    public void doWork(long time, TimeUnit unit) throws Exception {
        // 加锁,带有超时时间,超过超时时间未获取到锁抛出异常
        if (!lock.acquire(time, unit)) {
            throw new IllegalStateException(clientName + "加锁失败");
        }
        try {
            System.out.println(clientName + "加锁成功");
            // 应用程序的业务逻辑部分
            resource.use();
        } finally {
            System.out.println(clientName + "释放锁");
            // 释放锁
            try {
                lock.release();
```

```
            } catch (Exception e) {

            }
        }
    }
}
```

3. 单元测试

创建单元测试，其中使用 5 个线程并发对共享资源进行操作：

```
/**
 * @Author: zhouguanya
 * @Date: 2019/01/09
 * @Description: 单元测试
 */
@RunWith(SpringJUnit4ClassRunner.class)
@ContextConfiguration("classpath:spring-zookeeper.xml")
public class DistributeLockTest {
    private static final int QTY = 5;
    private static final String PATH = "/lock/lock-";
    @Autowired
    private ZookeeperClient zookeeperClient;
    @Test
    public void test() throws Exception {
        final SharedResource resource = new SharedResource();
        // 线程池
        ExecutorService service = Executors.newFixedThreadPool(QTY);
        try {
            // 5 个线程并发
            for (int i = 0; i < QTY; ++i) {
                final int index = i;
                Runnable task = () -> {
                    try {
                        // 每个线程都通过 DistributeClient 操作共享资源
                        final DistributeClient client = new DistributeClient
(zookeeperClient.getCuratorFramework(), PATH, resource, "Client " + index);
                        client.doWork(10, TimeUnit.SECONDS);
                    } catch (Throwable e) {
                        e.printStackTrace();
                    } finally {
                        CloseableUtils.closeQuietly(zookeeperClient.
getCuratorFramework());
                    }
                };
                service.submit(task);
```

```
            }
            service.shutdown();
            service.awaitTermination(10, TimeUnit.MINUTES);
        } finally {
            CloseableUtils.closeQuietly(zookeeperClient.
getCuratorFramework());
        }
    }
}
```

执行单元测试，运行结果如下：

```
Client 0 加锁成功
共享资源操作中
Client 0 释放锁
Client 3 加锁成功
共享资源操作中
Client 3 释放锁
java.lang.IllegalStateException: Client 2 加锁失败
    at com.test.zk.lock.DistributeClient.doWork(DistributeClient.java:30)
    at com.test.zk.lock.DistributeLockTest.lambda$test$0
(DistributeLockTest.java:41)
    at java.util.concurrent.Executors$RunnableAdapter.call
(Executors.java:511)
    at java.util.concurrent.FutureTask.run(FutureTask.java:266)
    at java.util.concurrent.ThreadPoolExecutor.runWorker
(ThreadPoolExecutor.java:114
    at java.lang.Thread.run(Thread.java:748)
java.lang.IllegalStateException: Client 4 加锁失败
    at com.test.zk.lock.DistributeClient.doWork(DistributeClient.java:30)
    at com.test.zk.lock.DistributeLockTest.lambda$test$0
(DistributeLockTest.java:41)
    at java.util.concurrent.Executors$RunnableAdapter.call
(Executors.java:511)
    at java.util.concurrent.FutureTask.run(FutureTask.java:266)
    at java.util.concurrent.ThreadPoolExecutor.runWorker
(ThreadPoolExecutor.java:114
    at java.lang.Thread.run(Thread.java:748)
java.lang.IllegalStateException: Client 1 加锁失败
    at com.test.zk.lock.DistributeClient.doWork(DistributeClient.java:30)
    at com.test.zk.lock.DistributeLockTest.lambda$test$0
(DistributeLockTest.java:41)
    at java.util.concurrent.Executors$RunnableAdapter.call
(Executors.java:511)
    at java.util.concurrent.FutureTask.run(FutureTask.java:266)
```

```
        at java.util.concurrent.ThreadPoolExecutor.runWorker
(ThreadPoolExecutor.java:114)
        at java.lang.Thread.run(Thread.java:748)
```

从测试结果可以看出，客户端 0 和客户端 3 加锁成功，其余客户端加锁失败。从以上案例可知，使用 ZooKeeper 分布式锁可以有效控制并发操作共享资源的问题。

15.6 小　　结

本章讲解了 ZooKeeper 集群的部署以及 Spring 与 ZooKeeper 的集成开发。ZooKeeper 在企业开发中是常用的分布式协调服务。熟练使用 ZooKeeper 对分布式环境中系统解耦和系统高可用性有很大帮助。

第 16 章

Spring 集成 Kafka

Kafka 是由 Apache 软件基金会开发的一个开源流处理平台,由 Scala 和 Java 编写。Kafka 是一种高吞吐量的分布式发布订阅消息系统。Kafka 具有高性能、持久化、多副本备份、横向扩展能力。生产者往队列里写消息,消费者从队列里取消息进行业务逻辑。一般在企业架构设计中起到解耦、削峰、异步处理的作用。

16.1 Kafka 集群安装

本书 Kafka 是在 Linux 环境下安装,如果读者使用的是 Windows 操作系统,可以在 Windows 上安装 Linux 虚拟机完成 Kafka 集群的安装。

Kafka 集群环境的搭建依赖于 ZooKeeper 集群。ZooKeeper 集群的安装步骤请参考 15.1 节相关内容。

1. 下载 Kafka

到 Kafka 官网 http://kafka.apache.org/ 下载需要的 Kafka 版本,本书使用的 Kafka 版本是 kafka_2.11-2.1.0.tgz。

2. 解压 Kafka

使用 tar 命令解压 Kafka 安装包:

```
tar -zxvf kafka_2.11-2.1.0.tgz
```

3. 创建配置文件

在解压后的 Kafka 目录下创建 servers 目录,在其中创建 3 个配置文件:

```
mkdir servers
cd servers
cp ../server.properties server1.properties
cp ../server.properties server2.properties
cp ../server.properties server3.properties
```

4. 修改配置文件

分别修改 server1.properties、server2.properties 和 server3.properties 文件。各文件的修改如下：

```
###################修改vim server1.properties###################
vim server1.properties
#当前机器在集群中的唯一标识
broker.id=1
#kafka实例broker监听端口
listeners=PLAINTEXT://127.0.0.1:9092
#消息存放的目录
log.dirs=../logs/server1
#topic的分区数
num.partitions=3
#Zookeeper的连接端口
zookeeper.connect=127.0.0.1:2182,127.0.0.1:2183,127.0.0.1:2184,127.0.0.1:2185,127.0.0.1:2186
#连接ZooKeeper超时时间
zookeeper.connection.timeout.ms=6000
###################修改vim server2.properties###################
vim server2.properties
#当前机器在集群中的唯一标识
broker.id=2
#Kafka实例broker监听端口
listeners=PLAINTEXT://127.0.0.1:9093
#消息存放的目录
log.dirs=../logs/server2
#topic的分区数
num.partitions=3
#Zookeeper的连接端口
zookeeper.connect=127.0.0.1:2182,127.0.0.1:2183,127.0.0.1:2184,127.0.0.1:2185,127.0.0.1:2186
#连接ZooKeeper超时时间
zookeeper.connection.timeout.ms=6000
###################修改vim server3.properties###################
vim server3.properties
#当前机器在集群中的唯一标识
broker.id=3
#Kafka实例broker监听端口
listeners=PLAINTEXT://127.0.0.1:9094
```

```
#消息存放的目录
log.dirs=../logs/server3
#topic 的分区数
num.partitions=3
#Zookeeper 的连接端口
zookeeper.connect=127.0.0.1:2182,127.0.0.1:2183,127.0.0.1:2184,127.0.0.1:
2185,127.0.0.1:2186
#连接 ZooKeeper 超时时间
zookeeper.connection.timeout.ms=6000
```

5. 启动 Kafka 集群

进入 bin 目录下执行启动脚本:

```
cd ../../bin/
#启动 Server1
./kafka-server-start.sh -daemon ../config/servers/server1.properties
./kafka-server-start.sh -daemon ../config/servers/server2.properties
./kafka-server-start.sh -daemon ../config/servers/server3.properties
```

6. 验证集群

Kafka 核心概念是 Topic、生产者和消费者。下面将创建一个名为 demo 的 Topic:

```
./kafka-topics.sh --create -zookeeper 127.0.0.1:2182,127.0.0.1:2183,
127.0.0.1:2184,127.0.0.1:2185,127.0.0.1:2186 --replication-factor 2
--partitions 3 --topic demo
```

创建的 Topic 详情如图 16-1 所示。

```
./kafka-topics.sh --describe --topic demo --zookeeper 127.0.0.1:2182
Topic:demo        PartitionCount:3        ReplicationFactor:2     Configs:
    Topic: demo         Partition: 0    Leader: 1       Replicas: 1,3   Isr: 1,3
    Topic: demo         Partition: 1    Leader: 2       Replicas: 2,1   Isr: 2,1
    Topic: demo         Partition: 2    Leader: 3       Replicas: 3,2   Isr: 3,2
```

图 16-1 查看 Topic 详情

各字段的含义如下:

- partiton: partion id。
- leader: 当前负责读写的 lead broker id。
- replicas: 当前 partition 的所有 replication broker list。
- isr: relicas 的子集,只包含处于活动状态的 broker。

使用 hello 这个 Topic 创建生产者:

```
./kafka-console-producer.sh --broker-list 127.0.0.1:9092 --topic demo
```

使用 hello 这个 Topic 创建消费者:

```
./kafka-console-consumer.sh --bootstrap-server 127.0.0.1:9092 --topic demo
--from-beginning
```

在生产者端输入"Hello Kafka",观察消费者端输出,发现消费者可以成功接收到生产者输入的消息"Hello Kafka"。

16.2 Kafka 总体架构

16.2.1 Kafka 的功能

1. 解耦

常见系统间的通信方式有 HTTP 或者 RPC 等,这种强依赖的系统互联方式的缺点是,如果其中某些部分出现故障,系统间将不能进行通信,系统功能将崩溃,如图 16-2 所示。

图 16-2　强依赖系统互联

消息系统在应用系统中间插入了一个隐含的、基于数据的接口层,各个应用系统的处理过程都要实现这一接口。这种架构下,即使系统中有部分功能故障,也不会造成整个功能不可用。如图 16-3 所示,此时引用系统 B 故障,并不会影响应用系统 A 与 Kafka 的通信。

图 16-3　Kafka 解耦系统

2. 冗余

在有些情况下,处理数据的过程会失败(如系统异常或者高并发场景下限流)。如果不对数据进行持久化,将造成丢失。Kafka 把数据进行持久化直到消息已经被完全处理,通过这一方式规避了数据丢失的风险。

3. 扩展性

因为 Kafka 对应用系统进行了解耦,所以增大消息入队和处理的频率是很容易的,只要另外增加处理过程即可,不需要改变代码,也不需要调节参数。

4. 峰值处理能力

在访问量剧增的情况下（如"秒杀"或者"双十一"），应用系统仍然需要保障高可用性，但是这样的突发流量并不常见。如果为了能处理这类峰值访问为标准来投入资源无疑是巨大的浪费。使用消息队列能够使关键组件顶住突发的访问压力，而不会因为突发的超负荷请求而导致应用系统完全崩溃。

5. 可恢复性

系统的一部分组件失效时，不会影响整个系统。Kafka 降低了应用系统之间的耦合度，所以即使一个处理消息的应用系统死机，已经加入队列中的消息仍然可以在应用系统恢复后被处理。

16.2.2 Kafka 的相关术语

Kafka 相关术语如表 16-1 所示。

表 16-1　Kafka 相关术语

字　段	描　述
Broker	Kafka 集群包含一个或多个服务器，集群中每个服务器被称为 Broker
Topic	每条发布到 Kafka 集群的消息都有一个类别，被称为 Topic
Partition	Partition 是物理上的概念，每个 Topic 分为一个或多个 Partition
Producer	负责发布消息到 Kafka Broker 的客户端
Consumer	消息消费者，从 Kafka Broker 读取消息的客户端
Consumer Group	每个 Consumer 属于一个特定的 Consumer Group

Kafka 拓扑结构如图 16-4 所示。

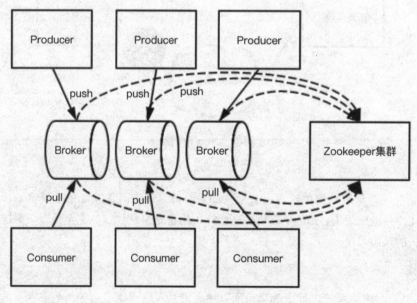

图 16-4　Kafka 拓扑结构

16.2.3 Topic 和 Partition

Topic 在逻辑上可以认为是一个队列,每条进入 Kafka 的消息都必须指定其 Topic。为了使 Kafka 的吞吐率可以线性提高,物理上把 Topic 分成一个或多个 Partition,每个 Partition 在物理上对应一个文件夹用来存储该 Partition 的所有消息和索引文件。

Kafka 是需要先写内存映射的文件,磁盘顺序读写的技术来提高性能的。Producer 生产的消息按照一定的分组策略被发送到 Broker 的 Partition 中的时候,这些消息如果在内存中放不下,就会放在 Partition 目录下的文件中,Partition 目录名是 Topic 的名称加上一个序号。在这个目录下有两类文件,一类是以 log 为后缀的文件,另一类是以 index 为后缀的文件。每个 log 文件和一个 index 文件相对应,这一对文件就是一个 Segment File,其中的 log 文件就是数据文件,里面存放的就是消息,而 index 文件是索引文件,记录了元数据信息,指向对应的数据文件中消息的物理偏移量。Segment File 示意图如图 16-5 所示。

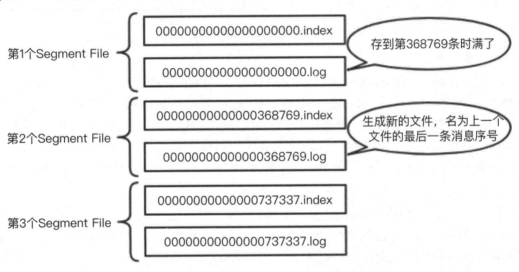

图 16-5　Segment File 命名规则

log 文件命名的规则是,Partition 全局的第一个 Segment 从 0(20 个 0)开始,后续的每一个文件的文件名是上一个文件最后一条消息的 offset 值。

index 文件里面存储的是 N 对 key-value,其中 key 是消息在 log 文件中的编号,比如 1,3,6,8……,表示第 1 条、第 3 条、第 6 条、第 8 条消息等。value 值表示该消息的物理偏移地址,如 0,497,1407 等。

索引文件存储大量元数据,数据文件存储大量消息,索引文件中元数据指向对应数据文件中 message 的物理偏移地址。log 文件和 index 文件是对应关系,如图 16-6 所示。

其中以索引文件中元数据 3 497 为例,在数据文件中表示第 3 条消息(在全局 partiton 表示第 368 772 条消息),以及该消息的物理偏移地址为 497。

例如读取 offset=368776 的 message,需要通过下面两个步骤查找。

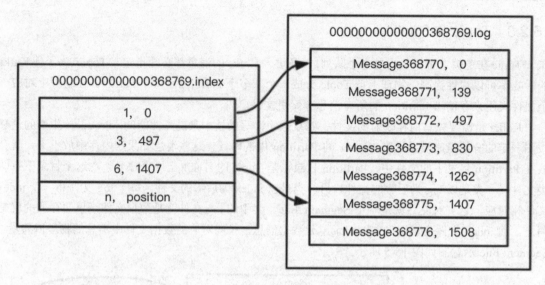

图 16-6　Segment File 命名规则

（1）查找 Segment File

其中 00000000000000000000.index 表示最开始的文件，起始偏移量（offset）为 0。第二个文件 00000000000000368769.index 的消息量起始偏移量为 368770=368769+1。同样，第三个文件 00000000000000737337.index 的起始偏移量为 737338=737337+1，其他后续文件依次类推，以起始偏移量命名并排序这些文件，只要根据 offset 对文件列表进行二分查找，就可以快速定位到具体文件。当 offset=368776 时定位到 00000000000000368769.index 文件和 00000000000000368769.log 文件。

（2）查找消息

通过步骤（1）中定位到的 Segment File 进行查找，当 offset=368776 时，一次定位到元数据 00000000000000368769.index 的物理位置和 00000000000000368769.log 的物理偏移地址。然后通过 00000000000000368769.log 顺序查找直到 offset=368776 为止。

16.2.4　消费组

消费组（Consumer Group）是 Kafka 提供的可扩展且具有容错性的消费者机制。组内可以有多个消费者或消费者实例（Consumer Instance），它们共享一个公共 ID，即 group ID。组内的所有消费者协调在一起来消费订阅主题（Subscribed Topics）的所有分区（Partition）。每个分区只能由同一个消费组内的一个消费者来消费消息，不同的 Consumer Group 可同时消费同一条消息。

消费组是 Kafka 用来实现一个 Topic 消息的广播（发给所有的 Consumer）和单播（发给某一个 Consumer）的手段。一个 Topic 可以对应多个 Consumer Group。如果需要实现广播，只要每个 Consumer 有一个独立的 Group 就可以了。要实现单播只要所有的 Consumer 在同一个 Group 里，消费组示意图如图 16-7 所示。

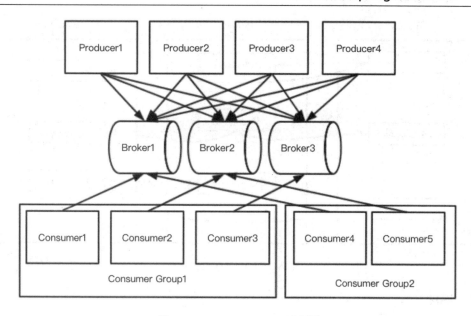

图 16-7　Consumer Group 示意图

16.2.5　Push 和 Pull

作为一个消息系统，Kafka 遵循了传统的方式，选择由 Producer 向 Broker Push 消息并由 Consumer 从 Broker Pull 消息。Push 模式和 Pull 模式各有优劣。

Push 模式很难适应消费速率不同的消费者，因为消息发送速率是由 Broker 决定的。Push 模式的目标是尽可能以最快速度传递消息，但是这样很容易造成 Consumer 来不及处理消息的后果，典型的表现就是拒绝服务以及网络拥塞。而 Pull 模式则可以根据 Consumer 的消费能力以适当的速率消费消息。

对于 Kafka 而言，Pull 模式更合适。Pull 模式可简化 Broker 的设计，Consumer 可自主控制消费消息的速率，同时 Consumer 可以自己控制消费方式——既可以批量消费也可逐条消费，同时还能选择不同的提交方式从而实现不同的传输语义。

16.2.6　复制原理

Kafka 中 Topic 的每个 Partition 有一个预写式的日志文件，虽然 Partition 可以继续细分为若干个 Segment File，但是对于上层应用来说可以将 Partition 看成最小的存储单元（一个含有多个 Segment 文件拼接的"巨型"文件），每个 Partition 都由不可变的消息组成，这些消息被连续的追加到 Partition 中。

为了提高消息的可靠性，Kafka 中每个 Topic 的 partition 有 N 个副本（replicas），其中 N（大于等于 1）是 Topic 的复制因子（replica fator）个数。Kafka 通过多副本机制实现故障自动转移。当 Kafka 集群中一个 Broker 失效情况下仍然保证服务可用。在 Kafka 中发生复制时确保 Partition 的日志能有序地写到其他节点上。当 N 个 replicas 中有一个为 Leader，其他都为 Follower，Leader 处理 Partition 的所有读写请求，与此同时，Follower 会被动定期地去复制 Leader 上的数据。Kafka 的复制原理如图 16-8 所示。

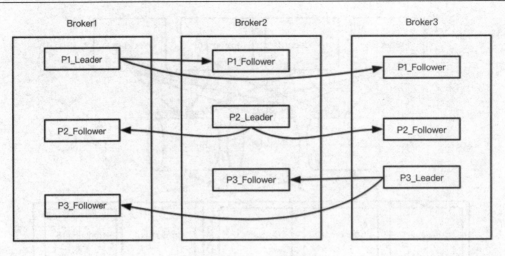

图 16-8　Kafka 的复制原理

16.2.7　ISR

如果 Leader 发生故障或挂掉，Kafka 将从同步副本列表中选举一个副本为 Leader，这个新 Leader 被选举出来并被接受客户端的消息成功写入。Leader 负责维护和跟踪 ISR（In-Sync Replicas 的缩写，表示副本同步队列，具体可参考下节）中所有 Follower 滞后的状态。当 Producer 发送一条消息到 Broker 后，Leader 写入消息并复制到所有 Follower 中。消息提交之后才被成功复制到所有的同步副本。消息复制延迟受最慢的 Follower 限制，对于那些"落后"太多或者失效的 Follower，Leader 将会把它从 ISR 中删除。

下面先介绍 LEO 和 HW 两个概念，如图 16-9 所示。

- LEO：LogEndOffset 的缩写，表示每个 Partition 的 log 文件中的最后一条消息的位置。
- HW 是 HighWatermark 的缩写，是指 Consumer 能够看到的 Partition 消息的位置。

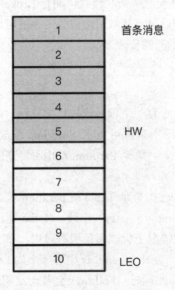

图 16-9　LEO 和 HW 示意图

Consumer 无法消费分区下 Leader 副本中（Follower）位移值大于分区 HW 的任何消息（即如图 16-9 中 6~10 部分消息）。这个涉及多副本的概念。

下面通过一个案例说明当 Producer 生产消息至 Broker 后，ISR、HW 和 LEO 的流转过程。

（1）初始状态下，HW 等于 LEO，Follower 将 Leader 中全部消息备份，此时有生产者向 Kafka 写入消息，如图 16-10 所示。

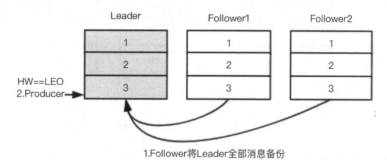

图 16-10　HW、LEO、Leader 和 Follower 初始状态

（2）生产者将消息写入 Leader 中，此时 Leader 将变更 LEO 的位置，Follower1 和 Follower2 将对 Leader 中的新增消息进行备份，如图 16-11 所示。

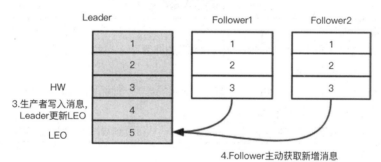

图 16-11　Leader 状态变更

（3）Follower1 完成 Leader 中所有消息的备份，Follower2 未完成备份，此时 HW 更新为 4，如图 16-12 所示。

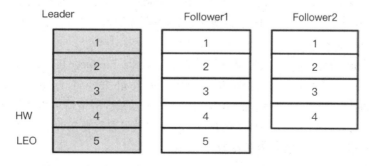

图 16-12　Follower1 完成对 Leader 的备份

(4)所有的 Follower 都将 Leader 中的消息备份完成,如图 16-13 所示。

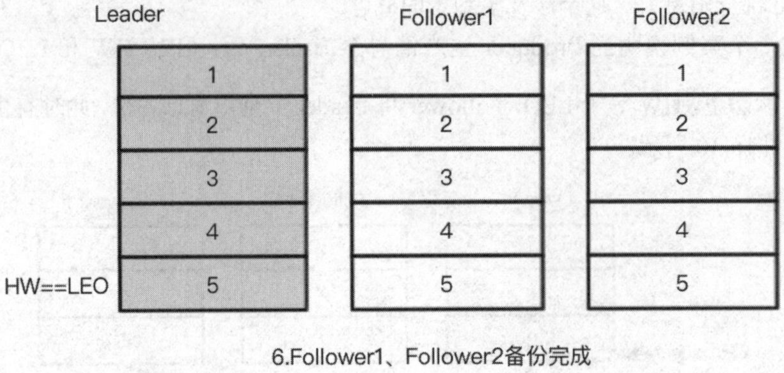

```
   Leader              Follower1            Follower2
┌─────────┐          ┌─────────┐          ┌─────────┐
│    1    │          │    1    │          │    1    │
├─────────┤          ├─────────┤          ├─────────┤
│    2    │          │    2    │          │    2    │
├─────────┤          ├─────────┤          ├─────────┤
│    3    │          │    3    │          │    3    │
├─────────┤          ├─────────┤          ├─────────┤
│    4    │          │    4    │          │    4    │
├─────────┤          ├─────────┤          ├─────────┤
│    5    │          │    5    │          │    5    │
└─────────┘          └─────────┘          └─────────┘
HW==LEO
```

6.Follower1、Follower2备份完成

图 16-13　所有 Follower 完成对 Leader 的备份

16.2.8　数据可靠性保障

当 Producer 向 Leader 发送数据时,可以通过 request.required.acks 参数来设置数据可靠性的级别,各个级别及其含义如表 16-2 所示。

表 16-2　Kafka 数据可靠性级别

参 数 值	描 述
1	默认级别。Producer 在 Leader 成功收到数据并得到确认后发送下一条消息。如果 Leader 死机了,则会丢失数据
0	Producer 无须等待来自 Broker 的确认而继续发送下一批消息。这种情况下数据传输效率最高,但是数据可靠性确是最低的
-1	Producer 需要等待 ISR 中的所有 Follower 都确认接收到数据后才算一次发送完成,可靠性最高

16.2.9　消息发送模式

Kafka 的发送模式由 Producer 端的配置参数 producer.type 来设置,这个参数指定了在后台线程中消息的发送方式是同步的还是异步的,默认是同步的方式,即 producer.type=sync。如果设置成异步的模式,即 producer.type=async,这种模式下 producer 以 batch 的形式 push(推送)数据,这样会极大地提高 Broker 的性能,同时也会增加丢失数据的风险。如果需要确保消息的可靠性,必须设置 producer.type=sync。

以 batch 方式推送数据可以极大地提高处理效率,Kafka Producer 可以将消息在内存中累计到一定数量后作为一个 batch 发送请求。batch 的数量大小可以通过 producer 的参数(batch.num.messages)控制。通过增加 batch 的大小,可以减少网络请求和磁盘 IO(写入和读出)的次数,但是随之而来的是数据丢失风险的增加。具体参数设置需要在效率和时效性方面做一个权衡。关于 Kafka 数据可靠性级别可参考表 16-3 所示。

表 16-3　Kafka 数据可靠性级别

参　数	描　述
queue.buffering.max.ms	默认值为 5000。启用异步模式时，Producer 缓存消息的时间。既默认每 5 秒发送一次缓存数据，这样可以极大提高 Broker 的吞吐量，同时也会造成时效性问题
queue.buffering.max.messages	默认值为 10000。启用异步模式时，Producer 缓存队列的最大消息数，如果超过这个值，Producer 会阻塞或者丢弃消息
queue.enqueue.timeout.ms	默认值为-1。达到上面参数时 Producer 阻塞的时间。如果设置为 0，Producer 缓存达到最大值后丢弃消息。如果设置为-1，Producer 不会丢弃消息，而是阻塞
batch.num.messages	默认值为 200。启用异步模式时，一个 batch 缓存的消息数量。达到这个值 Producer 将会发送消息

16.2.10　消息传输保障

Kafka 有以下 3 种可能的传输保障。

- at most once: 消息可能会丢，但绝不会重复传输。
- at least once: 消息绝不会丢，但可能会重复传输。
- exactly once: 每条消息肯定会被传输一次且仅传输一次。

当 Producer 向 Broker 发送消息时，一旦这条消息被 commit，由于副本机制（replication）的存在，消息就不会丢失。但是如果 Producer 发送数据给 Broker 后，遇到的网络问题而造成通信中断，那么 Producer 就无法判断该条消息是否已经提交（commit）。虽然 Kafka 无法确定网络故障期间发生了什么，但是 Producer 可以 retry（重试）多次，确保消息正确传输到 Broker 中，这样就实现了 at least once（至少一次的目标）。

Consumer 从 Broker 中读取消息后，可以选择何时进行 commit（提交）操作，commit 操作会在 Zookeeper 中保存该 Consumer 在该 Partition 下读取的消息的 offset。该 Consumer 下一次再读该 Partition 时会从下一条开始读取。如果此次 Consumer 没有提交（commit），下一次读取的开始位置会跟上一次 commit（提交）之后的开始位置相同。当然也可以将 Consumer 设置为自动提交，即 Consumer 一旦读取到数据立即自动提交。如果是自动提交，那 Kafka 确保了 exactly once（肯定且只有一次的目标）。但是如果 Producer 与 Broker 之间的某种原因导致消息重复发送，那么这里就是 at least once。

考虑这样一种情况，当 Consumer 读取消息之后先提交再处理消息，在这种模式下，如果 Consumer 在提交后还没来得及处理消息就下线了，下次 Consumer 重新开始工作后就无法读到上次已提交而未处理的消息，这就对应于 at most once（最多一次的目标）了。

读取消息先处理再提交（commit）。在这种模式下，如果处理完了消息在提交（commit）之前 Consumer 下线了，Consumer 重新开始工作时还会处理刚刚未提交（commit）的消息，实际上该消息已经被处理过了，这就对应于 at least once。

16.3 Spring 集成 Kafka 快速体验

本节使用 kafka-clients 结合 Spring 进行 Kafka 相关的开发，具体开发步骤如下。

1. 准备环境

在 pom.xml 文件中加入 Kafka 需要的依赖，Spring 集成 Kafka 需要以下 jar 包：

```xml
<dependency>
    <groupId>org.springframework.kafka</groupId>
    <artifactId>spring-kafka</artifactId>
    <version>2.2.3.RELEASE</version>
</dependency>
<dependency>
    <groupId>org.springframework.integration</groupId>
    <artifactId>spring-integration-kafka</artifactId>
    <version>3.1.0.RELEASE</version>
</dependency>
<dependency>
    <groupId>org.apache.kafka</groupId>
    <artifactId>kafka-clients</artifactId>
    <version>2.1.0</version>
</dependency>
```

2. 配置生产者和消费者

配置生产者和消费者的代码如下：

```xml
<!--kafka 生产者配置-->
<bean id="producerProperties" class="java.util.HashMap">
    <constructor-arg>
        <map>
            <!--kafka 集群-->
            <entry key="bootstrap.servers" value="127.0.0.1:9092,127.0.0.1:9093,127.0.0.1:9094"/>
            <entry key="retries" value="1"/>
            <entry key="batch.size" value="16384"/>
            <entry key="buffer.memory" value="10285760"/>
            <entry key="key.serializer" value="org.apache.kafka.common.serialization.StringSerializer"/>
            <entry key="value.serializer" value="org.apache.kafka.common.serialization.StringSerializer"/>
        </map>
    </constructor-arg>
</bean>
```

```xml
<!--配置 ProducerFactory-->
<bean id="producerFactory" class="org.springframework.kafka.core.DefaultKafkaProducerFactory">
    <constructor-arg>
        <ref bean="producerProperties"></ref>
    </constructor-arg>
</bean>
<!--KafkaTemplate 消息发送-->
<bean id="kafkaTemplate" class="org.springframework.kafka.core.KafkaTemplate">
    <constructor-arg ref="producerFactory"></constructor-arg>
    <constructor-arg name="autoFlush" value="true"></constructor-arg>
</bean>
<!--kafka 消费者配置-->
<bean id="consumerProperties" class="java.util.HashMap">
    <constructor-arg>
        <map>
            <!--kafka 集群-->
            <entry key="bootstrap.servers" value="127.0.0.1:9092,127.0.0.1:9093,127.0.0.1:9094"/>
            <entry key="group.id" value="kafka_consumer_group"/>
            <entry key="session.timeout.ms" value="30000"/>
            <entry key="key.deserializer" value="org.apache.kafka.common.serialization.StringDeserializer"/>
            <entry key="value.deserializer" value="org.apache.kafka.common.serialization.StringDeserializer"/>
        </map>
    </constructor-arg>
</bean>
<!--ConsumerFactory-->
<bean id="consumerFactory" class="org.springframework.kafka.core.DefaultKafkaConsumerFactory">
    <constructor-arg>
        <ref bean="consumerProperties"/>
    </constructor-arg>
</bean>
<!--实际执行消息消费的类(指向 kafka 的实际消费的类)-->
<bean id="messageConsumer" class="com.test.kafka.consumer.MessageConsumer"/>
<!--消费者容器配置信息-->
<bean id="containerProperties" class="org.springframework.kafka.listener.ContainerProperties">
    <constructor-arg value="spring-kafka-test"/>
    <property name="messageListener" ref="messageConsumer"/>
</bean>
```

```xml
<!-- 创建 messageListenerContainer bean，使用的时候，只需要注入这个 bean -->
<bean id="messageListenerContainer" class="org.springframework.
kafka.listener.KafkaMessageListenerContainer"
      init-method="doStart">
    <constructor-arg ref="consumerFactory"/>
    <constructor-arg ref="containerProperties"/>
</bean>
```

3. 创建生产者

创建生产者，使用 KafkaTemplate 发送消息至 Kafka 集群：

```java
/**
 * @Author: zhouguanya
 * @Date: 2019/01/14
 * @Description: 消息生产者
 */
@Component
public class MessageProducer {
    @Autowired
    private KafkaTemplate<String, String> kafkaTemplate;

    /**
     * 发送消息
     * @param topic 主题
     * @param value 消息
     */
    public void send(String topic, String value) {
        kafkaTemplate.send(topic, value);
    }
}
```

4. 创建消费者

实现 MessageListener 接口监听生产者发送到 Kafka 集群的消息：

```java
/**
 * @Author: zhouguanya
 * @Date: 2019/01/14
 * @Description: 消息消费者
 */
@Component
public class MessageConsumer implements MessageListener<String, String> {

    /**
     * 消费组监听消息
     */
    @Override
```

```
    public void onMessage(ConsumerRecord<String, String> data) {
        System.out.printf("监听到消息: topic=%s,value=%s%n", data.topic(),
data.value());
    }
}
```

5. 单元测试

创建单元测试,使用生产者发送 10 条消息到 Kafka 集群,消费者可以监听到这 10 条消息:

```
/**
 * @Author: zhouguanya
 * @Date: 2019/01/14
 * @Description: 单元测试
 */
@RunWith(SpringJUnit4ClassRunner.class)
@ContextConfiguration("classpath:spring-kafka.xml")
public class SpringKafkaTest {
    @Autowired
    private MessageProducer messageProducer;
    @Test
    public void sendMessage() throws InterruptedException {
        for (int i = 0; i < 10; i++) {
            messageProducer.send("spring-kafka-test", "Hello Kafka");
        }
        Thread.sleep(1000);
    }
}
```

运行单元测试,测试结果如下:

```
监听到消息: topic=spring-kafka-test,value=Hello Kafka
监听到消息: topic=spring-kafka-test,value=Hello Kafka
监听到消息: topic=spring-kafka-test,value=Hello Kafka
监听到消息: topic=spring-kafka-test,value=Hello Kafka
监听到消息: topic=spring-kafka-test,value=Hello Kafka
监听到消息: topic=spring-kafka-test,value=Hello Kafka
监听到消息: topic=spring-kafka-test,value=Hello Kafka
监听到消息: topic=spring-kafka-test,value=Hello Kafka
监听到消息: topic=spring-kafka-test,value=Hello Kafka
监听到消息: topic=spring-kafka-test,value=Hello Kafka
```

16.4 小　　结

Kafka 在企业开发中扮演着非常重要的角色,常见的使用场景如下。

（1）日志收集。企业开发中可以使用 Kafka 收集各种服务的日志，通过 Kafka 以统一接口服务的方式开放给各种消费者使用，如 Hadoop、Hbase 和 Solr 等。

（2）消息系统。将生产者和消费者解耦，常用于支付或订单场景系统解耦。

（3）用户行为跟踪。Kafka 可用于记录 Web 用户或者 App 用户的各种行为，如网页浏览、商品检索和"点击"等活动。这些活动信息被各个服务器发布到 Kafka 对应的 Topic 中，然后订阅者通过订阅这些 Topic 来做实时的监控分析，或者装载到 Hadoop 或数据仓库中做离线分析和挖掘。

（4）Kafka 也经常用来记录运维监控数据。包括收集各种分布式应用的数据，生产各种操作的集中反馈，比如报警和报告。

（5）流式处理。比如 Spark Streaming 和 Storm 等。

第 17 章

Spring 集成 Mycat

Mycat 是一个开源的、面向企业应用开发的数据库中间件产品,在企业开发中常常使用 Mycat 作为分库分表的组件。

17.1 Mycat 分库分表

随着时间和业务的发展,数据库中的数据量增长是不可控的,库和表中的数据会越来越大,随之带来的是更高的磁盘、IO、系统开销,甚至性能上的瓶颈。当数据库单表达到千万级别后,SQL 性能会开始下降,如果不进行优化,SQL 性能会继续下降。

一台服务的资源终究是有限的,因此需要对数据库和表进行拆分,从而更好地提供数据服务,提升 SQL 性能。

17.1.1 分库

分库的含义是根据业务需要,将原库拆分成多个库,通过降低单库大小来提高单库的性能。常见的分库方式有两种——垂直分库和水平分库,分库架构如图 17-1 所示。

(1)垂直分库。垂直分库根据业务进行划分,将同一类业务相关的数据表划分在同一个库中。如将原库中有关商品的数据表划分为一个数据库,将原库中有关订单相关的数据表划分为一个数据库。

(2)水平分库。水平分库是按照一定的规则对数据库进行划分的。每个数据库中各个表结构相同,数据存储在不同的数据库中,如根据年份划分不同的数据库。

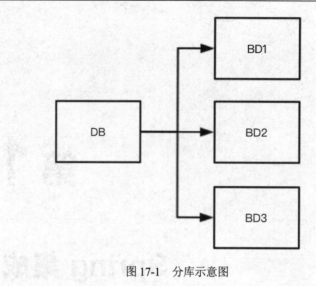

图 17-1　分库示意图

17.1.2　分表

分表的含义是根据业务需要，将大表拆分成多个子表，通过降低单表的大小来提高单表的性能。常见的分表方式有两种——垂直分表和水平分表，分库架构如图 17-2 所示。

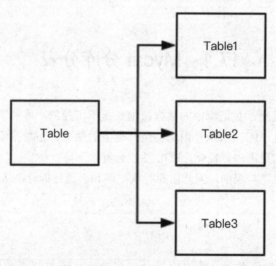

图 17-2　分表示意图

1. 垂直分表

垂直分表就是将一个大表根据业务功能拆分成多个分表，例如将原表可根据业务分成基本信息表和详细信息表等。

2. 水平分表

水平分表是按照一定的规则对数据表进行划分。每个数据表结构相同，数据存储在多个分表中。

17.2 Mycat 分库分表实战

本节讲解 Mycat 的安装和 Mycat 结合 MySQL 进行分库分表部署。

1. 下载 Mycat

可以在 Mycat 的官网 http://www.mycat.io/下载合适的 Mycat 版本，本书使用的 Mycat 版本是 Mycat-1.6.5。

2. 解压 Mycat

执行 tar 命令解压 Mycat 安装包：

```
tar -zxvf Mycat-server-1.6.5-release-20180122220033-linux.tar.gz
```

3. 配置 server.xml

进入 Mycat 解压目录。进入 conf 目录，对配置文件进行修改。

在 server.xml 中配置 Mycat 服务的端口和 Mycat 服务的用户：

```xml
<!-- Mycat 服务端口 8066 -->
<property name="serverPort">8066</property>
<!-- root 用户 -->
<user name="root" defaultAccount="true">
    <property name="password">123456</property>
    <property name="schemas">TESTDB</property>
</user>


<!-- 只读用户 -->
<user name="test">
    <property name="password">test</property>
        <property name="schemas">TESTDB</property>
    <property name="readOnly">true</property>
</user>
```

4. 配置 schema.xml

在 schema.xml 中配置逻辑库 TESTDB，该库中包含 customer 表、item 表和 customer_order 表，其中 customer 表是不使用分库也不使用分表的。item 表使用 Mycat 进行分库操作，分库规则是 mod-long。customer_order 表使用 Mycat 进行分表操作，分表规则是 mod-long，customer_order 分表为 customer_order1、customer_order2 和 customer_order3。除了配置逻辑库以外，还要配置 MySQL 的连接：

```xml
<?xml version="1.0"?>
<!DOCTYPE mycat:schema SYSTEM "schema.dtd">
<mycat:schema xmlns:mycat="http://io.mycat/">
```

```xml
        <!-- 定义一个 MyCat 的 schema，逻辑数据库名称 TestDB -->
        <!-- checkSQLschema: 描述的是当前的连接是否需要检测数据库的模式 -->
        <!-- sqlMaxLimit: 表示返回的最大的数据量的行数 -->
        <!-- dataNode: 该操作使用的数据节点的逻辑名称 -->
        <schema name="TESTDB" checkSQLschema="false" sqlMaxLimit="100">
            <!-- customer 客户表在 dn1 中不使用分库分表 -->
            <table name="customer" dataNode="dn1" />

            <!-- item 商品表在 dn2,dn3 上-->
            <table name="item" primaryKey="ID" dataNode="dn1,dn2,dn3"
rule="mod-long"/>

            <!-- order 订单表在 dn3 上 id 做分片-->
            <table name="customer_order" primaryKey="ID"
subTables="customer_order$1-3" dataNode="dn3" rule="mod-long" />

        </schema>
        <!-- 数据节点 -->
        <dataNode name="dn1" dataHost="localhost" database="mycat01" />
        <dataNode name="dn2" dataHost="localhost" database="mycat02" />
        <dataNode name="dn3" dataHost="localhost" database="mycat03" />
        <!-- 数据主机 -->
        <dataHost name="localhost" maxCon="1000" minCon="10" balance="0"
writeType="0" dbType="mysql" dbDriver="native" switchType="1"
slaveThreshold="100">
            <heartbeat>select user()</heartbeat>
            <!-- can have multi write hosts -->
            <writeHost host="hostM1" url="localhost:3306" user="root"
password="123456">
                <!-- can have multi read hosts -->
                <readHost host="hostS2" url="localhost:3306" user="root"
password="123456" />
            </writeHost>
        </dataHost>
    </mycat:schema>
```

5. 配置 rule.xml

本书使用的 mod-long 规则是根据 id 对 3 取模进行数据分片。mod-long 规则的实现如下：

```xml
    <tableRule name="mod-long">
        <rule>
            <columns>id</columns>
            <algorithm>mod-long</algorithm>
        </rule>
    </tableRule>
```

```xml
<function name="mod-long" class="io.mycat.route.function.PartitionByMod">
    <!-- how many data nodes -->
    <property name="count">3</property>
</function>
```

除了 mod-long 规则外，还有很多规则。如按照月份对数据进行分片，代码如下：

```xml
<tableRule name="sharding-by-month">
    <rule>
        <columns>create_time</columns>
        <algorithm>partbymonth</algorithm>
    </rule>
</tableRule>
<function name="partbymonth"
    class="io.mycat.route.function.PartitionByMonth">
    <property name="dateFormat">yyyy-MM-dd</property>
    <property name="sBeginDate">2015-01-01</property>
</function>
```

除了 Mycat 已定义好的分片规则，用户也可以自定义合适的分片规则。

6. 验证 Mycat 服务

进入 Mycat 解压目录下的 bin 目录，使用启动脚本启动 Mycat 服务：

```
./mycat start
```

查看 Mycat 服务，代码如下：

```
ps -ef | grep mycat | grep -v grep
```

7. 创建数据库和数据表

准备分库分表需要用到的创建脚本：

```sql
#####################不分库不分表############################
DROP DATABASE IF EXISTS mycat01;
CREATE DATABASE mycat01;
USE mycat01;
 CREATE TABLE customer (
  id INT NOT NULL AUTO_INCREMENT COMMENT '客户id',
  name VARCHAR(20) DEFAULT '' COMMENT '客户姓名',
  phone VARCHAR(11) DEFAULT '' COMMENT '客户手机号',
  adddate TIMESTAMP NOT NULL DEFAULT CURRENT_TIMESTAMP COMMENT '添加时间',
  updatedate TIMESTAMP NOT NULL DEFAULT CURRENT_TIMESTAMP ON UPDATE CURRENT_TIMESTAMP COMMENT '修改时间',
  PRIMARY KEY (`id`)
)ENGINE=InnoDB DEFAULT CHARSET=utf8;

 CREATE TABLE item (
```

```sql
    id INT NOT NULL AUTO_INCREMENT,
    value INT NOT NULL default 0,
    adddate TIMESTAMP NOT NULL DEFAULT CURRENT_TIMESTAMP COMMENT '添加时间',
    updatedate TIMESTAMP NOT NULL DEFAULT CURRENT_TIMESTAMP ON UPDATE CURRENT_TIMESTAMP COMMENT '修改时间',
    PRIMARY KEY (id)
)ENGINE=InnoDB DEFAULT CHARSET=utf8;

######################分库不分表#############################
DROP DATABASE IF EXISTS mycat02;
CREATE DATABASE mycat02;
USE mycat02;
 CREATE TABLE item (
    id INT NOT NULL AUTO_INCREMENT,
    value INT NOT NULL default 0,
    adddate TIMESTAMP NOT NULL DEFAULT CURRENT_TIMESTAMP COMMENT '添加时间',
    updatedate TIMESTAMP NOT NULL DEFAULT CURRENT_TIMESTAMP ON UPDATE CURRENT_TIMESTAMP COMMENT '修改时间',
    PRIMARY KEY (id)
)ENGINE=InnoDB DEFAULT CHARSET=utf8;


######################分库分表#############################
DROP DATABASE IF EXISTS mycat03;
CREATE DATABASE mycat03;
USE mycat03;
CREATE TABLE item (
    id INT NOT NULL AUTO_INCREMENT,
    value INT NOT NULL default 0,
    adddate TIMESTAMP NOT NULL DEFAULT CURRENT_TIMESTAMP COMMENT '添加时间',
    updatedate TIMESTAMP NOT NULL DEFAULT CURRENT_TIMESTAMP ON UPDATE CURRENT_TIMESTAMP COMMENT '修改时间',
    PRIMARY KEY (id)
)ENGINE=InnoDB DEFAULT CHARSET=utf8;

CREATE TABLE customer_order1 (
    id INT NOT NULL AUTO_INCREMENT,
    amount INT NOT NULL default 0,
    adddate TIMESTAMP NOT NULL DEFAULT CURRENT_TIMESTAMP COMMENT '添加时间',
    updatedate TIMESTAMP NOT NULL DEFAULT CURRENT_TIMESTAMP ON UPDATE CURRENT_TIMESTAMP COMMENT '修改时间',
    PRIMARY KEY (id)
)ENGINE=InnoDB DEFAULT CHARSET=utf8;
```

```sql
CREATE TABLE customer_order2 (
    id INT NOT NULL AUTO_INCREMENT,
    amount INT NOT NULL default 0,
    adddate TIMESTAMP NOT NULL DEFAULT CURRENT_TIMESTAMP COMMENT '添加时间',
    updatedate TIMESTAMP NOT NULL DEFAULT CURRENT_TIMESTAMP ON UPDATE CURRENT_TIMESTAMP COMMENT '修改时间',
    PRIMARY KEY (id)
)ENGINE=InnoDB DEFAULT CHARSET=utf8;

CREATE TABLE customer_order3 (
    id INT NOT NULL AUTO_INCREMENT,
    amount INT NOT NULL default 0,
    adddate TIMESTAMP NOT NULL DEFAULT CURRENT_TIMESTAMP COMMENT '添加时间',
    updatedate TIMESTAMP NOT NULL DEFAULT CURRENT_TIMESTAMP ON UPDATE CURRENT_TIMESTAMP COMMENT '修改时间',
    PRIMARY KEY (id)
)ENGINE=InnoDB DEFAULT CHARSET=utf8;
```

8. 登录 Mycat

使用以下命令查询登录 Mycat：

```
#登录 Mycat
mysql -uroot -p123456 -h127.0.0.1 -P8066 -DTESTDB;
```

查看创建的逻辑数据库 TESTDB：

```
show databases;
```

使用 TESTDB 数据库并查询 TESTDB 数据库中的数据表：

```
mysql> use TESTDB;
Database changed
mysql> show tables;
+------------------+
| Tables in TESTDB |
+------------------+
| customer         |
| customer_order   |
| item             |
+------------------+
3 rows in set (0.00 sec)
```

通过以上结果可以看出，从 Mycat 角度看，其实只维护了 TESTDB 这一个逻辑数据库，Mycat 屏蔽了分库分表的细节。

17.3　Spring+MyBatis+Mycat 快速体验

本节讲解使用 Spring 集成 MyBatis 作为持久化框架以及集成 Mycat 进行分库分表的实际应用。

1. 创建实体类

创建 Customer 用户类与 17.2 节 customer 表对应：

```java
/**
 * @Author: zhouguanya
 * @Date: 2019/01/04
 * @Description: 客户实体
 */
public class Customer {
    private int id;
    private String name;
    private String phone;
    private Date addDate;
    private Date updateDate;

    public int getId() {
        return id;
    }

    public void setId(int id) {
        this.id = id;
    }

    public String getName() {
        return name;
    }

    public void setName(String name) {
        this.name = name;
    }

    public String getPhone() {
        return phone;
    }

    public void setPhone(String phone) {
        this.phone = phone;
    }
```

```java
    public Date getAddDate() {
        return addDate;
    }

    public void setAddDate(Date addDate) {
        this.addDate = addDate;
    }

    public Date getUpdateDate() {
        return updateDate;
    }

    public void setUpdateDate(Date updateDate) {
        this.updateDate = updateDate;
    }
}
```

Item 商品类与 17.2 节中的 item 表相对应：

```java
/**
 * @Author: zhouguanya
 * @Date: 2019/01/04
 * @Description: 商品实体
 */
public class Item {
    private int id;
    private int value;
    private Date addDate;
    private Date updateDate;

    public int getId() {
        return id;
    }

    public void setId(int id) {
        this.id = id;
    }

    public int getValue() {
        return value;
    }

    public void setValue(int value) {
        this.value = value;
    }
```

```java
    public Date getAddDate() {
        return addDate;
    }

    public void setAddDate(Date addDate) {
        this.addDate = addDate;
    }

    public Date getUpdateDate() {
        return updateDate;
    }

    public void setUpdateDate(Date updateDate) {
        this.updateDate = updateDate;
    }
}
```

CustomerOrder 用户订单类，与 17.2 节 customer_order 表对应：

```java
/**
 * @Author: zhouguanya
 * @Date: 2019/01/04
 * @Description: 客户订单实体
 */
public class CustomerOrder {
    private int id;
    private int amount;
    private Date addDate;
    private Date updateDate;

    public int getId() {
        return id;
    }

    public void setId(int id) {
        this.id = id;
    }

    public int getAmount() {
        return amount;
    }

    public void setAmount(int amount) {
        this.amount = amount;
    }

    public Date getAddDate() {
        return addDate;
```

```
    }
    public void setAddDate(Date addDate) {
        this.addDate = addDate;
    }
    public Date getUpdateDate() {
        return updateDate;
    }
    public void setUpdateDate(Date updateDate) {
        this.updateDate = updateDate;
    }
}
```

2. 创建 DAO

创建 CustomerDao，用于 Customer 对象的数据库操作：

```
/**
 * @Author: zhouguanya
 * @Date: 2019/01/04
 * @Description:
 */
public interface CustomerDao {
    int save(Customer customer);
    Customer query(int id);
}
```

创建 CustomerOrderDao，用于 CustomerOrder 对象的数据库操作：

```
/**
 * @Author: zhouguanya
 * @Date: 2019/01/04
 * @Description:
 */
public interface CustomerDao {
    int save(Customer customer);
    Customer query(int id);
}
```

创建 ItemDao，用于 Item 对象的数据库操作：

```
/**
 * @Author: zhouguanya
 * @Date: 2019/01/04
 * @Description:
 */
public interface ItemDao {
    int save(Item customer);
```

```
    Item query(int id);
}
```

3. 创建 Mapper

创建 mybatis-customer-mapper.xml 文件，其中包含 customer 表的保存和查询操作：

```xml
<mapper namespace="com.test.mycat.dao.CustomerDao">
    <resultMap id="BaseResultMap" type="com.test.mycat.model.Customer">
        <id column="id" jdbcType="INTEGER" property="id" />
        <result column="name" jdbcType="VARCHAR" property="name" />
        <result column="phone" jdbcType="VARCHAR" property="phone" />
        <result column="adddate" jdbcType="TIMESTAMP" property="addDate" />
        <result column="updatedate" jdbcType="TIMESTAMP" property="updateDate" />
    </resultMap>
    <select id="query" parameterType="java.lang.Integer" resultMap="BaseResultMap">
        select
        *
        from customer
        where id = #{id,jdbcType=BIGINT}
    </select>

    <insert id="save" parameterType="com.test.mycat.model.Customer">
        insert into customer (id,name, phone)
        values (#{id,jdbcType=INTEGER}, #{name,jdbcType=VARCHAR}, #{phone,jdbcType=VARCHAR})
    </insert>
</mapper>
```

创建 mybatis-item-mapper.xml 文件，其中包含 item 表的保存和查询操作：

```xml
<mapper namespace="com.test.mycat.dao.ItemDao">
    <resultMap id="BaseResultMap" type="com.test.mycat.model.Item">
        <id column="id" jdbcType="INTEGER" property="id" />
        <result column="value" jdbcType="INTEGER" property="value" />
        <result column="adddate" jdbcType="TIMESTAMP" property="addDate" />
        <result column="updatedate" jdbcType="TIMESTAMP" property="updateDate" />
    </resultMap>
    <select id="query" parameterType="java.lang.Integer" resultMap="BaseResultMap">
        select
        *
        from item
        where id = #{id,jdbcType=BIGINT}
    </select>
```

```xml
<insert id="save" parameterType="com.test.mycat.model.Item">
    insert into item (id,value)
    values (#{id,jdbcType=INTEGER}, #{value,jdbcType=INTEGER})
</insert>
</mapper>
```

创建 mybatis-customer_order-mapper.xml 文件，其中包含 customer_order 表的保存和查询操作：

```xml
<mapper namespace="com.test.mycat.dao.CustomerOrderDao">
    <resultMap id="BaseResultMap" type="com.test.mycat.model.CustomerOrder">
        <id column="id" jdbcType="INTEGER" property="id" />
        <result column="amount" jdbcType="INTEGER" property="amount" />
        <result column="adddate" jdbcType="TIMESTAMP" property="addDate" />
        <result column="updatedate" jdbcType="TIMESTAMP" property="updateDate" />
    </resultMap>
    <select id="query" parameterType="java.lang.Integer" resultMap="BaseResultMap">
        select
        *
        from customer_order
        where id = #{id,jdbcType=BIGINT}
    </select>

    <insert id="save" parameterType="com.test.mycat.model.CustomerOrder">
        insert into customer_order (id,amount)
        values (#{id,jdbcType=INTEGER}, #{amount,jdbcType=INTEGER})
    </insert>
</mapper>
```

4. 创建 JDBC 配置文件

创建 jdbc.properties 文件，其中包含数据库驱动，数据库用户名和密码以及 Mycat 连接——在原 JDBC 连接的基础上修改为 Mycat 的 host、Mycat 端口和 Mycat 逻辑库。

```
driver=com.mysql.jdbc.Driver
#Mycat 连接
url=jdbc:mysql://127.0.0.1:8066/TESTDB
#MySQL 用户名
username=root
#MySQL 密码
password=123456
```

5. 在 Spring 中集成 Mycat

创建 spring-mycat.xml 文件，包含数据源和 MyBatis 相关配置：

```xml
<!-- 引入jdbc配置文件 -->
<bean id="propertyConfigurer" class="org.springframework.beans.factory.config.PropertyPlaceholderConfigurer">
    <property name="location" value="classpath:jdbc.properties" />
</bean>

<bean id="dataSource" class="org.springframework.jdbc.datasource.DriverManagerDataSource">
    <property name="driverClassName" value="${driver}" />
    <property name="url" value="${url}" />
    <property name="username" value="${username}" />
    <property name="password" value="${password}" />
</bean>

<!-- spring和MyBatis整合-->
<bean id="sqlSessionFactory" class="org.mybatis.spring.SqlSessionFactoryBean">
    <property name="dataSource" ref="dataSource" />
    <!-- 自动扫描mapping.xml文件，**表示迭代查找 -->
    <property name="mapperLocations">
        <array>
            <value>classpath:mapper/*.xml</value>
        </array>
    </property>
</bean>

<!-- DAO接口，Spring会自动查找其下的类，包下的类需要使用@MapperScan注解,否则容器注入会失败 -->
<bean class="org.mybatis.spring.mapper.MapperScannerConfigurer">
    <property name="basePackage" value="com.test.mycat.dao" />
    <property name="sqlSessionFactoryBeanName" value="sqlSessionFactory" />
</bean>
```

6. 验证不分库分表

创建 CustomerDaoTest 类用于测试 customer 表的保存和查询测试：

```java
/**
 * @Author: zhouguanya
 * @Date: 2019/01/04
 * @Description: CustomerDao 测试类
 */
@RunWith(SpringJUnit4ClassRunner.class)
@ContextConfiguration("classpath:spring-mycat.xml")
public class CustomerDaoTest {
    @Autowired
```

```java
    private CustomerDao customerDao;
    @Test
    public void testSave() {
        Customer customer_1 = new Customer();
        customer_1.setId(1);
        customer_1.setName("Michael");
        customer_1.setPhone("3344625292");
        customerDao.save(customer_1);
        Customer customer_2 = new Customer();
        customer_2.setId(2);
        customer_2.setName("Tom");
        customer_2.setPhone("3190976240");
        customerDao.save(customer_2);
    }
    @Test
    public void testQuery() {
        System.out.println("用户1=" + JSON.toJSONString (customerDao.query(1)));
        System.out.println("用户2=" + JSON.toJSONString (customerDao.query(2)));

    }
}
```

执行 testSave() 方法，然后执行 testQuery() 方法，发现此时数据写入成功，并且可以通过 Mycat 服务查询到写入的数据。证明在 Mycat 集成环境下，对不使用分表也不使用分库的数据可以得到很好地支持。

7. 验证分库功能

创建 ItemDaoTest 测试类，用于测试有关 item 商品表分库的测试：

```java
/**
 * @Author: zhouguanya
 * @Date: 2019/01/04
 * @Description: ItemDao 分库测试
 */
@RunWith(SpringJUnit4ClassRunner.class)
@ContextConfiguration("classpath:spring-mycat.xml")
public class ItemDaoTest {
    @Autowired
    private ItemDao itemDao;
    @Test
    public void testSave() {
        Item item_1 = new Item();
        item_1.setId(1);
```

```
        item_1.setValue(100);
        itemDao.save(item_1);
        Item item_2 = new Item();
        item_2.setId(2);
        item_2.setValue(200);
        itemDao.save(item_2);
        Item item_3 = new Item();
        item_3.setId(3);
        item_3.setValue(300);
        itemDao.save(item_3);
    }

    @Test
    public void testQuery() {
        System.out.println("商品1=" + JSON.toJSONString(itemDao.query(1)));
        System.out.println("商品2=" + JSON.toJSONString(itemDao.query(2)));
        System.out.println("商品3=" + JSON.toJSONString(itemDao.query(3)));
    }
}
```

执行 testSave() 方法，然后执行 testQuery() 方法，得到如下输出：

```
商品1={"addDate":1547772568000,"id":1,"updateDate":1547772568000,"value":100}
商品2={"addDate":1547772568000,"id":2,"updateDate":1547772568000,"value":200}
商品3={"addDate":1547772568000,"id":3,"updateDate":1547772568000,"value":300}
```

登录 MySQL 客户端，查看 item 表分库情况。

查看 mycat01 库 item 表的数据，如图 17-3 所示。

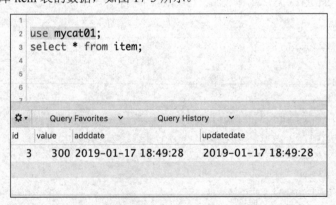

图 17-3 mycat01 库 item 表数据存储

查看 mycat02 库 item 表的数据，如图 17-4 所示。

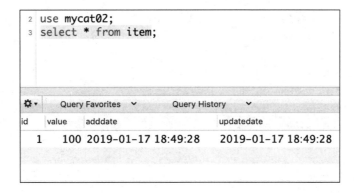

图 17-4　mycat02 库 item 表数据存储

查看 mycat03 库 item 表的数据，如图 17-5 所示。

图 17-5　mycat03 库 item 表数据存储

从以上测试结果可知，在 Mycat 集成环境下，对 item 表分库可以很好地支持。

8. 验证分表功能

创建 CustomerOrderDaoTest 类验证 customer_order 表的分表功能：

```java
/**
 * @Author: zhouguanya
 * @Date: 2019/01/04
 * @Description: CustomerOrderDao 分表测试
 */
@RunWith(SpringJUnit4ClassRunner.class)
@ContextConfiguration("classpath:spring-mycat.xml")
public class CustomerOrderDaoTest {
    @Autowired
    private CustomerOrderDao customerOrderDao;
    @Test
    public void testSave() {
        CustomerOrder customerOrder_1 = new CustomerOrder();
        customerOrder_1.setId(1);
        customerOrder_1.setAmount(100);
```

```
        customerOrderDao.save(customerOrder_1);
        CustomerOrder customerOrder_2 = new CustomerOrder();
        customerOrder_2.setId(2);
        customerOrder_2.setAmount(200);
        customerOrderDao.save(customerOrder_2);
        CustomerOrder customerOrder_3 = new CustomerOrder();
        customerOrder_3.setId(3);
        customerOrder_3.setAmount(300);
        customerOrderDao.save(customerOrder_3);
    }

    @Test
    public void testQuery() {
        System.out.println("订单1=" + JSON.toJSONString (customerOrderDao.query(1)));
        System.out.println("订单2=" + JSON.toJSONString (customerOrderDao.query(2)));
        System.out.println("订单3=" + JSON.toJSONString (customerOrderDao.query(3)));
    }
}
```

分别执行 testSave() 方法和 testQuery() 方法,得到如下输出:

```
订单1={"addDate":1547773668000,"amount":100,"id":1, "updateDate":1547773668000}
订单2={"addDate":1547773668000,"amount":200,"id":2, "updateDate":1547773668000}
订单3={"addDate":1547773668000,"amount":300,"id":3, "updateDate":1547773668000}
```

登录 MySQL 客户端验证 customer_order 分表情况。

查看 customer_order1 表的数据,如图 17-6 所示。

图 17-6 customer_order1 表数据存储

查看 customer_order2 表的数据,如图 17-7 所示。

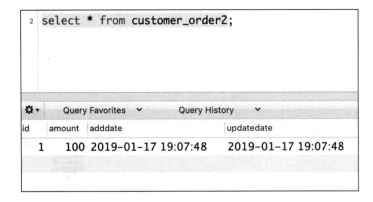

图 17-7 customer_order2 表数据存储

查看 customer_order3 表的数据，如图 17-8 所示。

图 17-8 customer_order3 表数据存储

从以上测试结果可知，在 Mycat 集成环境下，对 customer_order 表进行分表可以得到很好的支持。

17.4 小　　结

Mycat 隐藏了分库分表的细节，从开发人员的角度看，在无须知道具体哪个库的哪张分表进行操作的情况下，应用程序即可对数据库进行操作。对于一些老项目，引入 Mycat 进行分库分表，无须修改业务代码，只需要修改 JDBC 连接即可实现项目的升级。

第 18 章

Spring 集成 Sharding-JDBC

Sharding-JDBC 是开源的数据库中间件。Sharding-JDBC 定位为轻量级数据库驱动,由客户端直连数据库,以 jar 包形式提供服务,没有使用中间层,无须额外部署,无须其他依赖。Sharding-JDBC 可以实现旧代码迁移零成本的目标。Sharding-JDBC 与 MyCat 不同,MyCat 本质上是一种数据库代理。

18.1 Spring 集成 Sharding-JDBC 快速体验

Sharding-JDBC 直接封装 JDBC API,可以理解为增强版的 JDBC 驱动,旧代码迁移成本几乎为零。Sharding-JDBC 可适用于任何基于 Java 的 ORM 框架,如 JPA、Hibernate、Mybatis、Spring JDBC Template 或直接使用 JDBC。

下面通过案例说明使用 Sharding-JDBC 实现分库分表的功能。

1. 创建分库和分表

创建两个库 shop_0 和 shop_1。每个库中分别创建两个分表 shop_info_0 和 shop_info_1。创建脚本如下:

```
DROP DATABASE IF EXISTS `shop_0`;

CREATE DATABASE `shop_0`;

USE `shop_0`;

CREATE TABLE `shop_info_0` (
    `shop_id` bigint(19) NOT NULL,
```

```sql
  `shop_name` varchar(45) DEFAULT NULL,
  `account` varchar(45) NOT NULL,
  PRIMARY KEY (`shop_id`)
) ENGINE=InnoDB DEFAULT CHARSET=utf8;

CREATE TABLE `shop_info_1` (
  `shop_id` bigint(19) NOT NULL,
  `shop_name` varchar(45) DEFAULT NULL,
  `account` varchar(45) NOT NULL,
  PRIMARY KEY (`shop_id`)
) ENGINE=InnoDB DEFAULT CHARSET=utf8;

DROP DATABASE IF EXISTS `shop_1`;

CREATE DATABASE `shop_1`;

USE `shop_1`;

CREATE TABLE `shop_info_0` (
  `shop_id` bigint(19) NOT NULL,
  `shop_name` varchar(45) DEFAULT NULL,
  `account` varchar(45) NOT NULL,
  PRIMARY KEY (`shop_id`)
) ENGINE=InnoDB DEFAULT CHARSET=utf8;

CREATE TABLE `shop_info_1` (
  `shop_id` bigint(19) NOT NULL,
  `shop_name` varchar(45) DEFAULT NULL,
  `account` varchar(45) NOT NULL,
  PRIMARY KEY (`shop_id`)
) ENGINE=InnoDB DEFAULT CHARSET=utf8;
```

2. 创建商户实体

创建与数据库中 shop 表对应的实体类 ShopInfo：

```java
/**
 * @Author: zhouguanya
 * @Date: 2019/01/19
 * @Description: 商户信息
 */
public class ShopInfo {
    /**
     * 商户 id
     */
    private Long shopId;
```

```java
    /**
     * 商户名
     */
    private String shopName;

    /**
     * 商户账户
     */
    private String account;

    public Long getShopId() {
        return shopId;
    }

    public void setShopId(Long shopId) {
        this.shopId = shopId;
    }

    public String getUserName() {
        return shopName;
    }

    public void setUserName(String userName) {
        this.shopName = userName;
    }

    public String getAccount() {
        return account;
    }

    public void setAccount(String account) {
        this.account = account;
    }
}
```

3. 创建 DAO

创建 ShopInfoDao 用于 ShopInfo 的数据库操作：

```java
/**
 * @Author: zhouguanya
 * @Date: 2019/01/19
 * @Description: ShopInfo DAO 层
 */
public interface ShopInfoDao {
```

```
/**
 * 保存商户
 */
int insert(ShopInfo shopInfo);

/**
 * 查询商户
 */
ShopInfo selectByPrimaryKey(Long shopId);
}
```

4. 创建 ShopService

创建 ShopService 用于封装 DAO 操作并提供给调用方使用：

```
/**
 * @Author: zhouguanya
 * @Date: 2019/01/19
 * @Description: Service 层
 */
@Service
public class ShopService {

    @Resource
    ShopInfoDao shopInfoDao;

    /**
     * 保存商户
     */
    public void saveShop(ShopInfo userInfo) {
        shopInfoDao.insert(userInfo);
    }

    /**
     * 查询商户
     */
    public ShopInfo queryShop(Long userId) {
        return shopInfoDao.selectByPrimaryKey(userId);
    }
}
```

5. 集成 MyBatis

本例中使用 MyBatis 作为持久层框架，以下是本例中 Spring 集成 MyBatis 的配置：

```
<context:component-scan base-package="com.test"/>
```

```xml
<!-- spring 和 MyBatis 整合-->
<bean id="sqlSessionFactory" class="org.mybatis.spring.SqlSessionFactoryBean">
    <property name="dataSource" ref="shardingDataSource" />
    <!-- 自动扫描 mapping.xml 文件，**表示迭代查找 -->
    <property name="mapperLocations">
        <array>
            <value>classpath:mapper/ShopInfoMapper.xml</value>
        </array>
    </property>
</bean>
<!-- DAO 接口所在包名，Spring 会自动查找其下的类，包下的类需要使用@MapperScan 注解，否则
容器注入会失败 -->
<bean class="org.mybatis.spring.mapper.MapperScannerConfigurer">
    <property name="basePackage" value="com.test.sharding.dao" />
    <property name="sqlSessionFactoryBeanName" value="sqlSessionFactory" />
</bean>
```

数据源使用的是 Sharding-JDBC 数据源。Sharding-JDBC 数据源配置，请看下文。

6. 创建 Mapper 文件

创建 MyBatis Mapper 文件代码如下：

```xml
<mapper namespace="com.test.sharding.dao.ShopInfoDao">
  <resultMap id="BaseResultMap" type="com.test.sharding.entity.ShopInfo">
    <id column="shop_id" jdbcType="BIGINT" property="shopId" />
    <result column="shop_name" jdbcType="VARCHAR" property="shopName" />
    <result column="account" jdbcType="VARCHAR" property="account" />
  </resultMap>
  <sql id="Base_Column_List">
    shop_id, shop_name, account
  </sql>

  <select id="selectByPrimaryKey" parameterType="java.lang.Long" resultMap="BaseResultMap">
    select
    <include refid="Base_Column_List" />
    from shop_info
    where shop_id = #{shopId,jdbcType=BIGINT}
  </select>

  <insert id="insert" parameterType="com.test.sharding.entity.ShopInfo">
    insert into shop_info (shop_id, shop_name, account)
    values (#{shopId,jdbcType=BIGINT}, #{shopName,jdbcType=VARCHAR}, #{account,jdbcType=VARCHAR})
  </insert>
</mapper>
```

7. 配置数据源

使用 Sharding-JDBC 的数据源需要 MySQL 数据源和具体的分库和分表逻辑：

```java
/**
 * @Author: zhouguanya
 * @Date: 2019/01/19
 * @Description: sharding-jdbc 数据源配置
 */
@Configuration
public class DataSourceConfig {

    /**
     * sharding-jdbc 数据源
     */
    @Bean(name = "shardingDataSource")
    DataSource getShardingDataSource() throws SQLException {
        ShardingRuleConfiguration shardingRuleConfiguration;
        shardingRuleConfiguration = new ShardingRuleConfiguration();
        //表规则配置
        shardingRuleConfiguration.getTableRuleConfigs().add(getUserTableRuleConfiguration());
        //表的组 shop_info
        shardingRuleConfiguration.getBindingTableGroups().add("shop_info");
        //DataBase 的分片策略  配合 DemoDatabaseShardingAlgorithm 数据库分片逻辑
        shardingRuleConfiguration.setDefaultDatabaseShardingStrategyConfig(new StandardShardingStrategyConfiguration("shop_id", DemoDatabaseShardingAlgorithm.class.getName()));
        //Table 的分片策略
        shardingRuleConfiguration.setDefaultTableShardingStrategyConfig(new StandardShardingStrategyConfiguration("shop_id", DemoTableShardingAlgorithm.class.getName()));
        //根据配置实例化一个 ShardingDataSource bean
        return new ShardingDataSource(shardingRuleConfiguration.build(createDataSourceMap()));
    }

    /**
     * 商户表规则配置
     * @return
     */
    @Bean
    TableRuleConfiguration getUserTableRuleConfiguration() {
        TableRuleConfiguration orderTableRuleConfig = new TableRuleConfiguration();
```

```java
        //逻辑表是user_info
        orderTableRuleConfig.setLogicTable("shop_info");
        //实际的物理节点是user_${0..1}.user_info_${0..1} 即database.table
        orderTableRuleConfig.setActualDataNodes("shop_${0..1}.shop_info_${0..1}");
        orderTableRuleConfig.setKeyGeneratorColumnName("shop_id");
        return orderTableRuleConfig;
    }

    /**
     * 封装多个MySQL数据源
     */
    private Map<String, DataSource> createDataSourceMap() {
        Map<String, DataSource> result = new HashMap<>(2);
        result.put("shop_0", createDataSource("shop_0"));
        result.put("shop_1", createDataSource("shop_1"));
        return result;
    }

    /**
     * MySQL数据源
     */
    private DataSource createDataSource(final String dataSourceName) {
        BasicDataSource result = new BasicDataSource();
        result.setDriverClassName(com.mysql.jdbc.Driver.class.getName());
        result.setUrl(String.format("jdbc:mysql://localhost:3306/%s?characterEncoding=utf-8&useSSL=false", dataSourceName));
        result.setUsername("root");
        result.setPassword("123456");
        return result;
    }
}
```

8. 配置数据库分片规则

根据shop_id对2取模实现数据库分片。本例中shop_id为基数的商户信息会保存在尾号为基数的表中，shop_id为偶数的商户信息会保存在尾号为偶数的表中。

```java
/**
 * @Author: zhouguanya
 * @Date: 2019/01/19
 * @Description: 数据库分片的计算逻辑
 */
public class DemoDatabaseShardingAlgorithm implements PreciseShardingAlgorithm<Long> {
    @Override
```

```java
    public String doSharding(Collection<String> collection,
PreciseShardingValue<Long> preciseShardingValue) {
        for (String databaseName : collection) {
            //数据库后缀名
            String suffix = String.valueOf(preciseShardingValue.getValue() % 2);
            //如果数据库后缀 = suffix，则选择这个库
            if (databaseName.endsWith(suffix)) {
                return databaseName;
            }
        }
        throw new IllegalArgumentException("参数异常");
    }
}
```

9. 配置数据表分片规则

根据 shop_id 对 2 取模实现数据表分片。本例中 shop_id 为基数的商户信息会保存在尾数为奇数的表中，shop_id 为偶数的商户信息会保存在尾号为偶数的表中。

```java
/**
 * @Author: zhouguanya
 * @Date: 2019/01/19
 * @Description: 数据表的分片规则
 */
public class DemoTableShardingAlgorithm implements
PreciseShardingAlgorithm<Long> {
    @Override
    public String doSharding(Collection<String> collection,
PreciseShardingValue<Long> preciseShardingValue) {
        for (String tableName : collection) {
            //表的后缀名
            String suffix = String.valueOf(preciseShardingValue.getValue() % 2);
            //如果表的后缀 = suffix  则选择这个表
            if (tableName.endsWith(suffix)) {
                return tableName;
            }
        }
        throw new IllegalArgumentException("参数异常");
    }
}
```

10. 创建单元测试

创建单元测试，分别测试保存和查询商户：

```java
/**
 * @Author: zhouguanya
```

```java
 * @Date: 2019/01/19
 * @Description: 单元测试
 */
@RunWith(SpringJUnit4ClassRunner.class)
@ContextConfiguration("classpath:spring-sharding.xml")
public class ShopInfoTest {
    @Resource
    ShopService shopService;

    public static Long shopId = 0L;

    /**
     * 模拟保存商户
     */
    @Test
    public void saveShop() {
        //保存10个商户
        for (int i = 0; i < 10; i++) {
            ShopInfo shopInfo = new ShopInfo();
            shopInfo.setShopId(shopId++);
            shopInfo.setAccount("Account" + i);
            shopInfo.setUserName("name" + i);
            shopService.saveShop(shopInfo);
        }
    }

    /**
     * 模拟查询商户
     */
    @Test
    public void queryShop() {
        ShopInfo shopInfo = shopService.queryShop(1L);
        System.out.printf("商户信息是=%s%n", JSON.toJSONString(shopInfo));
    }
}
```

分别执行 saveShop()方法和 queryShop()方法，并观察 10 个商户的分片情况。

执行以下命令查询 shop_id 为偶数的商户的数据分片：

```
select * from shop_0.shop_info_0;
```

执行结果如图 18-1 所示。

执行以下命令查询 shop_id 为基数的商户的数据分片：

```
select * from shop_1.shop_info_1;
```

执行结果如图 18-2 所示。

读者可以通过 PreciseShardingAlgorithm 接口实现更加复杂、更加个性化的分库分表规则。

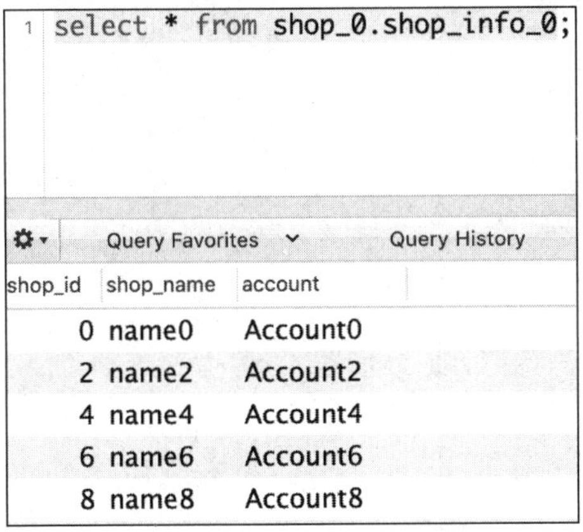

图 18-1 shop_0.shop_info_0 表数据存储

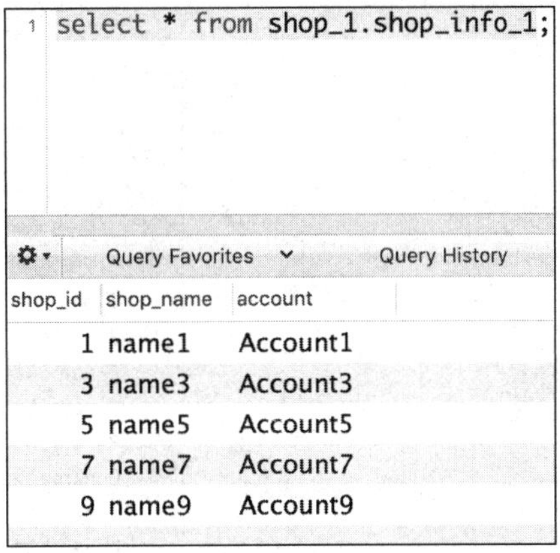

图 18-2 shop_1.shop_info_1 表数据存储

18.2　Sharding-JDBC 强制路由

当使用 Sharding-JDBC 配置好分库分表规则后，可以使用 HintManager 强制修改路由规则。

下面使用 HintManager 强制修改 shop_id=15 的商户路由规则，使 shop_id=15 的商户保存到 shop_0.shop_info_0 表中：

```
/**
 * 模拟保存商户
 */
@Test
public void saveShop2() {
    //保存 10 个商户
    for (int i = 10; i < 20; i++) {
        ShopInfo shopInfo = new ShopInfo();
        shopInfo.setShopId(Long.valueOf(i));
        shopInfo.setAccount("Account" + i);
        shopInfo.setUserName("name" + i);
        //强制修改路由规则
        if(i == 15) {
            HintManagerHolder.clear();
            HintManager hintManager = HintManager.getInstance();
            hintManager.addDatabaseShardingValue("shop_info", "shop_id", 2L);
            hintManager.addTableShardingValue("shop_info", "shop_id", 2L);
        }
        shopService.saveShop(shopInfo);
    }
}
```

执行单元测试后，验证商户信息保存，如图 18-3 所示。

```
1  select * from shop_0.shop_info_0;
```

shop_id	shop_name	account
0	name0	Account0
2	name2	Account2
4	name4	Account4
6	name6	Account6
8	name8	Account8
10	name10	Account10
12	name12	Account12
14	name14	Account14
15	name15	Account15
16	name16	Account16
18	name18	Account18

图 18-3 shop_0.shop_info_0 表数据存储

18.3 Sharding-JDBC 分布式主键

在传统企业软件开发中，主键自动生成技术是基本需求，各个数据库对于该自增主键的需求提供了相应的支持，如 MySQL 的自增键。对于 MySQL 而言，分库分表之后，不同库、不同表生成全局唯一的主键是非常麻烦的事情。因为同一个逻辑表内的不同物理表之间的自增主键是无法互相感知的，这样会生成重复的主键。

目前有许多第三方解决方案可以完美解决这个问题，比如 UUID 等依靠特定算法自生成不重复键，或者通过引入主键生成服务（Redis 或者 ZooKeeper）等。

Sharding-JDBC 提供了注解生成接口 KeyGenerator。各个实现类通过实现 generateKey() 方法即可对外提供生成主键的功能。下面通过案例阐述 Sharding-JDBC 分布式主键的使用。

1. 修改商户表规则

修改 getShopTableRuleConfiguration() 方法，配置主键生成实现类 DefaultKeyGenerator：

```java
/**
 * 商户表规则配置
 * @return
 */
@Bean
TableRuleConfiguration getShopTableRuleConfiguration() {
    TableRuleConfiguration orderTableRuleConfig = new TableRuleConfiguration();
    //逻辑表是 shop_info
    orderTableRuleConfig.setLogicTable("shop_info");
```

```
        //实际的物理节点是 shop_${0..1}.shop_info_${0..1} 即 database.table
        orderTableRuleConfig.setActualDataNodes ("shop_${0..1}.
shop_info_${0..1}");
        orderTableRuleConfig.setKeyGeneratorColumnName("shop_id");
        //主键生成类
        orderTableRuleConfig.setKeyGeneratorClass (DefaultKeyGenerator.class.
getName());
        return orderTableRuleConfig;
    }
```

2. 创建主键自增的 DAO 方法

创建主键自增的 DAO 方法，代码如下：

```
/**
 * 主键自增
 */
int insertAutoIncrement(ShopInfo shopInfo);
```

3. 创建 Mapper

创建 Mapper，代码如下：

```
<insert id="insertAutoIncrement" parameterType= "com.test.sharding.entity.
ShopInfo">
    insert into shop_info (shop_name, account)
    values (#{shopName,jdbcType=VARCHAR}, #{account,jdbcType=VARCHAR})
</insert>
```

4. 创建 service

创建 service，代码如下：

```
/**
 * 保存商户，主键自增
 */
public void saveShopAutoIncrement(ShopInfo userInfo) {
    shopInfoDao.insertAutoIncrement(userInfo);
}
```

5. 创建单元测试

单元测试中使用 10 个线程分别对商户信息进行写入操作，代码如下：

```
/**
 * 测试主键自增
 */
@Test
public void saveShop3() throws InterruptedException {
    //保存 10 个商户
```

```java
        for (int i = 0; i < 10; i++) {
            int finalI = i;
            //创建新的线程
            new Thread(new Runnable() {
                @Override
                public void run() {
                    ShopInfo shopInfo = new ShopInfo();
                    shopInfo.setAccount("Account" + finalI);
                    shopInfo.setUserName("name" + finalI);
                    try {
                        Thread.sleep(1);
                    } catch (InterruptedException e) {
                        e.printStackTrace();
                    }
                    //此处为使用 Sharding-JDBC 提供的主键
                    shopService.saveShopAutoIncrement(shopInfo);
                }
            }).start();
        }
        Thread.sleep(1000);
    }
```

执行单元测试，观察 10 个线程生成的主键情况。

6. 查询 shop_0.shop_info_0

查询 shop_0.shop_info_0 表保存的商户信息，如图 18-4 所示。

7. 查询 shop_1.shop_info_1

查询 shop_1.shop_info_1 表保存的商户信息，如图 18-5 所示。

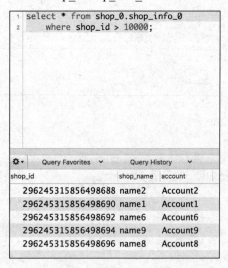

图 18-4　shop_0.shop_info_0 表数据存储

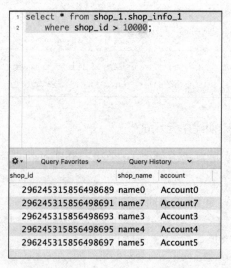

图 18-5　shop_1.shop_info_1 表数据存储

DefaultKeyGenerator 是 Sharding-JDBC 默认的主键生成器。该主键生成器采用 Twitter Snowflake 算法实现生成 64 位的 Long 型编号。国内很多大型互联网公司发号器服务基于该算法加部分改造实现。下面分析 DefaultKeyGenerator 产生的编号的组成。

DefaultKeyGenerator 生成的 64 位 Long 型编号的组成如图 18-6 所示。

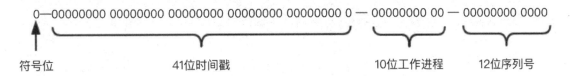

图 18-6 64 位 Long 型编号组成

64 位 Long 型编号中各个部分的如表 18-1 所示。

表 18-1 DefaultKeyGenerator 64 位编号组成

位 数	含 义	取值范围
1	符号位	0
41	时间戳	从 2016-11-01 00:00:00 000 开始的毫秒数,支持约 70 年
10	工作进程编号	最大进程编号 1024
12	序列号	每毫秒从 0 开始自增,每毫秒最多 4096 个编号,每秒最多 4096000 个编号

DefaultKeyGenerator 部分代码如下:

```
public final class DefaultKeyGenerator implements KeyGenerator {
    //2016-11-01 00:00:00 000 对应的毫秒数
    public static final long EPOCH;
    //12 位序列号
    private static final long SEQUENCE_BITS = 12L;
    //10 位工作进程号
    private static final long WORKER_ID_BITS = 10L;
    //12 位序列号自增量掩码(最大值)
    private static final long SEQUENCE_MASK = (1 << SEQUENCE_BITS) - 1;
    //工作进程 ID 左移比特数(位数)
    private static final long WORKER_ID_LEFT_SHIFT_BITS = SEQUENCE_BITS;
    //时间戳左移比特数(位数)
    private static final long TIMESTAMP_LEFT_SHIFT_BITS = WORKER_ID_LEFT_SHIFT_BITS + WORKER_ID_BITS;
    //工作进程 ID 最大值
    private static final long WORKER_ID_MAX_VALUE = 1L << WORKER_ID_BITS;
    //当前时间
    private static TimeService timeService = new TimeService();
    //工作进程 ID
    private static long workerId;
    // 初始化 EPOCH
    static {
```

```java
            Calendar calendar = Calendar.getInstance();
            calendar.set(2016, Calendar.NOVEMBER, 1);
            calendar.set(Calendar.HOUR_OF_DAY, 0);
            calendar.set(Calendar.MINUTE, 0);
            calendar.set(Calendar.SECOND, 0);
            calendar.set(Calendar.MILLISECOND, 0);
            EPOCH = calendar.getTimeInMillis();
    }
    //自增量
    private long sequence;
    //最后生成编号时间戳，单位：毫秒
    private long lastTime;
    /**
     * 设置工作进程
     */
    public static void setWorkerId(final long workerId) {
        Preconditions.checkArgument(workerId >= 0L && workerId < WORKER_ID_MAX_VALUE);
        DefaultKeyGenerator.workerId = workerId;
    }
    /**
     * 创建id
     */
    @Override
    public synchronized Number generateKey() {
        long currentMillis = timeService.getCurrentMillis();
        Preconditions.checkState(lastTime <= currentMillis, "Clock is moving backwards, last time is %d milliseconds, current time is %d milliseconds",lastTime, currentMillis);
        if (lastTime == currentMillis) {
            if (0L == (sequence = ++sequence & SEQUENCE_MASK)) {
                currentMillis = waitUntilNextTime(currentMillis);
            }
        } else {
            sequence = 0;
        }
        lastTime = currentMillis;
        ......省略代码......
        return ((currentMillis - EPOCH) << TIMESTAMP_LEFT_SHIFT_BITS) | (workerId << WORKER_ID_LEFT_SHIFT_BITS) | sequence;
    }

    private long waitUntilNextTime(final long lastTime) {
        long time = timeService.getCurrentMillis();
        while (time <= lastTime) {
```

```
            time = timeService.getCurrentMillis();
        }
        return time;
    }
}
```

18.4 小　　结

　　Sharding-JDBC 与 Mycat 类似，都是分库分表中间件。Mycat 以代理的形式提供数据库服务，对应用程序完全透明。Sharding-JDBC 采用在 JDBC 协议层扩展分库分表，是一个以 jar 包形式提供服务的轻量级组件。读者可以根据具体场景选择使用合适的分库分表组件。

第 19 章

Spring 集成 Dubbo

Dubbo 是阿里巴巴公司开源的一个高性能优秀的服务框架，Dubboo 使应用可通过高性能的 RPC 实现服务的输出和输入功能，并且可以与 Spring 框架无缝集成。

19.1 远程过程调用协议

远程过程调用协议（RPC，Remote Procedure Call）是一种通过网络从远程计算机程序上请求的服务，而不需要客户端程序了解底层网络技术的协议。RPC 协议使用某些传输协议，如 TCP 或 UDP，为通信程序之间传输信息数据。在 OSI 网络通信模型中，RPC 跨越了传输层和应用层。RPC 使开发分布式应用程序更加容易。

RPC 采用客户端/服务器模式。请求程序就是一个客户端，而服务提供程序就是一个服务器。首先，客户端调用进程发送一个有进程参数的调用信息到服务进程，然后等待应答信息。在服务器端，当一个调用信息到达，服务器获得调用参数，计算结果后返回响应信息，然后等待下一个调用信息。最后，客户端调用进程接收服务器端响应信息。RPC 使客户端像调用本地服务一样调用远程服务，调用者对网络通信透明。

RPC 调用流程如图 19-1 所示。

RPC 调用过程如下。

（1）Client 为服务调用方，以本地调用的方式调用服务。

（2）Client Stub 接受调用方的调用请求后，将方法和参数等序列化为消息体。

（3）Client Stub 寻找服务地址，并将消息发送到服务端。

（4）Server Stub 接受消息体，并将消息反序列化。

（5）Server Stub 使用反序列化的结果进行服务端本地调用。

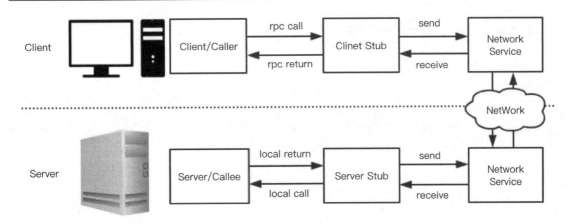

图 19-1　RPC 调用流程图

（6）服务端本地执行成功并返回结果给 Server Stub。
（7）Server Stub 将执行结果序列化，并通过网络传输给调用方。
（8）Client Stub 接收到响应消息体，并反序列化。
（9）调用方得到最终服务端响应结果。

19.2　Spring 集成 Dubbo 快速体验

本节将 Spring 与 Dubbo 整合，实现一个 RPC 快速应用。

1. 创建 ZooKeeper 集群

安装 ZooKeeper 集群详细步骤请参考本书 15.1 节。

2. 创建 API 模块

创建 API 模块，其中包含一个 HelloService 接口：

```
/**
 * @Author: zhouguanya
 * @Date: 2019/01/31
 * @Description: api
 */
public interface HelloService {
    /**
     * 抽象方法
     * @param name
     * @return
     */
    String sayHello(String name);
}
```

3. 创建服务提供方 provider 模块

在 pom.xml 中加入以下依赖：

```xml
<dependency>
    <groupId>com.test.spring5</groupId>
    <artifactId>dubbo-api</artifactId>
    <version>1.0-SNAPSHOT</version>
</dependency>
<dependency>
    <groupId>com.alibaba</groupId>
    <artifactId>dubbo</artifactId>
    <version>2.6.5</version>
</dependency>
<dependency>
    <groupId>org.apache.curator</groupId>
    <artifactId>curator-framework</artifactId>
    <version>4.0.0</version>
    <exclusions>
        <exclusion>
            <groupId>org.apache.zookeeper</groupId>
            <artifactId>zookeeper</artifactId>
        </exclusion>
    </exclusions>
</dependency>
<dependency>
    <groupId>org.apache.zookeeper</groupId>
    <artifactId>zookeeper</artifactId>
    <version>3.4.13</version>
</dependency>
```

创建 HelloService 接口实现类 HelloServiceImpl：

```java
/**
 * @Author: zhouguanya
 * @Date: 2019/01/31
 * @Description: 服务提供方实现接口
 */
public class HelloServiceImpl implements HelloService {
    /**
     * sayHello方法实现
     */
    @Override
    public String sayHello(String name) {
        System.out.println(LocalDateTime.now() + " hello" + name + ", response from provider: " + RpcContext.getContext().getLocalAddress());
```

```
        return "hello " + name;
    }
}
```

配置服务提供方。包括 ZooKeeper 集群的配置和通信端口等配置：

```xml
<!-- 服务提供方方应用名 -->
<dubbo:application name="dubbo-provider"/>
<!-- 注册中心暴露服务地址 -->
<dubbo:registry address="zookeeper://127.0.0.1:2182?backup=127.0.0.1:2183,127.0.0.1:2184,127.0.0.1:2185,127.0.0.1:2186" />
<!-- 用 dubbo 协议在 20881 端口暴露服务 -->
<dubbo:protocol name="dubbo" port="20881" />
<!-- Bean 管理 -->
<bean id="helloService" class="com.test.dubbo.provider.HelloServiceImpl"/>
<!-- 声明需要暴露的服务接口 -->
<dubbo:service interface="com.test.dubbo.api.HelloService" ref="helloService"/>
```

创建单元测试，启动服务提供者：

```java
/**
 * @Author: zhouguanya
 * @Date: 2019/01/31
 * @Description:
 */
@RunWith(SpringJUnit4ClassRunner.class)
@ContextConfiguration("classpath:dubbo-provider.xml")
public class DubboProviderTest {
    @Test
    public void startProvider() throws Exception {
        System.out.println("Dubbo Provider started successfully...");
        System.in.read();
    }
}
```

4. 创建服务调用方 consumer 模块

配置服务调用方，代码如下：

```xml
<!-- 服务消费方应用名 -->
<dubbo:application name="dubbo-consumer"/>
<!-- 注册中心地址 -->
<dubbo:registry address="zookeeper://127.0.0.1:2182?backup=127.0.0.1:2183,127.0.0.1:2184,127.0.0.1:2185,127.0.0.1:2186" />
<!-- 引用服务 -->
<dubbo:reference id="helloService" interface="com.test.dubbo.api.HelloService"/>
```

创建单元测试，测试服务调用者，代码如下：

```java
/**
 * @Author: zhouguanya
 * @Date: 2019/01/31
 * @Description: 测试服务调用者
 */
@RunWith(SpringJUnit4ClassRunner.class)
@ContextConfiguration("classpath:dubbo-consumer.xml")
public class DubboConsumerTest {
    @Autowired
    private HelloService helloService;
    @Test
    public void test() throws InterruptedException {
        for (int i = 0; i < 10; i++) {
            Thread.sleep(1000);
            System.out.println(helloService.sayHello("Michael"));
        }
    }
}
```

5. 验证

执行服务提供方的单元测试，结果如下所示：

```
Dubbo Provider started successfully...
```

执行服务调用方的单元测试，结果如下：

```
hello Michael
hello Michael
hello Michael
hello Michael
hello Michael
hello Michael
hello Michael
hello Michael
hello Michael
hello Michael
```

至此完成了一个简单的 Spring 集成 Dubbo 的开发环境的搭建。

19.3 Dubbo 代码分析

19.3.1 SPI

SPI（Service Provider Interface），是 Java 提供的一套用来被第三方实现或者扩展的 API，SPI 可以用来启用框架扩展和替换组件。Java SPI 是一种"基于接口的编程＋策略模式＋配置文件"组合实现的动态加载机制。下面通过一个案例说明 SPI 的使用。

（1）创建接口 SpiHelloService，接口中有一个 say()方法：

```
public interface SpiHelloService {
    void say();
}
```

（2）创建接口 SpiHelloService 实现类。

用 EnglishSpiHelloServiceImpl 实现类打印"Hello World"。

```
/**
 * @Author: zhouguanya
 * @Date: 2019-02-04
 * @Description:
 */
public class EnglishSpiHelloServiceImpl implements SpiHelloService {
    public void say() {
        System.out.println("Hello World");
    }
}
```

用 ChineseSpiHelloServiceImpl 实现类打印"世界，你好"。

```
/**
 * @Author: zhouguanya
 * @Date: 2019-02-04
 * @Description:
 */
public class ChineseSpiHelloServiceImpl implements SpiHelloService {
    public void say() {
        System.out.println("世界，你好");
    }
}
```

（3）创建配置文件。

在 resources 目录下创建"META-INF/services"文件夹，在文件夹下创建以 SpiHelloService 接口全路径名称命名的文件 com.test.SpiHelloService，在其中配置 SpiHelloService 接口的实现类。文件内容如下：

```
com.test.EnglishSpiHelloServiceImpl
com.test.ChineseSpiHelloServiceImpl
```

(4)编写单元测试。

使用 ServiceLoader 加载 SpiHelloService 的实现类,并执行每个实现类的方法:

```java
/**
 * @Author: zhouguanya
 * @Date: 2019-02-04
 * @Description:
 */
public class SpiDemo {
    public static void main(String[] args) {
        ServiceLoader<SpiHelloService> services = ServiceLoader.load
(SpiHelloService.class);
        for (SpiHelloService service : services) {
            service.say();
        }
    }
}
```

执行单元测试,结果如下所示:

```
Hello World
世界,你好
```

Dubbo 微内核和插件机制的核心是 ExtensionLoader,取代了 JDK 自带的 ServiceLoader。

JDK 提供的标准 SPI 会一次性实例化扩展点所有实现,即使没用到的实现类也将会被加载,如果有某个扩展类初始化很耗时,会很浪费资源。

ExtensionLoader 相比于 ServiceLoader 增加了对扩展点 IoC 和 AOP 的支持,一个扩展点可以直接使用 setter 注入其他扩展点。

以 LoadBalance 为例,com.alibaba.dubbo.rpc.cluster.LoadBalance 文件的内容如下:

```
random=com.alibaba.dubbo.rpc.cluster.loadbalance.RandomLoadBalance
roundrobin=com.alibaba.dubbo.rpc.cluster.loadbalance.RoundRobinLoadBalance
leastactive=com.alibaba.dubbo.rpc.cluster.loadbalance.
LeastActiveLoadBalance
consistenthash=com.alibaba.dubbo.rpc.cluster.loadbalance.
ConsistentHashLoadBalance
```

当用户配置 loadbalance="roundrobin" 时,Dubbo 仅加载 RoundRobinLoadBalance 类。

获取 ExtensionLoader 的 getExtensionLoader() 方法的代码如下:

```java
public static <T> ExtensionLoader<T> getExtensionLoader(Class<T> type) {
    if (type == null)
        throw new IllegalArgumentException("Extension type == null");
    if (!type.isInterface()) {
```

```
            throw new IllegalArgumentException("Extension type(" + type + ") is not
interface!");
        }
        if (!withExtensionAnnotation(type)) {
            throw new IllegalArgumentException("Extension type(" + type +") is not
extension, because WITHOUT @" + SPI.class.getSimpleName() + " Annotation!");
        }

        ExtensionLoader<T> loader = (ExtensionLoader<T>) EXTENSION_LOADERS.get
(type);
        if (loader == null) {
            EXTENSION_LOADERS.putIfAbsent(type, new ExtensionLoader<T>(type));
            loader = (ExtensionLoader<T>) EXTENSION_LOADERS.get(type);
        }
        return loader;
    }
```

此方法需要判断传入的 Class 是否为 interface 接口类型并且判断是否含有"@SPI"注解。然后创建 ExtensionLoader 对象后,在内存中缓存下来,保证每个类型的扩展点都有且仅有唯一的 ExtensionLoader 单例。

获取扩展点对象的 getExtension() 方法如下:

```
    public T getExtension(String name) {
        if (name == null || name.length() == 0)
            throw new IllegalArgumentException("Extension name == null");
        if ("true".equals(name)) {
            return getDefaultExtension();
        }
        Holder<Object> holder = cachedInstances.get(name);
        if (holder == null) {
            cachedInstances.putIfAbsent(name, new Holder<Object>());
            holder = cachedInstances.get(name);
        }
        Object instance = holder.get();
        if (instance == null) {
            synchronized (holder) {
                instance = holder.get();
                if (instance == null) {
                    instance = createExtension(name);
                    holder.set(instance);
                }
            }
        }
        return (T) instance;
    }
```

getExtension()方法通过调用createExtension()方法创建扩展点，createExtension()方法的代码如下：

```
private T createExtension(String name) {
    Class<?> clazz = getExtensionClasses().get(name);
    if (clazz == null) {
        throw findException(name);
    }
    try {
        T instance = (T) EXTENSION_INSTANCES.get(clazz);
        if (instance == null) {
            EXTENSION_INSTANCES.putIfAbsent(clazz, clazz.newInstance());
            instance = (T) EXTENSION_INSTANCES.get(clazz);
        }
        injectExtension(instance);
        Set<Class<?>> wrapperClasses = cachedWrapperClasses;
        if (wrapperClasses != null && !wrapperClasses.isEmpty()) {
            for (Class<?> wrapperClass : wrapperClasses) {
                instance = injectExtension((T) wrapperClass.getConstructor(type).newInstance(instance));
            }
        }
        return instance;
    } catch (Throwable t) {
        throw new IllegalStateException("Extension instance(name: " + name + ", class: " + type + ") could not be instantiated: " + t.getMessage(), t);
    }
}
```

createExtension()方法通过getExtensionClasses()方法查找与名称对应的Class对象：

```
private Map<String, Class<?>> getExtensionClasses() {
    Map<String, Class<?>> classes = cachedClasses.get();
    if (classes == null) {
        synchronized (cachedClasses) {
            classes = cachedClasses.get();
            if (classes == null) {
                classes = loadExtensionClasses();
                cachedClasses.set(classes);
            }
        }
    }
    return classes;
}
```

getExtensionClasses()方法通过调用 loadExtensionClasses()方法从配置文件中加载扩展点。Dubbo从以下3个路径中读取扩展点配置文件并加载：

- META-INF/services/
- META-INF/dubbo/
- META-INF/dubbo/internal/

找到 Class 对象后，实例化这个扩展点对象，并保存在 ConcurrentHashMap 缓存中。

createExtension()方法调用的 injectExtension()方法其实是通过 Setter 方法为扩展点注入属性的，代码如下：

```java
private T injectExtension(T instance) {
    try {
        if (objectFactory != null) {
            for (Method method : instance.getClass().getMethods()) {
                if (method.getName().startsWith("set")
                        && method.getParameterTypes().length == 1
                        && Modifier.isPublic(method.getModifiers())) {
                    /**
                     * Check {@link DisableInject} to see if we need auto injection for this property
                     */
                    if (method.getAnnotation(DisableInject.class) != null) {
                        continue;
                    }
                    Class<?> pt = method.getParameterTypes()[0];
                    try {
                        String property = method.getName().length() > 3 ?
method.getName().substring(3, 4).toLowerCase() + method.getName().substring(4) : "";
                        Object object = objectFactory.getExtension(pt, property);
                        if (object != null) {
                            method.invoke(instance, object);
                        }
                    } catch (Exception e) {
                        logger.error("fail to inject via method "
+method.getName()+ " of interface " + type.getName() + ": " + e.getMessage(), e);
                    }
                }
            }
        }
    } catch (Exception e) {
        logger.error(e.getMessage(), e);
    }
    return instance;
}
```

执行完 injectExtension()方法后，将扩展点实现类的对象通过 Wrapper 包装类封装，包装类可以将所有扩展点的公共逻辑封装起来。

Dubbo 还提供了"@Adaptive"和"@Activate"两个注解。

"@Adaptive"的作用是扩展点自适应，直到扩展点方法执行时才决定调用哪一个扩展点实现。扩展点的调用会有 URL 作为参数，通过"@Adaptive"注解可以提取约定 key 来决定调用哪个实现的方法。

以下代码段表示默认使用 Random 负载均衡策略，同时会根据用户在 XML 中配置的 loadbalance 参数来最终决定调用哪个扩展点实现类：

```
SPI(RandomLoadBalance.NAME)
public interface LoadBalance {

    @Adaptive("loadbalance")
    <T> Invoker<T> select(List<Invoker<T>> invokers, URL url, Invocation invocation) throws RpcException;

}
```

"@Activate"的作用是扩展点自动激活，指定 URL 中激活扩展点的 key，未指定 key 时表示无条件激活。

以下代码段表示只有在 Consumer 端才会激活：

```
@Activate(group = Constants.CONSUMER)
public class AsyncFilter implements Filter{

}
```

19.3.2 Dubbo 服务提供方代码分析

当 Spring 解析 19.2 节中服务提供方 Provider 中的配置文件 dubbo-provider.xml 时，解析器 DefaultBeanDefinitionDocumentReader 会将<dubbo:service>元素解析出来，如图 19-2 所示。

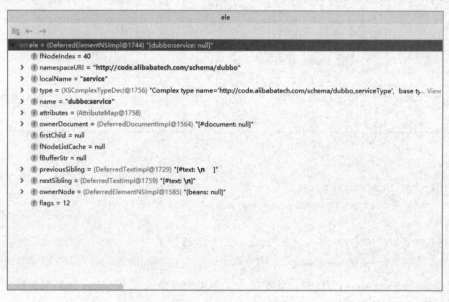

图 19-2　Spring 解析 dubbo-provider.xml 获取<dubbo:service>元素信息

Spring 将<dubbo:service>元素从 dubbo-provider.xml 解析出来后，会将<dubbo:service>元素解析成 ServiceBean 对象，如图 19-3 所示。

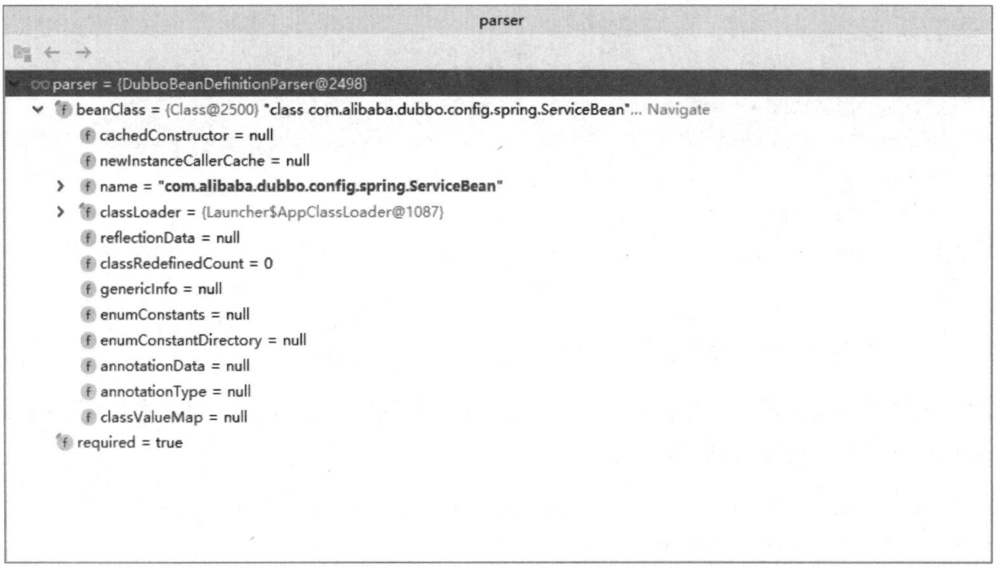

图 19-3 Spring 解析<dubbo:service>元素生成 ServiceBean

ServiceBean 实现了 ApplicationListener 接口，在 Spring 容器初始化的时候会调用 ServiceBean 的 onApplicationEvent()方法，其代码如下：

```
public void onApplicationEvent(ContextRefreshedEvent event) {
    if (isDelay() && !isExported() && !isUnexported()) {
        if (logger.isInfoEnabled()) {
            logger.info("The service ready on spring started. service: " +
getInterface());
        }
        export();
    }
}
```

onApplicationEvent()方法需要调用父类 ServiceConfig 中的 export()方法将提供服务，export() 方法代码如下：

```
public synchronized void export() {
    if (provider != null) {
        if (export == null) {
            export = provider.getExport();
        }
        if (delay == null) {
            delay = provider.getDelay();
        }
    }
    if (export != null && !export) {
```

```
            return;
        }

        if (delay != null && delay > 0) {
            delayExportExecutor.schedule(new Runnable() {
                public void run() {
                    doExport();
                }
            }, delay, TimeUnit.MILLISECONDS);
        } else {
            doExport();
        }
    }
```

如果配置了 delay 参数，Dubbo 将会启动一个守护线程，在守护线程休眠指定时段后调用 doExport()方法，守护线程的创建如下：

```
private static final ScheduledExecutorService delayExportExecutor =
Executors.newSingleThreadScheduledExecutor(new
NamedThreadFactory("DubboServiceDelayExporter", true));
```

如果没有配置 delay 参数，则直接调用 doExport()方法。

ServiceConfig 类的 doExport()方法的部分代码如下：

```
protected synchronized void doExport() {
    if (unexported) {
        throw new IllegalStateException("Already unexported!");
    }
    if (exported) {
        return;
    }
    exported = true;
    if (interfaceName == null || interfaceName.length() == 0) {
        throw new IllegalStateException("<dubbo:service interface=\"\" />
interface not allow null!");
    }
    checkDefault();
    ......省略代码......
    checkApplication();
    checkRegistry();
    checkProtocol();
    appendProperties(this);
    checkStubAndMock(interfaceClass);
    if (path == null || path.length() == 0) {
        path = interfaceName;
    }
    doExportUrls();
```

```
        ProviderModel providerModel = new ProviderModel(getUniqueServiceName(),
this, ref);
        ApplicationModel.initProviderModel(getUniqueServiceName(),
providerModel);
    }
```

doExport()方法调用doExportUrls()方法将提供服务,doExportUrls()方法如下:

```
private void doExportUrls() {
    List<URL> registryURLs = loadRegistries(true);
    for (ProtocolConfig protocolConfig : protocols) {
        doExportUrlsFor1Protocol(protocolConfig, registryURLs);
    }
}
```

doExportUrls()方法调用doExportUrlsFor1Protocol()方法对协议和注册中心等进行封装,将服务提供出去。

doExportUrlsFor1Protocol()方法部分代码如下:

```
    private void doExportUrlsFor1Protocol(ProtocolConfig protocolConfig,
List<URL> registryURLs) {
        String name = protocolConfig.getName();
        if (name == null || name.length() == 0) {
            name = "dubbo";
        }

        //处理host

        //处理port

        Map<String, String> map = new HashMap<String, String>();
        //设置参数到map

        // 导出服务
        String contextPath = protocolConfig.getContextpath();
        if ((contextPath == null || contextPath.length() == 0) && provider != null){
            contextPath = provider.getContextpath();
        }

        if (ExtensionLoader.getExtensionLoader(ConfiguratorFactory.class)
.hasExtension(url.getProtocol())) {
            url = ExtensionLoader.getExtensionLoader(ConfiguratorFactory.class)
.getExtension(url.getProtocol()).getConfigurator(url).configure(url);
        }

        //此处省略重要代码,具体分析请参考19.3.3节和19.3.4节
```

```
            this.urls.add(url);
        }
```

如果配置的 scope=none 则不提供服务。否则将分别对本地服务和远程服务进行处理。在上面省略的 doExportUrlsFor1Protocol()方法代码中有如下一段代码：

```
// don't export when none is configured
if (!Constants.SCOPE_NONE.toString().equalsIgnoreCase(scope)) {
    // export to local if the config is not
    //remote (export to remote only when config is remote)
    if (!Constants.SCOPE_REMOTE.toString().equalsIgnoreCase(scope)) {
        exportLocal(url);
    }
```

提供本地服务的 exportLocal()方法代码如下：

```
private void exportLocal(URL url) {
    if (!Constants.LOCAL_PROTOCOL.equalsIgnoreCase(url.getProtocol())) {
        URL local = URL.valueOf(url.toFullString())
                .setProtocol(Constants.LOCAL_PROTOCOL)
                .setHost(LOCALHOST)
                .setPort(0);
        ServiceClassHolder.getInstance().pushServiceClass
(getServiceClass(ref));
        Exporter<?> exporter = protocol.export(
                proxyFactory.getInvoker(ref, (Class) interfaceClass, local));
        exporters.add(exporter);
        logger.info("Export dubbo service " + interfaceClass.getName() + " to
 local registry");
    }
}
```

Dubbo 更常用的使用场景是提供远程服务，因此重点分析提供远程服务的场景。

19.3.3　Dubbo 服务提供方不使用注册中心

如果不使用注册中心，则直接调用对应协议的 export()方法提供远程服务，继续分析上面 doExportUrlsFor1Protocol()方法省略的部分，有如下部分代码段：

```
// export to remote if the config is not local (export to local only when config
 is local)
    if (!Constants.SCOPE_LOCAL.toString().equalsIgnoreCase(scope)) {
        if (logger.isInfoEnabled()) {
            logger.info("Export dubbo service " + interfaceClass.getName() + " to
 url " + url);
        }
        if (registryURLs != null && !registryURLs.isEmpty()) {
            //具体参考19.3.4 节
```

```
        } else {
            Invoker<?> invoker = proxyFactory.getInvoker(ref, (Class)
interfaceClass, url);
            DelegateProviderMetaDataInvoker wrapperInvoker = new
DelegateProviderMetaDataInvoker(invoker, this);
            //不使用注册中心的方式提供远程服务
            Exporter<?> exporter = protocol.export(wrapperInvoker);
            exporters.add(exporter);
        }
    }
```

以 Dubbo 协议为例，DubboProtocol 类的 export()方法将提供服务，其代码如下：

```
    public <T> Exporter<T> export(Invoker<T> invoker) throws RpcException {
        URL url = invoker.getUrl();

        // export service.
        String key = serviceKey(url);
        DubboExporter<T> exporter = new DubboExporter<T>(invoker, key,
exporterMap);
        exporterMap.put(key, exporter);

        //export an stub service for dispatching event
        Boolean isStubSupportEvent = url.getParameter(Constants.STUB_EVENT_KEY,
Constants.DEFAULT_STUB_EVENT);
        Boolean isCallbackservice =
url.getParameter(Constants.IS_CALLBACK_SERVICE, false);
        if (isStubSupportEvent && !isCallbackservice) {
            String stubServiceMethods =
url.getParameter(Constants.STUB_EVENT_METHODS_KEY);
            if (stubServiceMethods == null || stubServiceMethods.length() == 0) {
                if (logger.isWarnEnabled()) {
                    logger.warn(new IllegalStateException("consumer [" +
url.getParameter(Constants.INTERFACE_KEY) +"], has set stubproxy support
event ,but no stub methods founded."));
                }
            } else {
                stubServiceMethodsMap.put(url.getServiceKey(),
stubServiceMethods);
            }
        }

        openServer(url);
        optimizeSerialization(url);
        return exporter;
    }
```

export()方法会调用 openServer()方法创建服务端处理程序，openServer()方法如下：

```java
private void openServer(URL url) {
    // find server.
    String key = url.getAddress();
    //client can export a service which's only for server to invoke
    boolean isServer = url.getParameter(Constants.IS_SERVER_KEY, true);
    if (isServer) {
        ExchangeServer server = serverMap.get(key);
        if (server == null) {
            serverMap.put(key, createServer(url));
        } else {
            // server supports reset, use together with override
            server.reset(url);
        }
    }
}
```

openServer()方法调用 createServer()方法创建服务端处理程序，createServer()方法代码如下：

```java
private ExchangeServer createServer(URL url) {
    // send readonly event when server closes, it's enabled by default
    url = url.addParameterIfAbsent(Constants.CHANNEL_READONLYEVENT_SENT_KEY, Boolean.TRUE.toString());
    // enable heartbeat by default
    url = url.addParameterIfAbsent(Constants.HEARTBEAT_KEY, String.valueOf(Constants.DEFAULT_HEARTBEAT));
    String str = url.getParameter(Constants.SERVER_KEY, Constants.DEFAULT_REMOTING_SERVER);

    if (str != null && str.length() > 0 && !ExtensionLoader.getExtensionLoader(Transporter.class).hasExtension(str))
        throw new RpcException("Unsupported server type: " + str + ", url: " + url);

    url = url.addParameter(Constants.CODEC_KEY, DubboCodec.NAME);
    ExchangeServer server;
    try {
        server = Exchangers.bind(url, requestHandler);
    } catch (RemotingException e) {
        throw new RpcException("Fail to start server(url: " + url + ") " + e.getMessage(), e);
    }
    str = url.getParameter(Constants.CLIENT_KEY);
    if (str != null && str.length() > 0) {
```

```
            Set<String> supportedTypes = ExtensionLoader.getExtensionLoader
(Transporter.class).getSupportedExtensions();
            if (!supportedTypes.contains(str)) {
                throw new RpcException("Unsupported client type: " + str);
            }
        }
        return server;
    }
```

createServer()方法通过 Exchangers.bind()方法创建 Server，Exchangers 封装了请求和响应信息。Exchangers.bind()方法代码如下：

```
    public static ExchangeServer bind(URL url, ExchangeHandler handler) throws
RemotingException {
        if (url == null) {
            throw new IllegalArgumentException("url == null");
        }
        if (handler == null) {
            throw new IllegalArgumentException("handler == null");
        }
        url = url.addParameterIfAbsent(Constants.CODEC_KEY, "exchange");
        return getExchanger(url).bind(url, handler);
    }
```

Exchangers.bind()方法默认调用 HeaderExchanger 的 bind()方法，HeaderExchanger 的 bind()方法代码如下：

```
    public ExchangeServer bind(URL url, ExchangeHandler handler) throws
RemotingException {
        return new HeaderExchangeServer(Transporters.bind(url, new
DecodeHandler(new HeaderExchangeHandler(handler))));
    }
```

HeaderExchanger 的 bind()方法调用 Transporters 类的 bind()方法创建 Server，Transporters 的 bind()方法代码如下：

```
    public static Server bind(URL url, ChannelHandler... handlers) throws
RemotingException {
        if (url == null) {
            throw new IllegalArgumentException("url == null");
        }
        if (handlers == null || handlers.length == 0) {
            throw new IllegalArgumentException("handlers == null");
        }
        ChannelHandler handler;
        if (handlers.length == 1) {
            handler = handlers[0];
```

```
    } else {
        handler = new ChannelHandlerDispatcher(handlers);
    }
    return getTransporter().bind(url, handler);
}
```

Transporters 的 bind()方法调用了 NettyTransporter 类的 bind()方法，NettyTransporter 类的 bind()方法代码如下：

```
public Server bind(URL url, ChannelHandler listener) throws RemotingException{
    return new NettyServer(url, listener);
}
```

NettyTransporter 类的 bind()方法返回 NettyServer 对象，NettyServer 类对应的构造器代码如下：

```
public NettyServer(URL url, ChannelHandler handler) throws RemotingException{
    super(url, ChannelHandlers.wrap(handler, ExecutorUtil.setThreadName(url,
SERVER_THREAD_POOL_NAME)));
}
```

NettyServer 构造函数会调用父类 AbstractServer 类的构造器，AbstractServer 类构造器如下：

```
public AbstractServer(URL url, ChannelHandler handler) throws
RemotingException {
    super(url, handler);
    localAddress = getUrl().toInetSocketAddress();

    String bindIp = getUrl().getParameter(Constants.BIND_IP_KEY,
getUrl().getHost());
    int bindPort = getUrl().getParameter(Constants.BIND_PORT_KEY,
getUrl().getPort());
    if (url.getParameter(Constants.ANYHOST_KEY, false) ||
NetUtils.isInvalidLocalHost(bindIp)) {
        bindIp = NetUtils.ANYHOST;
    }
    bindAddress = new InetSocketAddress(bindIp, bindPort);
    this.accepts = url.getParameter(Constants.ACCEPTS_KEY,
Constants.DEFAULT_ACCEPTS);
    this.idleTimeout = url.getParameter(Constants.IDLE_TIMEOUT_KEY,
Constants.DEFAULT_IDLE_TIMEOUT);
    try {
        doOpen();
        if (logger.isInfoEnabled()) {
            logger.info("Start " + getClass().getSimpleName() + " bind " +
getBindAddress() + ", export " + getLocalAddress());
        }
    } catch (Throwable t) {
        throw new RemotingException(url.toInetSocketAddress(), null, "Failed to
```

```
bind " + getClass().getSimpleName() + " on " + getLocalAddress() + ", cause: " +
t.getMessage(), t);
        }
        //fixme replace this with better method
        DataStore dataStore =
ExtensionLoader.getExtensionLoader(DataStore.class).getDefaultExtension();
        executor = (ExecutorService)
dataStore.get(Constants.EXECUTOR_SERVICE_COMPONENT_KEY,
Integer.toString(url.getPort()));
    }
```

AbstractServer 构造器中包含 doOpen()抽象方法，由其子类 NettyServer 实现，其代码如下：

```
    protected void doOpen() throws Throwable {
        NettyHelper.setNettyLoggerFactory();
        ExecutorService boss = Executors.newCachedThreadPool(new
NamedThreadFactory("NettyServerBoss", true));
        ExecutorService worker = Executors.newCachedThreadPool(new
NamedThreadFactory("NettyServerWorker", true));
        ChannelFactory channelFactory = new NioServerSocketChannelFactory(boss,
worker, getUrl().getPositiveParameter(Constants.IO_THREADS_KEY,
Constants.DEFAULT_IO_THREADS));
        bootstrap = new ServerBootstrap(channelFactory);

        final NettyHandler nettyHandler = new NettyHandler(getUrl(), this);
        channels = nettyHandler.getChannels();
        // https://issues.jboss.org/browse/NETTY-365
        // https://issues.jboss.org/browse/NETTY-379
        // final Timer timer = new HashedWheelTimer(new
NamedThreadFactory("NettyIdleTimer", true));
        bootstrap.setOption("child.tcpNoDelay", true);
        bootstrap.setPipelineFactory(new ChannelPipelineFactory() {
            @Override
            public ChannelPipeline getPipeline() {
                NettyCodecAdapter adapter = new NettyCodecAdapter(getCodec(),
getUrl(), NettyServer.this);
                ChannelPipeline pipeline = Channels.pipeline();
                /*int idleTimeout = getIdleTimeout();
                if (idleTimeout > 10000) {
                    pipeline.addLast("timer", new IdleStateHandler(timer,
idleTimeout / 1000, 0, 0));
                }*/
                pipeline.addLast("decoder", adapter.getDecoder());
                pipeline.addLast("encoder", adapter.getEncoder());
                pipeline.addLast("handler", nettyHandler);
                return pipeline;
```

```
        });
        // bind
        channel = bootstrap.bind(getBindAddress());
    }
```

至此可以发现，Dubbo 的服务提供方通过 Netty 监听网络连接并对外提供服务。

19.3.4 Dubbo 服务提供方使用注册中心

如果使用了注册中心，则需要提供服务并且将服务注册到注册中心，这是 Dubbo 最常用的使用方式，下面将重点分析 doExportUrlsFor1Protocol()方法中的这部分逻辑，部分代码如下：

```
if (registryURLs != null && !registryURLs.isEmpty()) {
    for (URL registryURL : registryURLs) {
        url = url.addParameterIfAbsent(Constants.DYNAMIC_KEY,
registryURL.getParameter(Constants.DYNAMIC_KEY));
        URL monitorUrl = loadMonitor(registryURL);
        if (monitorUrl != null) {
            url = url.addParameterAndEncoded(Constants.MONITOR_KEY,
monitorUrl.toFullString());
        }
        if (logger.isInfoEnabled()) {
            logger.info("Register dubbo service " + interfaceClass.getName() +
" url " + url + " to registry " + registryURL);
        }

        // For providers, this is used to enable custom proxy to generate invoker
        String proxy = url.getParameter(Constants.PROXY_KEY);
        if (StringUtils.isNotEmpty(proxy)) {
            registryURL = registryURL.addParameter(Constants.PROXY_KEY,
proxy);
        }

        Invoker<?> invoker = proxyFactory.getInvoker(ref, (Class)
interfaceClass, registryURL.addParameterAndEncoded(Constants.EXPORT_KEY,
url.toFullString()));
        DelegateProviderMetaDataInvoker wrapperInvoker = new
DelegateProviderMetaDataInvoker(invoker, this);

        Exporter<?> exporter = protocol.export(wrapperInvoker);
        exporters.add(exporter);
    }
}
```

提供远程服务时获取 Invoker 的过程可以分为以下几个步骤：

（1）调用 ProxyFactory 类的 getInvoker()方法获取 Invoker 对象。
（2）调用 getInvoker()方法的过程需要将实现类做封装生成一个包装类 Wrapper。
（3）创建 Invoker 对象，其中包含生成的 Wrapper 类，该类含有具体的服务实现类。

Dubbo 获取 Invoke 的方式是通过调用 SPI 接口 ProxyFactory 来实现。上面代码段中 proxyFactory.getInvoker()方法其实是调用了 JavassistProxyFactory 类的 getInvoker()方法，JavassistProxyFactory 代码如下：

```
/**
 * JavaassistRpcProxyFactory
 */
public class JavassistProxyFactory extends AbstractProxyFactory {

    @Override
    @SuppressWarnings("unchecked")
    public <T> T getProxy(Invoker<T> invoker, Class<?>[] interfaces) {
        return (T) Proxy.getProxy(interfaces).newInstance(new InvokerInvocationHandler(invoker));
    }

    @Override
    public <T> Invoker<T> getInvoker(T proxy, Class<T> type, URL url) {

        final Wrapper wrapper = Wrapper.getWrapper(proxy.getClass()
            .getName().indexOf('$') < 0 ? proxy.getClass() : type);
        return new AbstractProxyInvoker<T>(proxy, type, url) {
            @Override
            protected Object doInvoke(T proxy, String methodName,Class<?>[] parameterTypes,Object[] arguments) throws Throwable {
                return wrapper.invokeMethod(proxy, methodName, parameterTypes, arguments);
            }
        };
    }

}
```

JavassistProxyFactory 类的 getInvoker()方法根据传入的 proxy 对象的类型信息创建对应的包装对象 Wrapper 并返回 Invoker 对象实例，Wrapper 类的 getWrapper()方法的代码如下：

```
public static Wrapper getWrapper(Class<?> c) {
    while (ClassGenerator.isDynamicClass(c)) // can not wrapper on dynamic class.
        c = c.getSuperclass();

    if (c == Object.class)
```

```
            return OBJECT_WRAPPER;
        Wrapper ret = WRAPPER_MAP.get(c);
        if (ret == null) {
            ret = makeWrapper(c);
            WRAPPER_MAP.put(c, ret);
        }
        return ret;
    }
```

getWrapper()方法调用 makeWrapper()方法的部分代码如下：

```
    private static Wrapper makeWrapper(Class<?> c) {
        if (c.isPrimitive())
            throw new IllegalArgumentException("Can not create wrapper for primitive type: " + c);

        String name = c.getName();
        ClassLoader cl = ClassHelper.getClassLoader(c);

        StringBuilder c1 = new StringBuilder("public void setPropertyValue(Object o, String n, Object v){ ");
        StringBuilder c2 = new StringBuilder("public Object getPropertyValue(Object o, String n){ ");
        StringBuilder c3 = new StringBuilder("public Object invokeMethod(Object o, String n, Class[] p, Object[] v) throws " + InvocationTargetException.class.getName() + "{ ");

        c1.append(name).append(" w; try{ w = ((").append(name).append(")$1); }catch(Throwable e){ throw new IllegalArgumentException(e); }");
        c2.append(name).append(" w; try{ w = ((").append(name).append(")$1); }catch(Throwable e){ throw new IllegalArgumentException(e); }");
        c3.append(name).append(" w; try{ w = ((").append(name).append(")$1); }catch(Throwable e){ throw new IllegalArgumentException(e); }");

        Map<String, Class<?>> pts = new HashMap<String, Class<?>>();
        Map<String, Method> ms = new LinkedHashMap<String, Method>();
        List<String> mns = new ArrayList<String>(); // method names.
        List<String> dmns = new ArrayList<String>(); // declaring method names.
```

这里其实就是在动态生成一个 Wrapper 类的对象，生成的 Wrapper 类中大概含有以下几个关键方法：

```
    public void setPropertyValue(Object o, String n, Object v) {
        com.test.dubbo.provider.HelloServiceImpl w;
        try {
            w = ((com.test.dubbo.provider.HelloServiceImpl) $1);
```

```
        } catch (Throwable e) {
            throw new IllegalArgumentException(e);
        }
    }

    public Object getPropertyValue(Object o, String n) {
        com.test.dubbo.provider.HelloServiceImpl w;
        try {
            w = ((com.test.dubbo.provider.HelloServiceImpl) $1);
        } catch (Throwable e) {
            throw new IllegalArgumentException(e);
        }
    }

    public Object invokeMethod(Object o, String n, Class[] p, Object[] v) throws
java.lang.reflect.InvocationTargetException {
        com.test.dubbo.provider.HelloServiceImpl w;
        try {
            w = ((com.test.dubbo.provider.HelloServiceImpl) $1);
        } catch (Throwable e) {
            throw new IllegalArgumentException(e);
        }
        try {
            if ("sayHello".equals($2) && $3.length == 0) {
                w.sayHello();
                return null;
            }
        } catch (Throwable e) {
            throw new java.lang.reflect.InvocationTargetException(e);
        }
    }
```

生成 Wrapper 以后，返回一个 AbstractProxyInvoker 实例。至此生成 Invoker 的步骤就完成了。可以看到 Invoker 对象执行方法的时候，会调用 Wrapper 的 invokeMethod()方法，这个方法中含有服务实现类的具体实现代码。

生成 Invoker 以后需要将服务提供出去，可以分为以下几个步骤：

（1）进入 RegistryProtocol 类的 export()方法。
（2）将服务提供出去，并返回 ExporterChangeableWrapper 对象，具体过程与 19.3.3 节类似。
（3）注册服务到注册中心。
（4）订阅注册中心的服务。
（5）生成一个 DestroyableExporter 对象，包含步骤 2 中 ExporterChangeableWrapper 对象的引用。

RegistryProtocol 类的 export()方法如下：

```java
public <T> Exporter<T> export(final Invoker<T> originInvoker) throws RpcException {
    //这里就交给了具体的协议去提供服务，与19.3.3节类似
    final ExporterChangeableWrapper<T> exporter = doLocalExport(originInvoker);
    //注册到注册中心的 URL
    URL registryUrl = getRegistryUrl(originInvoker);
    //根据 invoker 中的 url 获取 Registry 实例
    final Registry registry = getRegistry(originInvoker);
    final URL registeredProviderUrl = getRegisteredProviderUrl(originInvoker);

    //to judge to delay publish whether or not
    boolean register = registeredProviderUrl.getParameter("register", true);

    ProviderConsumerRegTable.registerProvider(originInvoker, registryUrl, registeredProviderUrl);

    if (register) {
    //调用远端注册中心的 register 方法进行服务注册
        register(registryUrl, registeredProviderUrl);
        ProviderConsumerRegTable.getProviderWrapper(originInvoker).setReg(true);
    }

    //FIXME 提供者订阅时，会影响同一 JVM 即暴露服务，又引用同一服务的场景
    //因为 subscribed 以服务名为缓存的 key，导致订阅信息覆盖。
    final URL overrideSubscribeUrl = getSubscribedOverrideUrl(registeredProviderUrl);
    final OverrideListener overrideSubscribeListener = new OverrideListener(overrideSubscribeUrl, originInvoker);
    overrideListeners.put(overrideSubscribeUrl, overrideSubscribeListener);
    //提供者向注册中心订阅所有注册服务的覆盖配置
    registry.subscribe(overrideSubscribeUrl, overrideSubscribeListener);
    //Ensure that a new exporter instance is returned every time export
    return new DestroyableExporter<T>(exporter, originInvoker, overrideSubscribeUrl, registeredProviderUrl);
}
```

export()方法调用的 register()方法如下：

```java
public void register(URL registryUrl, URL registedProviderUrl) {
    Registry registry = registryFactory.getRegistry(registryUrl);
    registry.register(registedProviderUrl);
}
```

register()方法调用的 registryFactory.getRegistry()方法是 AbstractRegistryFactory 类中定义的 getRegistry()方法，其代码如下：

```java
public Registry getRegistry(URL url) {
    url = url.setPath(RegistryService.class.getName())
            .addParameter(Constants.INTERFACE_KEY, RegistryService.class.getName())
            .removeParameters(Constants.EXPORT_KEY, Constants.REFER_KEY);
    String key = url.toServiceString();
    // 锁定注册中心获取过程，保证注册中心单一实例
    LOCK.lock();
    try {
        //缓存中获取Registry实例，缓存中存在就直接返回
        Registry registry = REGISTRIES.get(key);
        if (registry != null) {
            return registry;
        }
        //创建registry，会直接创建一个ZookeeperRegistry对象
        //具体创建实例是子类来实现的
        registry = createRegistry(url);
        if (registry == null) {
            throw new IllegalStateException("Can not create registry " + url);
        }
        //放到缓存中
        REGISTRIES.put(key, registry);
        return registry;
    } finally {
        //释放重入锁
        LOCK.unlock();
    }
}
```

createRegistry()方法由其子类实现，以 ZooKeeper 注册中心为例，createRegistry()方法的实现如下：

```java
public Registry createRegistry(URL url) {
    return new ZookeeperRegistry(url, zookeeperTransporter);
}
```

createRegistry()方法返回一个 ZookeeperRegistry 对象，从类的继承关系上可以看出，ZookeeperRegistry 继承 FailbackRegistry，FailbackRegistry 集成 AbstractRegistry，下面分析 AbstractRegistry 类的构造器，代码如下：

```java
public AbstractRegistry(URL url) {
    setUrl(url);
    // Start file save timer
```

```java
        syncSaveFile = url.getParameter(Constants.REGISTRY_FILESAVE_SYNC_KEY,
false);
        String filename = url.getParameter(Constants.FILE_KEY,System.getProperty
("user.home") + "/.dubbo/dubbo-registry-" + url.getParameter
(Constants.APPLICATION_KEY) + "-" + url.getAddress() + ".cache");
        File file = null;
        if (ConfigUtils.isNotEmpty(filename)) {
            file = new File(filename);
            if (!file.exists() && file.getParentFile() != null
&& !file.getParentFile().exists()) {
                if (!file.getParentFile().mkdirs()) {
                    throw new IllegalArgumentException("Invalid registry store file
" + file + ", cause: Failed to create directory " + file.getParentFile() + "!");
                }
            }
        }
        this.file = file;
        loadProperties();
        //通知订阅
        notify(url.getBackupUrls());
    }
```

AbstractRegistry 类构造器中调用的 notify() 方法代码如下：

```java
    protected void notify(List<URL> urls) {
        if (urls == null || urls.isEmpty()) return;

        for (Map.Entry<URL, Set<NotifyListener>> entry : getSubscribed().
entrySet()) {
            URL url = entry.getKey();

            if (!UrlUtils.isMatch(url, urls.get(0))) {
                continue;
            }

            Set<NotifyListener> listeners = entry.getValue();
            if (listeners != null) {
                for (NotifyListener listener : listeners) {
                    try {
                        notify(url, listener, filterEmpty(url, urls));
                    } catch (Throwable t) {
                        logger.error("Failed to notify registry event, urls: " + urls
+ ", cause: " + t.getMessage(), t);
                    }
                }
            }
        }
    }
```

这里的notify()方法会调用重载的notify()方法，其代码如下：

```java
protected void notify(URL url, NotifyListener listener, List<URL> urls) {
    if (url == null) {
        throw new IllegalArgumentException("notify url == null");
    }
    if (listener == null) {
        throw new IllegalArgumentException("notify listener == null");
    }
    if ((urls == null || urls.isEmpty())
            && !Constants.ANY_VALUE.equals(url.getServiceInterface())) {
        logger.warn("Ignore empty notify urls for subscribe url " + url);
        return;
    }
    if (logger.isInfoEnabled()) {
        logger.info("Notify urls for subscribe url " + url + ", urls: " + urls);
    }
    Map<String, List<URL>> result = new HashMap<String, List<URL>>();
    for (URL u : urls) {
        if (UrlUtils.isMatch(url, u)) {
            String category = u.getParameter(Constants.CATEGORY_KEY, Constants.DEFAULT_CATEGORY);
            List<URL> categoryList = result.get(category);
            if (categoryList == null) {
                categoryList = new ArrayList<URL>();
                result.put(category, categoryList);
            }
            categoryList.add(u);
        }
    }
    if (result.size() == 0) {
        return;
    }
    Map<String, List<URL>> categoryNotified = notified.get(url);
    if (categoryNotified == null) {
        notified.putIfAbsent(url, new ConcurrentHashMap<String,List<URL>>());
        categoryNotified = notified.get(url);
    }
    for (Map.Entry<String, List<URL>> entry : result.entrySet()) {
        String category = entry.getKey();
        List<URL> categoryList = entry.getValue();
        categoryNotified.put(category, categoryList);
        saveProperties(url);
        //通知监听器
```

```
            listener.notify(categoryList);
        }
    }
```

分析完 AbstractRegistry 构造器后，接着分析 FailbackRegistry 构造器，其代码如下：

```java
public FailbackRegistry(URL url) {
    super(url);
    //重试时间，默认 5000ms
    this.retryPeriod = url.getParameter(Constants.REGISTRY_RETRY_PERIOD_KEY,
Constants.DEFAULT_REGISTRY_RETRY_PERIOD);
    //启动失败重试定时器
    this.retryFuture = retryExecutor.scheduleWithFixedDelay(new Runnable() {
        @Override
        public void run() {
            // Check and connect to the registry
            try {
                //重试
                retry();
            } catch (Throwable t) { // Defensive fault tolerance
                logger.error("Unexpected error occur at failed retry, cause: "
+ t.getMessage(), t);
            }
        }
    }, retryPeriod, retryPeriod, TimeUnit.MILLISECONDS);
}
```

ZookeeperRegistry 类的构造器如下：

```java
public ZookeeperRegistry(URL url, ZookeeperTransporter zookeeperTransporter) {
    super(url);
    if (url.isAnyHost()) {
        throw new IllegalStateException("registry address == null");
    }
    String group = url.getParameter(Constants.GROUP_KEY, DEFAULT_ROOT);
    if (!group.startsWith(Constants.PATH_SEPARATOR)) {
        group = Constants.PATH_SEPARATOR + group;
    }
    this.root = group;
    zkClient = zookeeperTransporter.connect(url);
    zkClient.addStateListener(new StateListener() {
        @Override
        public void stateChanged(int state) {
            if (state == RECONNECTED) {
                try {
                    recover();
                } catch (Exception e) {
```

```
                logger.error(e.getMessage(), e);
            }
        }
    });
}
```

当执行到 register()方法调用 registry.register()方法时,将会进入子类的 register()方法,以 FailbackRegistry 为例,register()方法实现如下:

```
public void register(URL url) {
    super.register(url);
    failedRegistered.remove(url);
    failedUnregistered.remove(url);
    try {
        // Sending a registration request to the server side
        doRegister(url);
    } catch (Exception e) {
        Throwable t = e;
        // If the startup detection is opened, the Exception is thrown directly.
        boolean check = getUrl().getParameter(Constants.CHECK_KEY, true)&&
url.getParameter(Constants.CHECK_KEY,true)&& !Constants.CONSUMER_PROTOCOL.equa
ls(url.getProtocol());
        boolean skipFailback = t instanceof SkipFailbackWrapperException;
        if (check || skipFailback) {
            if (skipFailback) {
                t = t.getCause();
            }
            throw new IllegalStateException("Failed to register " + url + " to 
registry " + getUrl().getAddress() + ", cause: " + t.getMessage(), t);
        } else {
            logger.error("Failed to register " + url + ", waiting for retry, 
cause: " + t.getMessage(), t);
        }

        // Record a failed registration request to a failed list, retry regularly
        failedRegistered.add(url);
    }
}
```

这里调用的 doRegister()方法由其子类实现,以 ZookeeperRegistry 为例,doRegister()方法的实现如下:

```
protected void doRegister(URL url) {
    try {
        zkClient.create(toUrlPath(url),url.getParameter(Constants.DYNAMIC_KEY,
true));
```

```
        } catch (Throwable e) {
            throw new RpcException("Failed to register " + url + " to zookeeper "
+ getUrl() + ", cause: " + e.getMessage(), e);
        }
    }
```

doRegister()方法会在 ZooKeeper 上创建一个临时节点完成服务注册。

19.3.5　Dubbo 服务调用方代码分析

与 Dubbo 服务提供方类似，在 Dubbo 的服务调用方配置文件中的<dubbo:reference>标签将会被解析为一个 ReferenceBean 对象，ReferenceBean 对象实现了 FactoryBean 接口，因此重写的 getObject()方法将会进行节点发现和服务发现流程，getObject()方法代码如下：

```
public Object getObject() throws Exception {
    return get();
}
```

getObject()方法会调用 get()方法，get()方法代码如下：

```
public synchronized T get() {
    if (destroyed) {
        throw new IllegalStateException("Already destroyed!");
    }
    if (ref == null) {
        init();
    }
    return ref;
}
```

get()方法会调用 init()方法，完成对代理对象 ref 的创建。init()方法的部分代码如下：

```
private void init() {
    ......省略代码......
    StaticContext.getSystemContext().putAll(attributes);
    ref = createProxy(map);
    ConsumerModel consumerModel = new ConsumerModel(getUniqueServiceName(),
this, ref, interfaceClass.getMethods());
    ApplicationModel.initConsumerModel(getUniqueServiceName(),
consumerModel);
}
```

进入 createProxy()方法，其部分代码如下：

```
private T createProxy(Map<String, String> map) {
    ......省略代码......
    if (urls.size() == 1) {
        //服务发现
```

```
            invoker = refprotocol.refer(interfaceClass, urls.get(0));
    } else {
        List<Invoker<?>> invokers = new ArrayList<Invoker<?>>();
        URL registryURL = null;
        for (URL url : urls) {
            invokers.add(refprotocol.refer(interfaceClass, url));
            if (Constants.REGISTRY_PROTOCOL.equals(url.getProtocol())) {
                registryURL = url; // use last registry url
            }
        }
        if (registryURL != null) { // registry url is available
            // use AvailableCluster only when register's cluster is available
            URL u = registryURL.addParameter(Constants.CLUSTER_KEY,
AvailableCluster.NAME);
            invoker = cluster.join(new StaticDirectory(u, invokers));
        } else { // not a registry url
            invoker = cluster.join(new StaticDirectory(invokers));
        }
    }
    ......省略代码......
//创建服务代理
    return (T) proxyFactory.getProxy(invoker);
}
```

createProxy()方法中做了两个重要操作。

（1）通过 refprotocol.refer()方法进行服务发现。

（2）通过 proxyFactory.getProxy()方法创建服务代理。

下面将分别对这两个方法进行分析。

19.3.6 Dubbo 服务调用方服务发现

进入 RegistryProtocol 类的 refer()方法，其代码如下：

```
public <T> Invoker<T> refer(Class<T> type, URL url) throws RpcException {
    url = url.setProtocol(url.getParameter(Constants.REGISTRY_KEY,
Constants.DEFAULT_REGISTRY)).removeParameter(Constants.REGISTRY_KEY);
    Registry registry = registryFactory.getRegistry(url);
    if (RegistryService.class.equals(type)) {
        return proxyFactory.getInvoker((T) registry, type, url);
    }

    // group="a,b" or group="*"
    Map<String, String> qs = StringUtils.parseQueryString(url.
getParameterAndDecoded(Constants.REFER_KEY));
    String group = qs.get(Constants.GROUP_KEY);
```

```
        if (group != null && group.length() > 0) {
            if ((Constants.COMMA_SPLIT_PATTERN.split(group)).length > 1||
"*".equals(group)) {
                return doRefer(getMergeableCluster(), registry, type, url);
            }
        }
        return doRefer(cluster, registry, type, url);
    }
```

refer()方法会调用 doRefer()方法，doRefer()方法的代码如下：

```
    private <T> Invoker<T> doRefer(Cluster cluster, Registry registry, Class<T> type, URL url) {
        RegistryDirectory<T> directory = new RegistryDirectory<T>(type, url);
        directory.setRegistry(registry);
        directory.setProtocol(protocol);
        // all attributes of REFER_KEY
        Map<String, String> parameters = new HashMap<String, String>(directory.getUrl().getParameters());
        URL subscribeUrl = new URL(Constants.CONSUMER_PROTOCOL, parameters.remove(Constants.REGISTER_IP_KEY), 0, type.getName(), parameters);
        if (!Constants.ANY_VALUE.equals(url.getServiceInterface())
                && url.getParameter(Constants.REGISTER_KEY, true)) {
            registry.register(subscribeUrl.addParameters(Constants.CATEGORY_KEY, Constants.CONSUMERS_CATEGORY,
                    Constants.CHECK_KEY, String.valueOf(false)));
        }
        directory.subscribe(subscribeUrl.addParameter(Constants.CATEGORY_KEY, Constants.PROVIDERS_CATEGORY+ "," + Constants.CONFIGURATORS_CATEGORY+ "," + Constants.ROUTERS_CATEGORY));

        Invoker invoker = cluster.join(directory);
        ProviderConsumerRegTable.registerConsumer(invoker, url, subscribeUrl, directory);
        return invoker;
    }
```

这里调用的 RegistryDirectory 的 subscribe()方法向 ZooKeeper 订阅 subscribeUrl 的信息并监听变更，这样就实现了服务自动发现，subscribe()方法的代码如下：

```
    public void subscribe(URL url) {
        setConsumerUrl(url);
        registry.subscribe(url, this);
    }
```

subscribe()方法调用的 Registry 的 subscribe()方法有多个实现，下面以 FailbackRegistry 类的实现为例说明：

```java
public void subscribe(URL url, NotifyListener listener) {
    super.subscribe(url, listener);
    removeFailedSubscribed(url, listener);
    try {
        // Sending a subscription request to the server side
        doSubscribe(url, listener);
    } catch (Exception e) {
        Throwable t = e;

        List<URL> urls = getCacheUrls(url);
        if (urls != null && !urls.isEmpty()) {
            notify(url, listener, urls);
            logger.error("Failed to subscribe " + url + ", Using cached list: " + urls + " from cache file: " + getUrl().getParameter(Constants.FILE_KEY, System.getProperty("user.home") + "/dubbo-registry-" + url.getHost() + ".cache") + ", cause: " + t.getMessage(), t);
        } else {
            // If the startup detection is opened, the Exception is thrown directly.
            boolean check = getUrl().getParameter(Constants.CHECK_KEY, true)
                    && url.getParameter(Constants.CHECK_KEY, true);
            boolean skipFailback = t instanceof SkipFailbackWrapperException;
            if (check || skipFailback) {
                if (skipFailback) {
                    t = t.getCause();
                }
                throw new IllegalStateException("Failed to subscribe " + url + ", cause: " + t.getMessage(), t);
            } else {
                logger.error("Failed to subscribe " + url + ", waiting for retry, cause: " + t.getMessage(), t);
            }
        }
        // Record a failed registration request to a failed list, retry regularly
        addFailedSubscribed(url, listener);
    }
}
```

FailbackRegistry 类的 subscribe()方法会调用 doSubscribe()方法，这个方法由其子类实现。下面以 ZookeeperRegistry 类的 doSubscribe()方法为例说明：

```java
protected void doSubscribe(final URL url, final NotifyListener listener) {
    try {
        if (Constants.ANY_VALUE.equals(url.getServiceInterface())) {
            ......省略代码......
        } else {
```

```
                List<URL> urls = new ArrayList<URL>();
                for (String path : toCategoriesPath(url)) {
                    ConcurrentMap<NotifyListener, ChildListener> listeners = 
zkListeners.get(url);
                    if (listeners == null) {
                        zkListeners.putIfAbsent(url, new 
ConcurrentHashMap<NotifyListener, ChildListener>());
                        listeners = zkListeners.get(url);
                    }
                    ChildListener zkListener = listeners.get(listener);
                    if (zkListener == null) {
                        listeners.putIfAbsent(listener, new ChildListener() {
                            @Override
                            public void childChanged(String parentPath, List<String> 
currentChilds) {
                                ZookeeperRegistry.this.notify(url, listener, 
toUrlsWithEmpty(url, parentPath, currentChilds));
                            }
                        });
                        zkListener = listeners.get(listener);
                    }
                    zkClient.create(path, false);
                    List<String> children = zkClient.addChildListener(path, 
zkListener);
                    if (children != null) {
                        urls.addAll(toUrlsWithEmpty(url, path, children));
                    }
                }
                notify(url, listener, urls);
            }
        } catch (Throwable e) {
            throw new RpcException("Failed to subscribe " + url + " to zookeeper " 
+ getUrl() + ", cause: " + e.getMessage(), e);
        }
    }
```

19.3.7　Dubbo 服务调用方服务代理

下面进行 createProxy()方法调用 proxyFactory.getProxy()获取代理对象的分析。

createProxy()方法会通过 SPI 的方式获取 MockClusterInvoker 对象，将这个对象传入 JavassistProxyFactory 类的 getProxy()方法，并创建代理对象。

JavassistProxyFactory 的 getProxy()方法代码如下：

```
public <T> T getProxy(Invoker<T> invoker, Class<?>[] interfaces) {
```

```
            return (T) Proxy.getProxy(interfaces).newInstance(new
InvokerInvocationHandler(invoker));
    }
```

getProxy()方法将 Invoker 对象封装在 InvokerInvocationHandler 对象中。

当代理对象执行方法调用的时候，将会进入 InvokerInvocationHandler 类的 invoke()方法，最终会进入 MockClusterInvoker 类的 invoke()方法中。MockClusterInvoker 类的 invoke()方法实现如下：

```
    public Result invoke(Invocation invocation) throws RpcException {
        Result result = null;

        String value = directory.getUrl().getMethodParameter
(invocation.getMethodName(),Constants.MOCK_KEY,
Boolean.FALSE.toString()).trim();
        if (value.length() == 0 || value.equalsIgnoreCase("false")) {
            //no mock
            result = this.invoker.invoke(invocation);
        } else if (value.startsWith("force")) {
            if (logger.isWarnEnabled()) {
                logger.info("force-mock: " + invocation.getMethodName() + "
force-mock enabled , url : " + directory.getUrl());
            }
            //force:direct mock
            result = doMockInvoke(invocation, null);
        } else {
            //fail-mock
            try {
                result = this.invoker.invoke(invocation);
            } catch (RpcException e) {
                if (e.isBiz()) {
                    throw e;
                } else {
                    if (logger.isWarnEnabled()) {
                        logger.warn("fail-mock: " + invocation.getMethodName() + "
fail-mock enabled , url : " + directory.getUrl(), e);
                    }
                    result = doMockInvoke(invocation, e);
                }
            }
        }
        return result;
    }
```

MockClusterInvoker 类的 invoke()方法会调用 AbstractClusterInvoker 的 invoke()方法，其代码如下：

```java
    public Result invoke(final Invocation invocation) throws RpcException {
        checkWhetherDestroyed();
        LoadBalance loadbalance = null;

        // binding attachments into invocation.
        Map<String, String> contextAttachments = RpcContext.getContext().
getAttachments();
        if (contextAttachments != null && contextAttachments.size() != 0) {
            ((RpcInvocation) invocation).addAttachments(contextAttachments);
        }

        List<Invoker<T>> invokers = list(invocation);
        if (invokers != null && !invokers.isEmpty()) {
            loadbalance = ExtensionLoader.getExtensionLoader(LoadBalance.class).
getExtension(invokers.get(0).getUrl().getMethodParameter(RpcUtils.getMethodNam
e(invocation),Constants.LOADBALANCE_KEY, Constants.DEFAULT_LOADBALANCE));
        }
        RpcUtils.attachInvocationIdIfAsync(getUrl(), invocation);
        return doInvoke(invocation, invokers, loadbalance);
    }
```

这里的doInvoke()方法由其子类实现，以FailoverClusterInvoker为例，其实现如下：

```java
    public Result doInvoke(Invocation invocation, final List<Invoker<T>> invokers,
LoadBalance loadbalance) throws RpcException {
        List<Invoker<T>> copyinvokers = invokers;
        checkInvokers(copyinvokers, invocation);
        int len = getUrl().getMethodParameter(invocation.getMethodName(),
Constants.RETRIES_KEY, Constants.DEFAULT_RETRIES) + 1;
        if (len <= 0) {
            len = 1;
        }
        // retry loop.
        RpcException le = null; // last exception.
        List<Invoker<T>> invoked = new ArrayList<Invoker<T>>(copyinvokers.size());
        Set<String> providers = new HashSet<String>(len);
        for (int i = 0; i < len; i++) {
            //Reselect before retry to avoid a change of candidate `invokers`.
            //NOTE: if `invokers` changed, then `invoked` also lose accuracy.
            if (i > 0) {
                checkWhetherDestroyed();
                copyinvokers = list(invocation);
                // check again
                checkInvokers(copyinvokers, invocation);
            }
```

```
        Invoker<T> invoker = select(loadbalance, invocation, copyinvokers,
invoked);
        invoked.add(invoker);
        RpcContext.getContext().setInvokers((List) invoked);
        try {
            Result result = invoker.invoke(invocation);
......省略代码......
```

在 doInvoke()方法中调用的 select()方法在每次调用或重试时，都通过负载均衡算法选出一个 Invoker 进行调用：

```
    protected Invoker<T> select(LoadBalance loadbalance, Invocation invocation,
List<Invoker<T>> invokers, List<Invoker<T>> selected) throws RpcException {
        if (invokers == null || invokers.isEmpty())
            return null;
        String methodName = invocation == null ? "" : invocation.getMethodName();

        boolean sticky = invokers.get(0).getUrl().getMethodParameter(methodName,
Constants.CLUSTER_STICKY_KEY, Constants.DEFAULT_CLUSTER_STICKY);
        {
            //ignore overloaded method
            if (stickyInvoker != null && !invokers.contains(stickyInvoker)) {
                stickyInvoker = null;
            }
            //ignore concurrency problem
            if (sticky && stickyInvoker != null && (selected == null
|| !selected.contains(stickyInvoker))) {
                if (availablecheck && stickyInvoker.isAvailable()) {
                    return stickyInvoker;
                }
            }
        }
        Invoker<T> invoker = doSelect(loadbalance, invocation, invokers,
selected);

        if (sticky) {
            stickyInvoker = invoker;
        }
        return invoker;
    }
```

select()方法调用的 doSelect()方法的代码如下：

```
    private Invoker<T> doSelect(LoadBalance loadbalance, Invocation invocation,
List<Invoker<T>> invokers, List<Invoker<T>> selected) throws RpcException {
        if (invokers == null || invokers.isEmpty())
            return null;
```

```
            if (invokers.size() == 1)
                return invokers.get(0);
            if (loadbalance == null) {
                loadbalance = ExtensionLoader.getExtensionLoader(LoadBalance.class).
getExtension(Constants.DEFAULT_LOADBALANCE);
            }
            Invoker<T> invoker = loadbalance.select(invokers, getUrl(), invocation);

            //If the `invoker` is in the `selected` or invoker is unavailable &&
availablecheck is true, reselect.
            if ((selected != null && selected.contains(invoker))
                    || (!invoker.isAvailable() && getUrl() != null && availablecheck)){
                try {
                    Invoker<T> rinvoker = reselect(loadbalance, invocation, invokers,
selected, availablecheck);
                    if (rinvoker != null) {
                        invoker = rinvoker;
                    } else {

                        int index = invokers.indexOf(invoker);
                        try {
                            //Avoid collision
                            invoker = index < invokers.size() - 1 ? invokers.get(index +
1) : invokers.get(0);
                        } catch (Exception e) {

                        }
                    }
                } catch (Throwable t) {
                    logger.error("cluster reselect fail reason is :" + t.getMessage()
+ " if can not solve, you can set cluster.availablecheck=false in url", t);
                }
            }
            return invoker;
    }
```

得到 Invoker 以后，通过调用 Invoker 的 invoke()方法得到远程调用的结果。

19.4 小　　结

　　Dubbo 是互联网公司微服务开发的利器，熟练掌握 Dubbo 对企业微服务架构演进和升级有重要的意义。

附录 A

设 计 模 式

设计模式（Design pattern）是软件工程领域的最佳实践，本书讲解的 Spring 代码解析部分涉及大量的设计模式，设计模式也是面试中经常被问到的。设计模式不是高超的技术，而是众多软件开发人员经过长时间的试验和改正错误中总结出来的。

A.1 工厂模式

工厂模式（Factory Pattern）是 Spring 中最常用的设计模式之一。工厂模式属于创建型模式，它提供了一种创建对象的最佳方式。在工厂模式中，隐藏了创建对象的逻辑，并通过使用一个共同的接口来供调用方使用。工厂设计模式如图 A-1 所示。

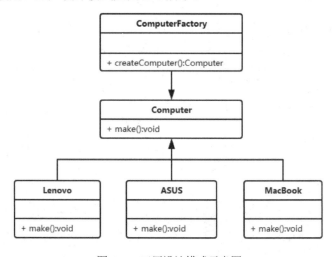

图 A-1　工厂设计模式示意图

通过以下案例说明工厂设计模式的使用。

（1）创建 Computer 接口，代码如下：

```java
/**
 * @Author: zhouguanya
 * @Date: 2019/02/05
 * @Description: 电脑接口
 */
public interface Computer {
    /**
     * 制造电脑的方法
     */
    void make();
}
```

（2）创建 ASUS 实现类，代码如下：

```java
/**
 * @Author: zhouguaya
 * @Date: 2019/02/05
 * @Description: ASUS 类
 */
public class ASUS implements Computer {
    /**
     * 制造 ASUS 电脑
     */
    @Override
    public void make() {
        System.out.println("produce a ASUS Computer");
    }
}
```

（3）创建 Lenovo 实现类，代码如下：

```java
/**
 * @Author: zhouguanya
 * @Date: 2019/02/05
 * @Description: Lenovo 电脑类
 */
public class Lenovo implements Computer {
    /**
     * 制造 Lenovo 电脑
     */
    @Override
    public void make() {
```

```
        System.out.println("produce a Lenovo Computer");
    }
}
```

（4）创建 MacBook 实现类，代码如下：

```
/**
 * @Author: zhouguanya
 * @Date: 2019/02/05
 * @Description: MacBook 类
 */
public class MacBook implements Computer {
    /**
     * 制造 MacBook 电脑
     */
    @Override
    public void make() {
        System.out.println("produce a MacBook Computer");
    }
}
```

（5）创建电脑工厂，代码如下：

```
/**
 * @Author: zhouguanya
 * @Date: 2019/02/05
 * @Description: 电脑工厂
 */
public class ComputerFactory {
    /**
     * createComputer 方法返回不同品牌的电脑
     */
    public Computer createComputer(String type) {
        if (type == null || type.equals("")) {
            return null;
        }
        switch (type) {
            case "ASUS":
                return new ASUS();
            case "Lenovo":
                return new Lenovo();
            case "MacBook":
                return new MacBook();
            default:
```

```
            return null;
        }
    }
}
```

（6）创建单元测试，代码如下：

```
/**
 * @Author: zhouguanya
 * @Date: 2019/02/05
 * @Description: 测试类
 */
public class ComputerFactoryDemo {
    public static void main(String[] args) {
        ComputerFactory computerFactory = new ComputerFactory();
        //创建 ASUS
        Computer asus = computerFactory.createComputer("ASUS");
        asus.make();
        //创建 Lenovo
        Computer lenovo = computerFactory.createComputer("Lenovo");
        lenovo.make();
        //创建 MacBook
        Computer macBook = computerFactory.createComputer("MacBook");
        macBook.make();
    }
}
```

执行测试类，得到如下的执行结果：

```
produce a ASUS Computer
produce a Lenovo Computer
produce a MacBook Computer
```

A.2 抽象工厂模式

抽象工厂围绕一个超级工厂创建其他工厂。在抽象工厂模式中，接口负责创建一个相关对象的工厂，不需要显示指定它们的类型。每个从抽象工厂中生成的工厂都能按照工厂模式提供对象。抽象工厂设计模式如图 A-2 所示。

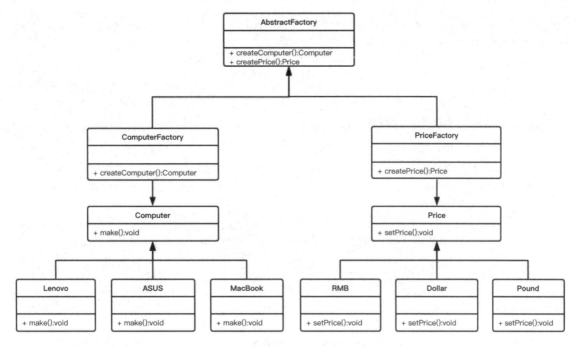

图 A-2 抽象工厂设计模式示意图

通过以下案例说明抽象工厂设计模式的使用。

（1）创建 Price 接口，代码如下：

```java
/**
 * @Author: zhouguanya
 * @Date: 2019/02/05
 * @Description: 价格接口
 */
public interface Price {
    /**
     * 设置价格
     */
    void setPrice();
}
```

（2）创建美元实现类，代码如下：

```java
/**
 * @Author: zhouguanya
 * @Date: 2019/02/05
 * @Description:
 */
public class Dollar implements Price {
    /**
     * 设置价格
     */
```

```java
    @Override
    public void setPrice() {
        System.out.println("制定电脑的美元售价");
    }
}
```

（3）创建英镑实现类，代码如下：

```java
/**
 * @Author: zhouguanya
 * @Date: 2019/02/05
 * @Description:
 */
public class Pound implements Price {
    /**
     * 设置价格
     */
    @Override
    public void setPrice() {
        System.out.println("制定电脑的英镑售价");
    }
}
```

（4）创建人民币实现类，代码如下：

```java
/**
 * @Author: zhouguanya
 * @Date: 2019/02/05
 * @Description:
 */
public class RMB implements Price {
    /**
     * 设置价格
     */
    @Override
    public void setPrice() {
        System.out.println("制定电脑的人民币售价");
    }
}
```

（5）创建抽象工厂，代码如下：

```java
/**
 * @Author: zhouguanya
 * @Date: 2019/02/05
 * @Description: 抽象工厂
 */
public abstract class AbstractFactory {
```

```java
    /**
     * 创建电脑
     */
    abstract Computer createComputer(String type);
    /**
     * 制定电脑价格
     */
    abstract Price createPrice(String currency);
}
```

（6）创建电脑工厂，代码如下：

```java
/**
 * @Author: zhouguanya
 * @Date: 2019/02/05
 * @Description: 电脑工厂
 */
public class ComputerFactory extends AbstractFactory{
    /**
     * createComputer 方法返回不同品牌的电脑
     */
    public Computer createComputer(String type) {
        if (type == null || type.equals("")) {
            return null;
        }
        switch (type) {
            case "ASUS":
                return new ASUS();
            case "Lenovo":
                return new Lenovo();
            case "MacBook":
                return new MacBook();
            default:
                return null;
        }
    }

    /**
     * 制定电脑价格
     *
     * @param currency
     */
    @Override
    Price createPrice(String currency) {
```

```
        return null;
    }
}
```

(7) 创建价格工厂,代码如下:

```
/**
 * @Author: zhouguanya
 * @Date: 2019/02/05
 * @Description: 价格工厂
 */
public class PriceFactory extends AbstractFactory{
    /**
     * 创建电脑
     */
    @Override
    Computer createComputer(String type) {
        return null;
    }

    /**
     * 制定电脑价格
     */
    @Override
    Price createPrice(String currency) {
        if (currency == null || currency.equals("")) {
            return null;
        }
        switch (currency) {
            case "RMB":
                return new RMB();
            case "Dollar":
                return new Dollar();
            case "Pound":
                return new Pound();
            default:
                return null;
        }
    }
}
```

(8) 创建工厂生成器,代码如下:

```
/**
 * @Author: zhouguanya
 * @Date: 2019/02/05
 * @Description: 工厂生成器
```

```java
 */
public class FactoryProducer {
    public static AbstractFactory getFactory(String factoryType) {
        if ("Computer".equals(factoryType)) {
            return new ComputerFactory();
        } else if ("Price".equals(factoryType)) {
            return new PriceFactory();
        }
        return null;
    }
}
```

（9）创建抽象工厂测试类，代码如下：

```java
/**
 * @Author: zhouguanya
 * @Date: 2019/02/05
 * @Description: 抽象工厂测试类
 */
public class AbstractFactoryDemo {
    public static void main(String[] args) {
        // 电脑工厂
        AbstractFactory computerFactory = FactoryProducer.getFactory("Computer");
        Computer asus = computerFactory.createComputer("ASUS");
        asus.make();
        Computer lenovo = computerFactory.createComputer("Lenovo");
        lenovo.make();
        Computer macBook = computerFactory.createComputer("MacBook");
        macBook.make();
        // 价格工厂
        AbstractFactory priceFactory = FactoryProducer.getFactory("Price");
        Price rmb = priceFactory.createPrice("RMB");
        rmb.setPrice();
        Price dollar = priceFactory.createPrice("Dollar");
        dollar.setPrice();
        Price pound = priceFactory.createPrice("Pound");
        pound.setPrice();
    }
}
```

执行单元测试，得到如下执行结果：

```
produce a ASUS Computer
produce a Lenovo Computer
produce a MacBook Computer
制定电脑的人民币售价
```

制定电脑的美元售价
制定电脑的英镑售价

A.3 单例模式

单例模式涉及一个单一的类，该类负责创建自己的对象，同时确保这个类只有单个对象被创建。Spring 管理的 Bean 默认都是单例的。单例工厂设计模式如图 A-3 所示。

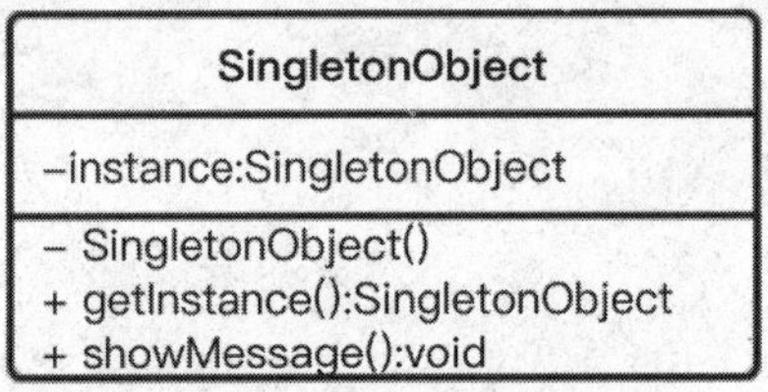

```
           SingletonObject
–instance:SingletonObject
– SingletonObject()
+ getInstance():SingletonObject
+ showMessage():void
```

图 A-3　单例工厂设计模式示意图

通过以下案例说明单例设计模式的使用。

（1）创建单例类，代码如下：

```java
/**
 * @Author: zhouguanya
 * @Date: 2019/02/05
 * @Description: 单例
 */
public class SingletonObject {
    //创建 SingletonObject 的一个对象
    private static SingletonObject instance = new SingletonObject();
    //让构造函数为 private，这样该类就不会被实例化
    private SingletonObject(){

    }
    //获取唯一可用的对象
    public static SingletonObject getInstance(){
        return instance;
    }
    public void showMessage(){
        System.out.println("Hello World!");
    }
}
```

（2）创建单例测试类，代码如下：

```java
/**
 * @Author: zhouguanya
 * @Date: 2019/02/05
```

```
 * @Description：单例测试类
 */
public class SingletonObjectDemo {
    public static void main(String[] args) {
        SingletonObject singletonObject = SingletonObject.getInstance();
        singletonObject.showMessage();
    }
}
```

执行单例测试类，得到如下所示结果：

```
Hello World!
```

A.4 建造者模式

建造者模式将多个简单对象构建成一个复杂的对象。一个 Builder 类会一步一步构造最终的对象。该 Builder 类是独立于其他对象的。

假设以下场景，每台电脑都由 CPU 和显示器组成。现在 CPU 常见的厂商有 Intel 和 AMD，显示器常见的厂商有 DELL 和 PHILIPS。每台电脑由一个 CPU 和显示器组成。建造者设计模式如图 A-4 所示。

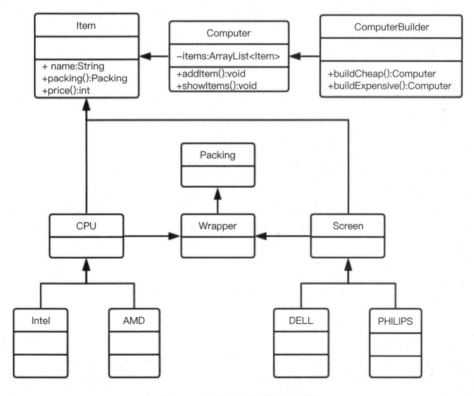

图 A-4　建造者设计模式示意图

通过以下案例说明建造者模式的使用。

(1) 创建电脑配件接口，代码如下：

```java
/**
 * @Author: zhouguanya
 * @Date: 2019/02/06
 * @Description: 电脑配件接口
 */
public interface Item {
    String name();
    Packing packing();
    int price();
}
```

(2) 创建组装接口 Packing，代码如下：

```java
/**
 * @Author: zhouguanya
 * @Date: 2019/02/06
 * @Description: 组装接口
 */
public interface Packing {
    String pack();
}
```

(3) 创建组装接口实现类 Wrapper，代码如下：

```java
/**
 * @Author: zhouguanya
 * @Date: 2019/02/06
 * @Description: 组装
 */
public class Wrapper implements Packing {
    @Override
    public String pack() {
        return "组装";
    }
}
```

(4) 创建电脑配件实现类 CPU，代码如下：

```java
/**
 * @Author: zhouguanya
 * @Date: 2019/02/06
 * @Description: CPU
 */
public abstract class CPU implements Item {
```

```java
    @Override
    public Packing packing() {
        return new Wrapper();
    }
}
```

（5）创建 Intel 处理器，代码如下：

```java
/**
 * @Author: zhouguanya
 * @Date: 2019/02/06
 * @Description:
 */
public class Intel extends CPU {
    @Override
    public String name() {
        return "Intel 处理器";
    }

    @Override
    public int price() {
        return 3000;
    }
}
```

（6）创建 AMD 处理器，代码如下：

```java
/**
 * @Author: zhouguanya
 * @Date: 2019/02/06
 * @Description:
 */
public class AMD extends CPU {
    @Override
    public String name() {
        return "AMD 处理器";
    }

    @Override
    public int price() {
        return 2000;
    }
}
```

（7）创建电脑配件实现类 Screen，代码如下：

```java
/**
 * @Author: zhouguanya
```

```
 * @Date: 2019/02/06
 * @Description: 显示器
 */
public abstract class Screen implements Item {

    @Override
    public Packing packing() {
        return new Wrapper();
    }
}
```

(8) 创建 DELL 处理器,代码如下:

```
/**
 * @Author: zhouguanya
 * @Date: 2019/02/06
 * @Description:
 */
public class DELL extends Screen {
    @Override
    public String name() {
        return "DELL 显示器";
    }

    @Override
    public int price() {
        return 2000;
    }
}
```

(9) 创建 PHILIPS 处理器,代码如下:

```
/**
 * @Author: zhouguanya
 * @Date: 2019/02/06
 * @Description:
 */
public class PHILIPS extends Screen {
    @Override
    public String name() {
        return "PHILIPS 显示器";
    }

    @Override
    public int price() {
        return 1000;
    }
}
```

（10）创建电脑类 Computer，代码如下：

```java
/**
 * @Author: zhouguanya
 * @Date: 2019/02/06
 * @Description: 电脑类
 */
public class Computer {
    private List<Item> itemList = new ArrayList<>();

    public void addItem(Item item){
        itemList.add(item);
    }

    public void showItems() {
        int total = 0;
        for (Item item : itemList) {
            System.out.print(item.packing().pack() + item.name() + ",价格=" + item.price() + "\t");
            total += item.price();
        }
        System.out.println("电脑总价=" + total);
    }
}
```

（11）创建电脑建造者 ComputerBuilder，代码如下：

```java
/**
 * @Author: zhouguanya
 * @Date: 2019/02/06
 * @Description: 电脑建造者
 */
public class ComputerBuilder {

    /**
     * 创建廉价电脑
     */
    public Computer buildCheap() {
        Computer computer = new Computer();
        computer.addItem(new AMD());
        computer.addItem(new PHILIPS());
        return computer;
    }

    /**
     * 创建高价电脑
```

```java
     */
    public Computer buildExpensive() {
        Computer computer = new Computer();
        computer.addItem(new Intel());
        computer.addItem(new DELL());
        return computer;
    }
}
```

（12）创建建造者模式测试类，代码如下：

```java
/**
 * @Author: zhouguanya
 * @Date: 2019/02/06
 * @Description: 建造者模式测试
 */
public class BuilderPattenDemo {
    public static void main(String[] args) {
        ComputerBuilder computerBuilder = new ComputerBuilder();
        Computer cheapComputer = computerBuilder.buildCheap();
        cheapComputer.showItems();
        ComputerBuilder expensiveBuilder = new ComputerBuilder();
        Computer expensiveComputer = expensiveBuilder.buildExpensive();
        expensiveComputer.showItems();
    }
}
```

执行建造者测试类，执行结果如下所示：

组装 AMD 处理器,价格=2000	组装 PHILIPS 显示器,价格=1000	电脑总价=3000
组装 Intel 处理器,价格=3000	组装 DELL 显示器,价格=2000	电脑总价=5000

A.5 原型模式

原型模式用于创建重复的对象，这种模式实现了一个原型接口，该接口用于创建当前对象的克隆对象。当创建对象的代价较大时，比较适合使用这种设计模式。如创建一个对象需要高代价的数据库操作和远程调用，这时可以将该对象缓存。当下一个请求到来时返回该对象的克隆对象，在需要的时候缓存该对象，以此来减少数据库调用和远程调用。原型设计模式如图 A-5 所示。

通过以下案例说明原型设计模式的使用。

（1）创建 Computer 类实现 Cloneable 接口，代码如下：

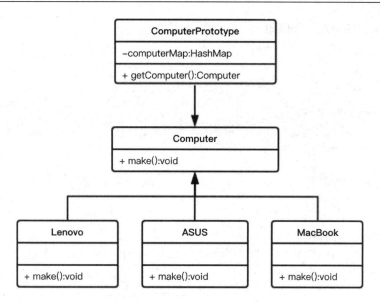

图 A-5　原型设计模式示意图

```
/**
 * @Author: zhouguanya
 * @Date: 2019/02/05
 * @Description: 电脑接口
 */
public abstract class Computer implements Cloneable{
    protected String type;
    /**
     * 制造电脑的方法
     */
    void make() {

    }

    /**
     * 克隆方法
     */
    public Object clone() {
        Object clone = null;
        try {
            clone = super.clone();
        } catch (CloneNotSupportedException e) {
            e.printStackTrace();
        }
        return clone;
    }
}
```

（2）创建 ASUS 电脑，代码如下：

```java
/**
 * @Author: zhouguaya
 * @Date: 2019/02/05
 * @Description: ASUS 类
 */
public class ASUS extends Computer {
    /**
     * 制造 ASUS 电脑
     */
    @Override
    public void make() {
        System.out.println("produce a ASUS Computer");
    }
}
```

（3）创建 Lenovo 电脑，代码如下：

```java
/**
 * @Author: zhouguanya
 * @Date: 2019/02/05
 * @Description: Lenovo 电脑类
 */
public class Lenovo extends Computer {
    /**
     * 制造 Lenovo 电脑
     */
    @Override
    public void make() {
        System.out.println("produce a Lenovo Computer");
    }
}
```

（4）创建 MacBook 电脑，代码如下：

```java
/**
 * @Author: zhouguanya
 * @Date: 2019/02/05
 * @Description: MacBook
 */
public class MacBook extends Computer {
    /**
     * 制造 MacBook 电脑
     */
    @Override
    public void make() {
```

```
        System.out.println("produce a MacBook Computer");
    }
}
```

（5）创建电脑原型类，代码如下：

```
/**
 * @Author: zhouguanya
 * @Date: 2019/02/06
 * @Description: 电脑原型类
 */
public class ComputerPrototype {
    private static Map<String, Computer> computerMap = new HashMap<>();
    static {
        computerMap.put("ASUS", new ASUS());
        computerMap.put("Lenovo", new Lenovo());
        computerMap.put("MacBook", new MacBook());
    }

    public static Computer getComputer(String type) {
        Computer computer = computerMap.get(type);
        if (computer != null) {
            //返回克隆对象
            return (Computer) computer.clone();
        }
        return null;
    }
}
```

（6）创建原型模式测试类，代码如下：

```
/**
 * @Author: zhouguanya
 * @Date: 2019/02/06
 * @Description: 原型模式测试
 */
public class PrototypePatternDemo {
    public static void main(String[] args) {
        Computer asus = ComputerPrototype.getComputer("ASUS");
        asus.make();
        Computer lenovo = ComputerPrototype.getComputer("Lenovo");
        lenovo.make();
        Computer macBook = ComputerPrototype.getComputer("MacBook");
        macBook.make();
    }
}
```

执行原型模式测试类,执行结果如下:

```
Produce a ASUS Computer
produce a Lenovo Computer
produce a MacBook Computer
```

A.6 适配器模式

适配器模式作为不兼容接口之前的桥梁。适配器模式涉及到一个类,该类负责加入独立的或者不兼容的接口功能。

下面通过案例演示适配器模式的使用。其中音频播放器只能播放 MP3 格式的文件,视频播放器可以播放 MP4 格式的文件和 RMVB 格式的文件。现在想要通过适配器模式使音频播放器不仅可以播放 MP3 格式的文件,还可以播放其他格式的文件。适配器设计模式如图 A-6 所示。

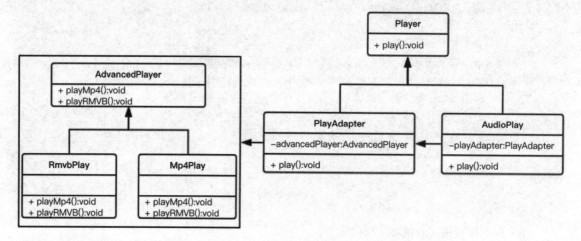

图 A-6 适配器设计模式示意图

通过以下案例说明适配器设计模式的使用。

(1)创建高级播放器接口 AdvancedPlayer,代码如下:

```
/**
 * @Author: zhouguanya
 * @Date: 2019/02/06
 * @Description: 高级播放器接口
 */
public interface AdvancedPlayer {
    /**
     * 播放 MP4
     */
    void playMp4();
    /**
     * 播放 RMVB
```

```
    */
    void playRMVB();
}
```

（2）创建 MP4 格式播放器，代码如下：

```
/**
 * @Author: zhouguanya
 * @Date: 2019/02/06
 * @Description: MP4 格式播放器
 */
public class Mp4Play implements AdvancedPlayer {
    /**
     * 播放 MP4
     */
    @Override
    public void playMp4() {
        System.out.println("播放 MP4 格式的文件");
    }

    /**
     * 播放 RMVB
     */
    @Override
    public void playRMVB() {

    }
}
```

（3）创建 RMVB 格式播放器，代码如下：

```
/**
 * @Author: zhouguanya
 * @Date: 2019/02/06
 * @Description: RMVB 格式播放器
 */
public class RmvbPlay implements AdvancedPlayer {
    /**
     * 播放 MP4
     */
    @Override
    public void playMp4() {

    }

    /**
     * 播放 RMVB
```

```java
     */
    @Override
    public void playRMVB() {
        System.out.println("播放 RMVB 格式的文件");
    }
}
```

（4）创建播放器适配器 PlayAdapter，代码如下：

```java
/**
 * @Author: zhouguanya
 * @Date: 2019/02/06
 * @Description: 播放器适配器
 */
public class PlayAdapter implements Player {
    private AdvancedPlayer advancedPlayer;

    public PlayAdapter (String type) {
        if ("MP4".equals(type)) {
            advancedPlayer = new Mp4Play();
        } else if ("RMVB".equals(type)) {
            advancedPlayer = new RmvbPlay();
        }
    }

    @Override
    public void play(String type) {
        if ("MP4".equals(type)) {
            advancedPlayer.playMp4();
        } else if ("RMVB".equals(type)) {
            advancedPlayer.playRMVB();
        }
    }
}
```

（5）创建播放器接口 Player，代码如下：

```java
/**
 * @Author: zhouguanya
 * @Date: 2019/02/06
 * @Description: 播放器接口
 */
public interface Player {
    void play(String type);
}
```

（6）创建音频播放器 AudioPlay，代码如下：

```java
/**
 * @Author: zhouguanya
 * @Date: 2019/02/06
 * @Description: 音频播放器
 */
public class AudioPlay implements Player {
    PlayAdapter playAdapter;

    @Override
    public void play(String type) {
        if ("MP3".equals(type)) {
            System.out.println("播放MP3格式的文件");
        } else if ("MP4".equals(type) || "RMVB".equals(type)) {
            playAdapter = new PlayAdapter(type);
            playAdapter.play(type);
        }
    }
}
```

（7）创建适配器模式测试类，代码如下：

```java
/**
 * @Author: zhouguanya
 * @Date: 2019/02/06
 * @Description: 适配器模式测试类
 */
public class AdapterPatternDemo {
    public static void main(String[] args) {
        AudioPlay audioPlay = new AudioPlay();
        audioPlay.play("MP3");
        audioPlay.play("MP4");
        audioPlay.play("RMVB");
    }
}
```

执行适配器模式测试类代码，执行结果如下：

```
播放MP3格式的文件
播放MP4格式的文件
播放RMVB格式的文件
```

A.7 桥接模式

如果软件系统中某个类存在两个独立变化的维度，通过桥接模式可以将这两个维度分离出来，使两者可以独立扩展，让系统更加符合"单一职责原则"。与多层继承方案不同，桥接模式将两个独立变化的维度设计为两个独立的继承结构，并且在抽象层建立一个抽象关联，该关联关系类似一条连接两个独立继承结构的桥，因此称作桥接模式。桥接设计模式如图 A-7 所示。

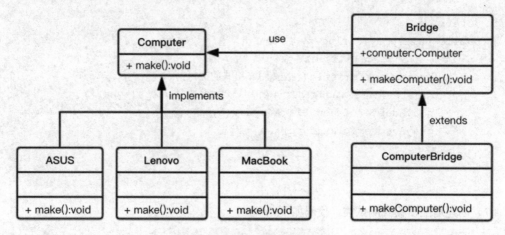

图 A-7 桥接设计模式示意图

通过以下案例说明桥接设计模式的使用。本例中使用的 ASUS、Lenovo 和 MacBook 类参考 A.1 节。

（1）创建抽象类 Bridge，代码如下：

```
/**
 * @Author: zhouguanya
 * @Date: 2019/02/06
 * @Description: 桥接
 */
public abstract class Bridge {
   protected Computer computer;
   public Bridge(Computer computer) {
       this.computer = computer;
   }

   public abstract void makeComputer();
}
```

（2）创建 ComputerBridge 类，集成 Bridge，实现抽象方法，代码如下：

```
/**
 * @Author: zhouguanya
```

```
 * @Date: 2019/02/06
 * @Description:
 */
public class ComputerBridge extends Bridge {

    public ComputerBridge(Computer computer) {
        super(computer);
    }

    @Override
    public void makeComputer() {
        computer.make();
    }
}
```

（3）创建桥接设计模式测试类，代码如下：

```
/**
 * @Author: zhouguanya
 * @Date: 2019/02/06
 * @Description: 桥接设计模式测试
 */
public class BridgePatternDemo {
    public static void main(String[] args) {
        ComputerBridge asus = new ComputerBridge(new ASUS());
        asus.makeComputer();
        ComputerBridge lenovo = new ComputerBridge(new Lenovo());
        lenovo.makeComputer();
        ComputerBridge macBook = new ComputerBridge(new MacBook());
        macBook.makeComputer();
    }
}
```

执行单元测试，执行结果如下：

```
produce a ASUS Computer
produce a Lenovo Computer
produce a MacBook Computer
```

A.8 标准模式

标准模式允许开发人员使用不同的标准来过滤一组对象，可以通过标准模式结合多个标准来获得单一标准。标准模式如图 A-8 所示。

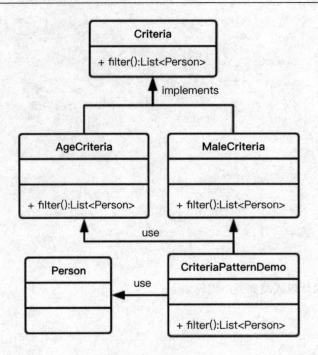

图 A-8 标准模式示意图

通过以下案例说明标准模式的使用。

（1）创建实体类 Person，代码如下：

```
/**
 * @Author: zhouguanya
 * @Date: 2019/02/06
 * @Description:
 */
public class Person {
    private String name;
    private int age;
    private String gender;
    public Person(String name, int age, String gender){
        this.name = name;
        this.age = age;
        this.gender = gender;
    }
    public String getName() {
        return name;
    }
    public int getAge() {
        return age;
    }
    public String getGender() {
```

```
        return gender;
    }
}
```

（2）创建 Criteria 接口，代码如下：

```
/**
 * @Author: zhouguanya
 * @Date: 2019/02/06
 * @Description:
 */
public interface Criteria {
    List<Person> filter(List<Person> personList);
}
```

（3）创建 MaleCriteria 过滤男性，代码如下：

```
/**
 * @Author: zhouguanya
 * @Date: 2019/02/06
 * @Description: 过滤男性
 */
public class MaleCriteria implements Criteria {
    @Override
    public List<Person> filter(List<Person> personList) {
        List<Person> filtered = new ArrayList<>();
        for (Person person : personList) {
            if ("Male".equals(person.getGender())) {
                filtered.add(person);
            }
        }
        return filtered;
    }
}
```

（4）创建 AgeCriteria 过滤年龄，代码如下：

```
/**
 * @Author: zhouguanya
 * @Date: 2019/02/06
 * @Description: 过滤年龄
 */
public class AgeCriteria implements Criteria {
    @Override
    public List<Person> filter(List<Person> personList) {
        List<Person> filtered = new ArrayList<>();
        for (Person person : personList) {
            if (person.getAge() > 20) {
```

```
                filtered.add(person);
            }
        }
        return filtered;
    }
}
```

（5）创建单元测试类，代码如下：

```java
/**
 * @Author: zhouguanya
 * @Date: 2019/02/06
 * @Description:
 */
public class CriteriaPatternDemo {
    public static void main(String[] args) {
        List<Person> persons = new ArrayList<Person>();
        persons.add(new Person("Michael", 18, "Male"));
        persons.add(new Person("Tom", 24, "Female"));
        persons.add(new Person("Robert", 22, "Male"));
        persons.add(new Person("John", 19, "Female"));
        persons.add(new Person("Bobby", 25, "Male"));
        Criteria maleCriteria = new MaleCriteria();
        printPersons(maleCriteria.filter(persons));
        System.out.println("-------------分割线-------------");
        Criteria ageCriteria = new AgeCriteria();
        printPersons(ageCriteria.filter(persons));
    }

    public static void printPersons(List<Person> persons){
        for (Person person : persons) {
            System.out.println("Person : [ Name : " + person.getName()
                    +", Gender : " + person.getGender()
                    +", Age : " + person.getAge()
                    +" ]");
        }
    }
}
```

执行单元测试，得到如下执行结果：

```
Person : [ Name : Michael, Gender : Male, Age : 18 ]
Person : [ Name : Robert, Gender : Male, Age : 22 ]
Person : [ Name : Bobby, Gender : Male, Age : 25 ]
-------------分割线-------------
Person : [ Name : Tom, Gender : Female, Age : 24 ]
Person : [ Name : Robert, Gender : Male, Age : 22 ]
```

```
Person : [ Name : Bobby, Gender : Male, Age : 25 ]
```

A.9 组合模式

组合模式是把一组相关的对象当作一个单一的对象对待的模式。组合模式依据树形结构来组合对象,用来表示部分以及整体层次。组合模式如图 A-9 所示。

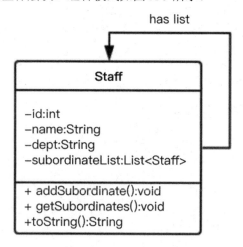

图 A-9 组合模式示意图

通过以下案例说明组合模式的使用。

(1) 创建员工类,代码如下:

```java
/**
 * @Author: zhouguanya
 * @Date: 2019/02/07
 * @Description: 员工类
 */
public class Staff {
    private int id;
    private String name;
    private String dept;
    private List<Staff> subordinateList;
    public Staff(int id, String name, String dept) {
        this.id = id;
        this.name = name;
        this.dept = dept;
        subordinateList = new ArrayList<>();
    }

    public void addSubordinate(Staff staff) {
```

```java
        subordinateList.add(staff);
    }

    public List<Staff> getSubordinates(){
        return subordinateList;
    }

    public String toString(){
        return ("Employee :[ id : "+ id
            +", name : "+ name + ", dept : "
            + dept+" ]");
    }
}
```

（2）创建组合模式测试类，代码如下：

```java
/**
 * @Author: zhouguanya
 * @Date: 2019/02/07
 * @Description: 组合模式测试类
 */
public class CompositePatternDemo {
    public static void main(String[] args) {
        Staff boss = new Staff(1, "Tom", "Boss");
        Staff cto = new Staff(2,"John", "IT");
        Staff salesDirector = new Staff(3,"Robert", "Marketing");
        boss.addSubordinate(cto);
        boss.addSubordinate(salesDirector);
        Staff engineer1 = new Staff(4,"Bob", "IT");
        Staff engineer2 = new Staff(5,"Michael", "IT");
        cto.addSubordinate(engineer1);
        cto.addSubordinate(engineer2);
        Staff salesExecutive1 = new Staff(6,"Richard", "Marketing");
        Staff salesExecutive2 = new Staff(7,"Rob", "Marketing");
        salesDirector.addSubordinate(salesExecutive1);
        salesDirector.addSubordinate(salesExecutive2);
        //打印该组织的所有员工
        System.out.println(boss);
        for (Staff manager : boss.getSubordinates()) {
            System.out.println(manager);
            for (Staff staff : manager.getSubordinates()) {
                System.out.println(staff);
            }
        }
    }
}
```

执行组合模式测试类,执行结果如下:

```
Employee :[ id : 1, name : Tom, dept : Boss ]
Employee :[ id : 2, name : John, dept : IT ]
Employee :[ id : 4, name : Bob, dept : IT ]
Employee :[ id : 5, name : Michael, dept : IT ]
Employee :[ id : 3, name : Robert, dept : Marketing ]
Employee :[ id : 6, name : Richard, dept : Marketing ]
Employee :[ id : 7, name : Rob, dept : Marketing ]
```

A.10 装饰器模式

装饰器模式向现有的对象添加新的功能,并且不破坏对象的结构。用装饰器模式创建一个装饰类,用于包装原有的类。组合模式如图 A-10 所示。

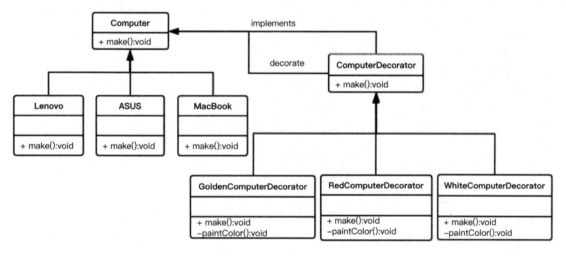

图 A-10 装饰器模式示意图

通过以下案例说明桥接设计模式的使用。本例中使用的 ASUS、Lenovo 和 MacBook 类参考 A.1 节。

(1) 创建抽象类 ComputerDecorator,代码如下:

```
/**
 * @Author: zhouguanya
 * @Date: 2019/02/07
 * @Description:
 */
public abstract class ComputerDecorator implements Computer {
    Computer computer;
    public ComputerDecorator (Computer computer) {
        this.computer = computer;
```

```java
    }

    public void make() {
        computer.make();
    }
}
```

(2) 创建金色电脑装饰器，代码如下：

```java
/**
 * @Author: zhouguanya
 * @Date: 2019/02/07
 * @Description: 金色电脑装饰器
 */
public class GoldenComputerDecorator extends ComputerDecorator {
    public GoldenComputerDecorator(Computer computer) {
        super(computer);
    }

    @Override
    public void make() {
        super.make();
        paintColor();
    }

    private void paintColor() {
        System.out.println("给电脑涂上金色");
    }
}
```

(3) 创建红色电脑装饰器，代码如下：

```java
/**
 * @Author: zhouguanya
 * @Date: 2019/02/07
 * @Description: 红色电脑装饰器
 */
public class RedComputerDecorator extends ComputerDecorator {
    public RedComputerDecorator(Computer computer) {
        super(computer);
    }

    @Override
    public void make() {
        super.make();
        paintColor();
    }
```

```java
    private void paintColor() {
        System.out.println("给电脑涂上红色");
    }
}
```

(4) 创建白色电脑装饰器，代码如下：

```java
/**
 * @Author: zhouguanya
 * @Date: 2019/02/07
 * @Description: 白色电脑装饰器
 */
public class WhiteComputerDecorator extends ComputerDecorator {
    public WhiteComputerDecorator(Computer computer) {
        super(computer);
    }

    @Override
    public void make() {
        super.make();
        paintColor();
    }

    private void paintColor() {
        System.out.println("给电脑涂上白色");
    }
}
```

(5) 创建装饰器模式测试类，代码如下：

```java
/**
 * @Author: zhouguanya
 * @Date: 2019/02/07
 * @Description: 装饰器模式测试类
 */
public class DecoratorPatternDemo {
    public static void main(String[] args) {
        Computer whiteComputer = new WhiteComputerDecorator(new ASUS());
        whiteComputer.make();
        Computer goldenComputer = new GoldenComputerDecorator(new Lenovo());
        goldenComputer.make();
        Computer redComputer = new RedComputerDecorator(new MacBook());
        redComputer.make();
    }
}
```

执行测试类,执行结果如下:

```
produce a ASUS Computer
给电脑涂上白色
produce a Lenovo Computer
给电脑涂上金色
produce a MacBook Computer
给电脑涂上红色
```

A.11 外观模式

外观模式隐藏了系统的复杂性,并为子系统中的一组接口提供了一个统一的高层访问接口,这个接口使得子系统更容易被访问或者使用。外观模式的优点是用户使用方便,把过度拆分的分散功能,组合成一个整体,对外提供一个统一的接口,隐藏了底层实现。

以医院看病为例,把医院作为一个系统,按照部门职能,该系统可以划分为挂号、门诊、化验、缴费、取药等部门。病人要与这些部门打交道,就如同一个系统的客户端与一个系统的各个不同的类打交道,客户端是需要处理很多逻辑,以确定何时应该调用某个类,如图 A-11 所示。

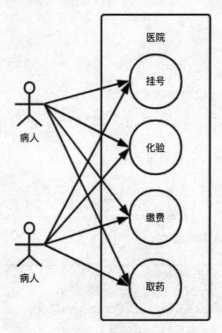

图 A-11 看病场景示意图

解决这种客户端复杂度的方法是使用门面模式,医院可以设置一个接待员的岗位,由接待员负责引导病人挂号、划价、缴费、取药等。这个接待员就是门面模式的体现,病人只接触接待员,由接待员与各个部门打交道,如图 A-12 所示。

通过以下案例说明外观设计模式的使用。本例中使用的 ASUS、Lenovo 和 MacBook 类参考 A.1 节。外观模式示意图如图 A-13 所示。

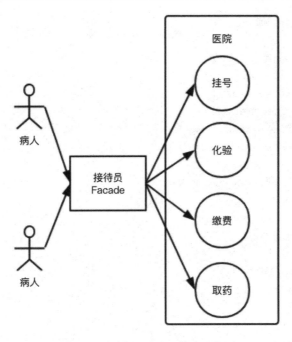

图 A-12　看病场景外观模式示意图

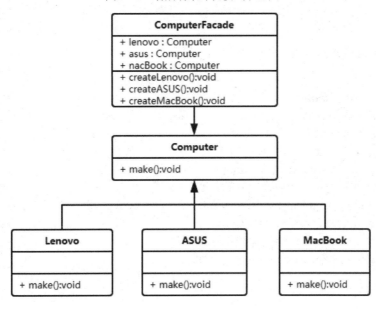

图 A-13　外观模式示意图

通过以下案例说明外观设计模式的使用。本例中使用的 ASUS、Lenovo 和 MacBook 类参考 A.1 节。

（1）创建电脑门面，代码如下：

```
/**
 * @Author: zhouguanya
 * @Date: 2019/02/07
```

```java
 * @Description: 门面
 */
public class ComputerFacade {
    private Computer lenovo;
    private Computer asus;
    private Computer macBook;
    public ComputerFacade() {
        lenovo = new Lenovo();
        asus = new ASUS();
        macBook = new MacBook();
    }
    public void makeLenovo(){
        lenovo.make();
    }
    public void makeASUS(){
        asus.make();
    }
    public void makeMacBook(){
        macBook.make();
    }
}
```

（2）创建外观模式测试类，代码如下：

```java
/**
 * @Author: zhouguanya
 * @Date: 2019/02/07
 * @Description: 测试类
 */
public class FacadePatternDemo {
    public static void main(String[] args) {
        ComputerFacade computerFacade = new ComputerFacade();
        computerFacade.makeLenovo();
        computerFacade.makeASUS();
        computerFacade.makeMacBook();
    }
}
```

运行外观模式测试类，执行结果如下：

```
produce a Lenovo Computer
produce a ASUS Computer
produce a MacBook Computer
```

A.12 享元模式

享元模式主要通过减少创建对象的数量，从而达到提高性能的目的。享元模式尝试重用现有的同类对象，如果现有的对象未匹配，则创建新对象。享元模式示意图如图 A-14 所示。

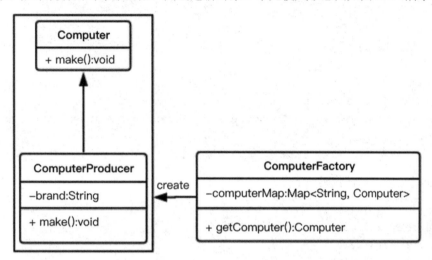

图 A-14　外观模式示意图

通过以下案例说明外观设计模式的使用。

（1）创建 Computer 接口，代码如下：

```
/**
 * @Author: zhouguanya
 * @Date: 2019/02/05
 * @Description: 电脑接口
 */
public interface Computer {
    /**
     * 制造电脑的方法
     */
    void make();
}
```

（2）创建 ComputerProducer 类，代码如下：

```
/**
 * @Author: zhouguaya
 * @Date: 2019/02/05
 * @Description: 电脑制造类
 */
```

```java
public class ComputerProducer implements Computer {

    private String brand;

    public ComputerProducer(String brand) {
        this.brand = brand;
    }

    /**
     * 制造电脑
     */
    @Override
    public void make() {
        System.out.println("produce a " + brand + " Computer");
    }
}
```

（3）创建电脑工厂，代码如下：

```java
/**
 * @Author: zhouguanya
 * @Date: 2019/02/07
 * @Description: 电脑工厂
 */
public class ComputerFactory {
    private static final Map<String, Computer> computerMap = new HashMap<>();

    public static Computer getComputer(String brand) {
        Computer computer = computerMap.get(brand);
        if (computer == null) {
            computer = new ComputerProducer(brand);
            computerMap.put(brand, computer);

        }
        return computer;
    }
}
```

（4）创建享原模式测试类，代码如下：

```java
/**
 * @Author: zhouguanya
 * @Date: 2019/02/07
 * @Description: 享原模式测试类
 */
public class FlyweightPatternDemo {
    static String brands[] = { "ASUA", "Lenovo", "MacBook"};
```

```java
    public static void main(String[] args) {

        for (int i = 0; i < 20; i++) {
            Computer computer = ComputerFactory.getComputer(getRandomBrand());
            computer.make();
        }
    }

    private static String getRandomBrand() {
        return brands[(int)(Math.random() * brands.length)];
    }
}
```

运行享原模式测试类,执行结果如下:

```
produce a ASUA Computer
produce a MacBook Computer
produce a Lenovo Computer
produce a Lenovo Computer
produce a Lenovo Computer
produce a Lenovo Computer
produce a Lenovo Computer
produce a MacBook Computer
produce a MacBook Computer
produce a ASUA Computer
produce a Lenovo Computer
produce a ASUA Computer
produce a Lenovo Computer
produce a Lenovo Computer
produce a MacBook Computer
produce a MacBook Computer
produce a ASUA Computer
produce a Lenovo Computer
produce a Lenovo Computer
produce a ASUA Computer
```

A.13 代理模式

代理模式使用一个类代表另一个类的功能,通过代理可以控制对这个对象的访问。代理模式示意图如图 A-15 所示。

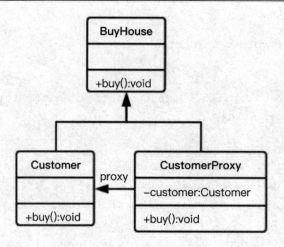

图 A-15　外观模式示意图

通过以下案例说明代理设计模式的使用。

（1）创建 BuyHouse 接口，代码如下：

```
/**
 * @Author: zhouguanya
 * @Date: 2019/02/07
 * @Description: 买房接口
 */
public interface BuyHouse {
    void buy();
}
```

（2）创建 Customer 类，代码如下：

```
/**
 * @Author: zhouguanya
 * @Date: 2019/02/07
 * @Description: 客户
 */
public class Customer implements BuyHouse {
    @Override
    public void buy() {
        System.out.println("我是客户，我想买房");
    }
}
```

（3）创建客户代理类 CustomerProxy，代码如下：

```
/**
 * @Author: zhouguanya
 * @Date: 2019/02/07
 * @Description: 客户代理
```

```java
 */
public class CustomerProxy implements BuyHouse {
    private Customer customer;

    public CustomerProxy(Customer customer) {
        this.customer = customer;
    }

    @Override
    public void buy() {
        customer.buy();
    }
}
```

（4）创建代理模式测试类，代码如下：

```java
/**
 * @Author: zhouguanya
 * @Date: 2019/02/07
 * @Description: 代理模式测试类
 */
public class ProxyPatternDemo {
    public static void main(String[] args) {
        CustomerProxy customerProxy = new CustomerProxy(new Customer());
        customerProxy.buy();
    }
}
```

运行代理模式测试类，执行结果如下：

我是客户，我想买房

A.14 责任链模式

责任链模式为请求创建了一系列处理对象，这些处理对象的形成链条。责任模式将请求的发送者和处理对象进行解耦。在这种模式中，如果一个处理对象不能处理该请求，那么该处理对象将会把相同的请求传给下一个接收者，以此类推。责任链的使用场景有 Struts 中的拦截器和 Servlet 中的过滤器等。责任链模式示意图如图 A-16 所示。

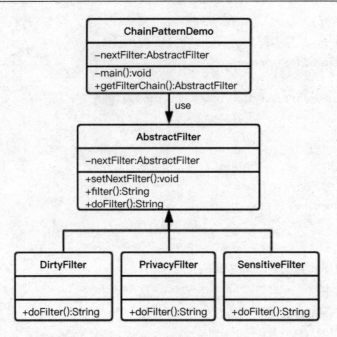

图 A-16 责任链模式示意图

通过以下案例说明责任链设计模式的使用。

（1）创建抽象类 AbstractFilter，代码如下：

```java
/**
 * @Author: zhouguanya
 * @Date: 2019/02/07
 * @Description:
 */
public abstract class AbstractFilter {
    AbstractFilter nextFilter;
    //责任链中的下一个元素
    public void setNextFilter(AbstractFilter nextFilter) {
        this.nextFilter = nextFilter;
    }

    public String filter(String content) {
        String filtered = doFilter(content);
        if (nextFilter != null) {
            return nextFilter.filter(filtered);
        }
        return filtered;
    }

    protected abstract String doFilter(String content);

}
```

（2）创建 DirtyFilter 过滤器，代码如下：

```java
/**
 * @Author: zhouguanya
 * @Date: 2019/02/07
 * @Description:
 */
public class DirtyFilter extends AbstractFilter {
    @Override
    protected String doFilter(String content) {
        return content.replace("Dirty Word","dw");
    }
}
```

（3）创建 PrivacyFilter 过滤器，代码如下：

```java
/**
 * @Author: zhouguanya
 * @Date: 2019/02/07
 * @Description:
 */
public class PrivacyFilter extends AbstractFilter {
    @Override
    protected String doFilter(String content) {
        return content.replace("Privacy Word", "pw");
    }
}
```

（4）创建 SensitiveFilter 过滤器，代码如下：

```java
/**
 * @Author: zhouguanya
 * @Date: 2019/02/07
 * @Description:
 */
public class SensitiveFilter extends AbstractFilter {
    @Override
    protected String doFilter(String content) {
        return content.replace("Sensitive Word", "sw");
    }
}
```

（5）创建责任链模式测试类，代码如下：

```java
/**
 * @Author: zhouguanya
 * @Date: 2019/02/07
 * @Description: 责任链模式测试类
```

```
 */
public class ChainPatternDemo {
    static String content = "Dirty Word, Privacy Word, Sensitive Word";
    public static void main(String[] args) {
        AbstractFilter filterChain = getFilterChain();
        System.out.println(filterChain.filter(content));
    }
    private static AbstractFilter getFilterChain() {
        DirtyFilter dirtyFilter = new DirtyFilter();
        PrivacyFilter privacyFilter = new PrivacyFilter();
        SensitiveFilter sensitiveFilter = new SensitiveFilter();
        dirtyFilter.setNextFilter(privacyFilter);
        privacyFilter.setNextFilter(sensitiveFilter);
        sensitiveFilter.setNextFilter(null);
        return dirtyFilter;
    }
}
```

执行责任链模式测试类，运行结果如下：

```
dw, pw, sw
```

A.15 命 令 模 式

命令模式是一种数据驱动的设计模式。命令模式请求将以命令的形式封装在对象中，并传递给调用对象，调用对象将寻找可以处理该命令的合适的对象，并把该命令传给相应的处理对象进行处理。命令模式示意图如图 A-17 所示。

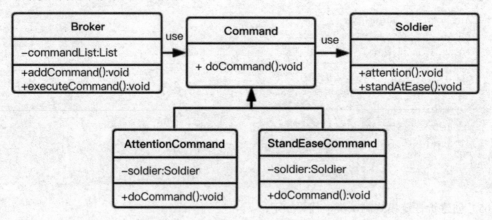

图 A-17 命令模式示意图

通过以下案例说明命令设计模式的使用。

（1）创建 Command 接口，代码如下：

```java
/**
 * @Author: zhouguanya
 * @Date: 2019/02/07
 * @Description:
 */
public interface Command {
    void doCommand();
}
```

（2）创建 Soldier 类，代码如下：

```java
/**
 * @Author: zhouguanya
 * @Date: 2019/02/07
 * @Description: 士兵
 */
public class Soldier {

    public void attention() {
        System.out.println("立正");
    }

    public void standAtEase() {
        System.out.println("稍息");
    }
}
```

（3）创建 AttentionCommand 命令，代码如下：

```java
/**
 * @Author: zhouguanya
 * @Date: 2019/02/07
 * @Description: 立正命令
 */
public class AttentionCommand implements Command {
    private Soldier soldier;

    public AttentionCommand(Soldier soldier) {
        this.soldier = soldier;
    }

    @Override
    public void doCommand() {
        soldier.attention();
    }
}
```

（4）创建 StandEaseCommand 命令，代码如下：

```java
/**
 * @Author: zhouguanya
 * @Date: 2019/02/07
 * @Description: 稍息命令
 */
public class StandEaseCommand implements Command {
    private Soldier soldier;

    public StandEaseCommand(Soldier soldier) {
        this.soldier = soldier;
    }

    @Override
    public void doCommand() {
        soldier.standAtEase();
    }
}
```

（5）创建命令调用类，代码如下：

```java
/**
 * @Author: zhouguanya
 * @Date: 2019/02/07
 * @Description: 命令的调用类
 */
public class Broker {
    private List<Command> commandList = new ArrayList<>();

    /**
     * 添加命令
     */
    public void addCommand(Command command) {
        commandList.add(command);
    }

    /**
     * 执行命令
     */
    public void executeCommand() {
        for (Command command : commandList) {
            command.doCommand();
        }
    }
}
```

（6）命令模式测试类，代码如下：

```java
/**
 * @Author: zhouguanya
 * @Date: 2019/02/07
 * @Description: 命令模式测试类
 */
public class CommandPatternDemo {
    public static void main(String[] args) {
        Soldier soldier = new Soldier();
        AttentionCommand attentionCommand = new AttentionCommand(soldier);
        StandEaseCommand standEaseCommand = new StandEaseCommand(soldier);
        Broker broker = new Broker();
        broker.addCommand(attentionCommand);
        broker.addCommand(standEaseCommand);
        broker.executeCommand();
    }
}
```

执行命令模式测试类，运行结果如下：

立正
稍息

A.16 解释器模式

解释器模式提供了解析语法或表达式的功能。解释器模式被用在 SQL 解析、符号处理引擎等。解释器模式示意图如图 A-18 所示。

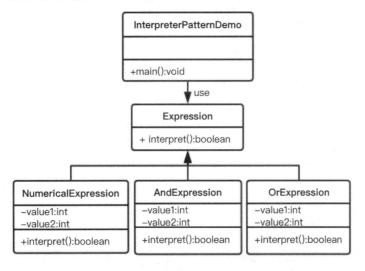

图 A-18 解释器模式示意图

通过以下案例说明解释器设计模式的使用。

（1）创建 Expression 接口，代码如下：

```java
/**
 * @Author: zhouguanya
 * @Date: 2019/02/07
 * @Description: 解释器接口
 */
public interface Expression {
    boolean interpret();
}
```

（2）创建 NumericalExpression 数值解释器，代码如下：

```java
/**
 * @Author: zhouguanya
 * @Date: 2019/02/07
 * @Description:
 */
public class NumericalExpression implements Expression {
    private int value1;
    private int value2;

    public NumericalExpression(int value1, int value2) {
        this.value1 = value1;
        this.value2 = value2;
    }

    @Override
    public boolean interpret() {
        return (value1 - value2) > 0;
    }
}
```

（3）创建 AndExpression 与表达式解释器，代码如下：

```java
/**
 * @Author: zhouguanya
 * @Date: 2019/02/07
 * @Description: 与表达式
 */
public class AndExpression implements Expression {
    private Expression expression1 = null;
    private Expression expression2 = null;
    public AndExpression(Expression expression1, Expression expression2) {
        this.expression1 = expression1;
        this.expression2 = expression2;
```

```java
    }

    @Override
    public boolean interpret() {
        return expression1.interpret() && expression2.interpret();
    }
}
```

(4) 创建 OrExpression 或表达式解释器，代码如下：

```java
/**
 * @Author: zhouguanya
 * @Date: 2019/02/07
 * @Description: 或表达式解释器
 */
public class OrExpression implements Expression {
    private Expression expression1 = null;
    private Expression expression2 = null;

    public OrExpression(Expression expression1, Expression expression2) {
        this.expression1 = expression1;
        this.expression2 = expression2;
    }

    @Override
    public boolean interpret() {
        return expression1.interpret() || expression2.interpret();
    }
}
```

(5) 创建解释器模式测试类，代码如下：

```java
/**
 * @Author: zhouguanya
 * @Date: 2019/02/07
 * @Description: 解释器模式测试类
 */
public class InterpreterPatternDemo {
    public static void main(String[] args) {
        NumericalExpression expression1 = new NumericalExpression(10, 8);
        NumericalExpression expression2 = new NumericalExpression(10, 20);
        AndExpression andExpression = new AndExpression(expression1,expression2);
        OrExpression orExpression = new OrExpression(expression1, expression2);
        System.out.println("10 > 8 && 10 > 20 ? " + andExpression.interpret());
        System.out.println("10 > 8 || 10 > 20 ? " + orExpression.interpret());
    }
}
```

执行解释器模式测试类，运行结果如下：

```
10 > 8 && 10 > 20 ? false
10 > 8 || 10 > 20 ? true
```

A.17 迭代器模式

迭代器模式是 JDK 中常用的设计模式。使用这种模式会顺序访问集合对象的元素，不需要知道集合对象的底层存储情况。迭代器模式示意图如图 A-19 所示。

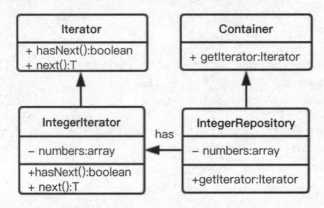

图 A-19 解释器模式示意图

通过以下案例说明迭代器设计模式的使用。

（1）创建迭代器接口，代码如下：

```
/**
 * @Author: zhouguanya
 * @Date: 2019/02/07
 * @Description: 迭代器接口
 */
public interface Iterator<T> {
    boolean hasNext();
    T next();
}
```

（2）建迭代器实现类，代码如下：

```
/**
 * @Author: zhouguanya
 * @Date: 2019/02/07
 * @Description:
 */
public class IntegerIterator implements Iterator<Integer> {
    private Integer[] numbers;
```

```java
        int index;
        public IntegerIterator (Integer[] numbers) {
            this.numbers = numbers;
        }

        @Override
        public boolean hasNext() {
            if (index < numbers.length) {
                return true;
            }
            return false;
        }

        @Override
        public Integer next() {
            if (hasNext()) {
                return numbers[index++];
            }
            return null;
        }
    }
```

（3）创建 Container 接口，代码如下：

```java
/**
 * @Author: zhouguanya
 * @Date: 2019/02/07
 * @Description:
 */
public interface Container {
    Iterator getIterator();
}
```

（4）创建 Container 接口实现类，代码如下：

```java
/**
 * @Author: zhouguanya
 * @Date: 2019/02/07
 * @Description:
 */
public class IntegerRepository implements Container {
    private Integer[] numbers;

    public IntegerRepository (Integer[] numbers) {
        this.numbers = numbers;
    }
```

```java
    @Override
    public Iterator getIterator() {
        return new IntegerIterator(numbers);
    }
}
```

（5）建迭代器模式测试类，代码如下：

```java
/**
 * @Author: zhouguanya
 * @Date: 2019/02/07
 * @Description: 迭代器模式测试类
 */
public class IteratorPatternDemo {
    public static void main(String[] args) {
        Integer numbers[] = {1, 2, 3, 4, 5};
        IntegerRepository repository = new IntegerRepository(numbers);
        Iterator iterator = repository.getIterator();
        while (iterator.hasNext()) {
            System.out.println(iterator.next());
        }
    }
}
```

执行迭代器模式测试类，运行结果如下：

```
1
2
3
4
5
```

A.18 中介者模式

中介者模式通常用来降低多个对象间沟通的复杂度。中介者模式提供了一个中介类，这个类处理不同对象之间的沟通，是多个对象之间保持松耦合，使代码易于维护。中介者模式示意图如图A-20所示。

通过以下案例说明中介者设计模式的使用。通过论坛实例来演示中介者模式。实例中，多个用户可以在论坛留言，论坛显示所有的用户留言。

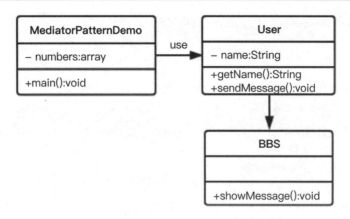

图 A-20 中介者模式示意图

（1）创建 BBS 类，代码如下：

```
/**
 * @Author: zhouguanya
 * @Date: 2019/02/07
 * @Description: 论坛
 */
public class BBS {
    public static void showMessage(User user, String message){
        System.out.println(new Date().toString() + " [" + user.getName() +"] : " + message);
    }
}
```

（2）创建用户实体，代码如下：

```
/**
 * @Author: zhouguanya
 * @Date: 2019/02/07
 * @Description: 用户
 */
public class User {
    private String name;

    public String getName() {
        return name;
    }

    public User(String name){
        this.name = name;
    }

    public void sendMessage(String message){
```

```
        BBS.showMessage(this,message);
    }
}
```

（3）创建中介者模式测试类，代码如下：

```
/**
 * @Author: zhouguanya
 * @Date: 2019/02/07
 * @Description: 中介者模式测试类
 */
public class MediatorPatternDemo {
    public static void main(String[] args) {
        User robert = new User("Tom");
        User john = new User("John");
        robert.sendMessage("Headline : celestial dog devouring the sun !");
        john.sendMessage("Reply : no, it is Solar Eclipse ~");
    }
}
```

运行中介者模式测试类，执行结果如下：

```
Thu Feb 07 21:51:32 CST 2019 [Tom] : Headline : celestial dog devouring the sun !
Thu Feb 07 21:51:32 CST 2019 [John] : Reply : no, it is Solar Eclipse ~
```

A.19　备忘录模式

备忘录模式用于保存一个对象的状态，以便在适当的时候恢复对象。备忘录模式在不破坏封装的前提下，捕获一个对象的内部状态，并在该对象之外保存这个状态，这样可以在以后将对象恢复到原先保存的状态。备忘录模式示意图如图 A-21 所示。

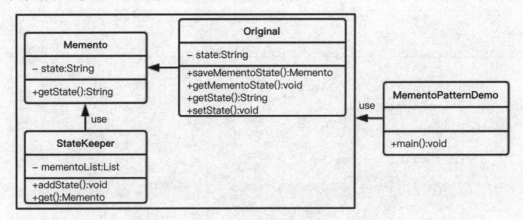

图 A-21　备忘录模式示意图

通过以下案例说明备忘录设计模式的使用。

（1）创建 Memento 类，代码如下：

```
/**
 * @Author: zhouguanya
 * @Date: 2019/02/07
 * @Description:
 */
public class Memento {
    private String state;
    public Memento(String state){
        this.state = state;
    }
    public String getState(){
        return state;
    }
}
```

（2）创建原始类，代码如下：

```
/**
 * @Author: zhouguanya
 * @Date: 2019/02/07
 * @Description: 原始类
 */
public class Original {
    private String state;
    public String getState() {
        return state;
    }
    public void setState(String state) {
        this.state = state;
    }
    public Memento saveMementoState() {
        return new Memento(state);
    }
    public void getMementoState(Memento memento) {
        state = memento.getState();
    }
}
```

（3）创建状态管理类，代码如下：

```
/**
 * @Author: zhouguanya
```

```java
 * @Date: 2019/02/07
 * @Description: 状态管理类
 */
public class StateKeeper {
    private List<Memento> mementoList = new ArrayList<>();

    public void addState(Memento state){
        mementoList.add(state);
    }

    public Memento get(int index){
        return mementoList.get(index);
    }
}
```

（4）创建备忘录模式测试类，代码如下：

```java
/**
 * @Author: zhouguanya
 * @Date: 2019/02/07
 * @Description: 备忘录模式测试类
 */
public class MementoPatternDemo {
    public static void main(String[] args) {
        Original original = new Original();
        StateKeeper keeper = new StateKeeper();
        original.setState("State 1");
        keeper.addState(original.saveMementoState());
        //状态变更
        original.setState("State 2");
        keeper.addState(original.saveMementoState());
        //状态变更
        original.setState("State 3");
        System.out.println("Current State is :" + original.getState());
        //第1次保存的状态
        original.getMementoState(keeper.get(0));
        System.out.println("Initial State: " + original.getState());
        //第2次保存的状态
        original.getMementoState(keeper.get(1));
        System.out.println("Second State: " + original.getState());
    }
}
```

运行备忘录模式测试类，执行结果如下：

```
Current State is :State 3
Initial State: State 1
Second State: State 2
```

A.20　观察者模式

当对象间存在一对多关系时，则使用观察者模式。观察者模式的主要作用是当一个对象的状态发生改变时，所有依赖于它的对象都得到通知并被自动更新，备忘录模式示意图如图 A-22 所示。

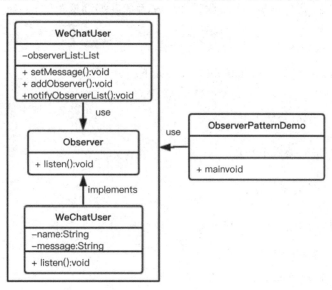

图 A-22　观察者模式示意图

通过以下案例说明观察者设计模式的使用。

（1）创建观察接口 Observer，代码如下：

```
/**
 * @Author: zhouguanya
 * @Date: 2019/02/08
 * @Description: 观察接口
 */
public interface Observer {
    void listen(String message);
}
```

（2）创建微信用户类 WeChatUser，实现 Observer 接口，代码如下：

```
/**
 * @Author: zhouguanya
 * @Date: 2019/02/08
 * @Description: 微信用户
 */
public class WeChatUser implements Observer {
    private String name;
```

```java
    private String message;
    public WeChatUser(String name) {
        this.name = name;
    }
    @Override
    public void listen(String message) {
        this.message = message;
        System.out.printf("%s 收到微信公众号的消息：%s%n", name, this.message);
    }
}
```

（3）创建微信公众号类 WeChatPulic，代码如下：

```java
/**
 * @Author: zhouguanya
 * @Date: 2019/02/08
 * @Description: 微信公众号
 */
public class WeChatPulic {
    //观察者列表
    private List<Observer> observerList;

    private String message;

    public WeChatPulic() {
        observerList = new ArrayList<>();
    }
    // 状态变更通知观察者
    public void setMessage(String message) {
        this.message = message;
        System.out.println("微信公众号更新消息: " + message);
        notifyObserverList();
    }

    public void addObserver(Observer observer) {
        observerList.add(observer);
    }

    public void notifyObserverList() {
        for (Observer observer : observerList) {
            observer.listen(message);
        }
    }
}
```

（4）创建观察者模式测试类，代码如下：

```java
/**
 * @Author: zhouguanya
```

```
 * @Date: 2019/02/08
 * @Description: 观察者模式测试类
 */
public class ObserverPatternDemo {
    public static void main(String[] args) {
        WeChatPulic weChatPulic = new WeChatPulic();
        WeChatUser weChatUser1 = new WeChatUser("Jack");
        WeChatUser weChatUser2 = new WeChatUser("Tom");
        WeChatUser weChatUser3 = new WeChatUser("John");
        weChatPulic.addObserver(weChatUser1);
        weChatPulic.addObserver(weChatUser2);
        weChatPulic.addObserver(weChatUser3);
        weChatPulic.setMessage("Hello World");
    }
}
```

执行观察者模式测试类，运行结果如下：

```
微信公众号更新消息：Hello World
Jack 收到微信公众号的消息：Hello World
Tom 收到微信公众号的消息：Hello World
John 收到微信公众号的消息：Hello World
```

A.21 状 态 模 式

在状态模式中，类的行为是基于状态改变的，可以创建表示各种状态的对象和一个行为随着状态对象改变而改变的 Context 对象。状态模式示意图如图 A-23 所示。

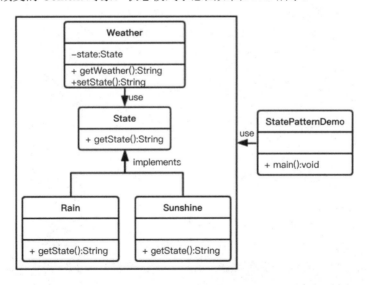

图 A-23 状态模式示意图

通过以下案例说明状态设计模式的使用。

（1）创建状态接口 State，代码如下：

```
/**
 * @Author: zhouguanya
 * @Date: 2019/02/08
 * @Description: 状态
 */
public interface State {
    //获取天气情况
    String getState();
}
```

（2）创建 Rain 类实现 State 接口，代码如下：

```
/**
 * @Author: zhouguanya
 * @Date: 2019/02/08
 * @Description: 下雨
 */
public class Rain implements State {
    @Override
    public String getState() {
        return "今天的天气：下雨";
    }
}
```

（3）创建 Sunshine 实现 State 接口，代码如下：

```
/**
 * @Author: zhouguanya
 * @Date: 2019/02/08
 * @Description: 晴天
 */
public class Sunshine implements State {
    @Override
    public String getState() {
        return "今天的天气：晴天";
    }
}
```

（4）创建天气类，代码如下：

```
/**
 * @Author: zhouguanya
 * @Date: 2019/02/08
 * @Description: 天气类
 */
```

```java
public class Weather {
    private State state;
    public void setState(State state) {
        this.state = state;
    }
    public String getWeather() {
        return state.getState();
    }
}
```

（5）创建状态模式测试类，代码如下：

```java
/**
 * @Author: zhouguanya
 * @Date: 2019/02/08
 * @Description: 状态模式测试类
 */
public class StatePatternDemo {
    public static void main(String[] args) {
        Weather weather = new Weather();
        weather.setState(new Rain());
        System.out.println(weather.getWeather());
        weather.setState(new Sunshine());
        System.out.println(weather.getWeather());
    }
}
```

执行状态模式测试类，运行结果如下：

```
今天的天气：下雨
今天的天气：晴天
```

A.22 空对象模式

在空对象模式中，使用一个空对象取代 null。空对象可以加强系统的稳固性，能有效地防止空指针报错对整个系统的影响，使系统更加稳定。空对象模式示意图如图 A-24 所示。

通过以下案例说明空对象设计模式的使用。

（1）创建空对象接口 AbstractObject，代码如下：

```java
/**
 * @Author: zhouguanya
 * @Date: 2019/02/08
 * @Description: 空对象接口
 */
```

```java
public interface AbstractObject {
    boolean isNull();
    void show();
}
```

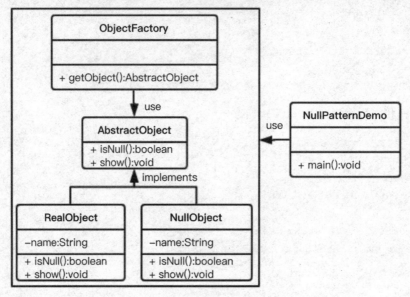

图 A-24　空对象模式示意图

（2）创建真实对象 RealObject 实现 AbstractObject 接口，代码如下：

```java
/**
 * @Author: zhouguanya
 * @Date: 2019/02/08
 * @Description:
 */
public class RealObject implements AbstractObject {
    private String name;
    public RealObject(String name) {
        this.name = name;
    }
    @Override
    public boolean isNull() {
        return false;
    }

    @Override
    public void show() {
        System.out.println("real object " + name + " shows now");
    }
}
```

（3）创建 NullObject 实现 AbstractObject 接口，代码如下：

```java
/**
 * @Author: zhouguanya
 * @Date: 2019/02/08
 * @Description:
 */
public class NullObject implements AbstractObject {
    private String name;
    public NullObject(String name) {
        this.name = name;
    }
    @Override
    public boolean isNull() {
        return true;
    }

    @Override
    public void show() {
        // do nothing
        System.out.println(name + " Object not exist");
    }
}
```

（4）创建对象工厂，代码如下：

```java
/**
 * @Author: zhouguanya
 * @Date: 2019/02/08
 * @Description: 对象工厂
 */
public class ObjectFactory {
    public static final String[] names = {"table", "light", "bed"};
    public static AbstractObject getObject(String name){
        for (int i = 0; i < names.length; i++) {
            if (names[i].equalsIgnoreCase(name)){
                return new RealObject(name);
            } else {
                return new NullObject(name);
            }
        }
        return new NullObject("");
    }
}
```

（5）创建空对象模式实现类，代码如下：

```java
/**
 * @Author: zhouguanya
 * @Date: 2019/02/08
 * @Description: 空对象模式实现类
 */
public class NullPatternDemo {
    public static void main(String[] args) {
        AbstractObject object1 = ObjectFactory.getObject("light");
        AbstractObject object2 = ObjectFactory.getObject("bed");
        AbstractObject object3 = ObjectFactory.getObject("table");
        AbstractObject object4 = ObjectFactory.getObject("sun");
        object1.show();
        object2.show();
        object3.show();
        object4.show();
    }
}
```

运行空对象模式实现类，执行结果如下：

```
light Object not exist
bed Object not exist
real object table shows now
sun Object not exist
```

A.23 策略模式

策略模式定义了一系列的算法，并将每一个算法封装起来，使每个算法可以相互替代，将算法和使用算法的客户端分割开来，相互独立。策略模式示意图如图 A-25 所示。

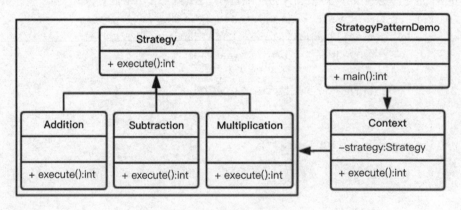

图 A-25　策略模式示意图

通过以下案例说明策略设计模式的使用。

（1）创建 Strategy 接口，代码如下：

```
/**
 * @Author: zhouguanya
 * @Date: 2019/02/08
 * @Description: 策略类
 */
public interface Strategy {
    int execute(int num1, int num2);
}
```

（2）创建 Strategy 接口实现类 Addition，代码如下：

```
/**
 * @Author: zhouguanya
 * @Date: 2019/02/08
 * @Description: 加法
 */
public class Addition implements Strategy {

    @Override
    public int execute(int num1, int num2) {
        return num1 + num2;
    }
}
```

（3）创建 Strategy 接口实现类 Subtraction，代码如下：

```
/**
 * @Author: zhouguanya
 * @Date: 2019/02/08
 * @Description: 减法
 */
public class Subtraction implements Strategy {
    @Override
    public int execute(int num1, int num2) {
        return num1 - num2;
    }
}
```

（4）创建 Strategy 接口实现类 Multiplication，代码如下：

```
/**
 * @Author: zhouguanya
 * @Date: 2019/02/08
 * @Description:
 */
```

```java
public class Multiplication implements Strategy {
    @Override
    public int execute(int num1, int num2) {
        return num1 * num2;
    }
}
```

(5)创建 Context，代码如下：

```java
/**
 * @Author: zhouguanya
 * @Date: 2019/02/08
 * @Description:
 */
public class Context {
    private Strategy strategy;
    public Context(Strategy strategy) {
        this.strategy = strategy;
    }

    public int execute(int num1, int num2) {
        return strategy.execute(num1, num2);
    }
}
```

(6)创建策略模式测试类，代码如下：

```java
/**
 * @Author: zhouguanya
 * @Date: 2019/02/08
 * @Description: 策略模式测试类
 */
public class StrategyPatternDemo {
    public static void main(String[] args) {
        Context context = new Context(new Addition());
        System.out.println("10 + 20 = " + context.execute(10, 20));
        context = new Context(new Subtraction());
        System.out.println("10 - 20 = " + context.execute(10, 20));
        context = new Context(new Multiplication());
        System.out.println("10 * 20 = " + context.execute(10, 20));
    }
}
```

运行策略模式测试类，执行结果如下：

```
10 + 20 = 30
10 - 20 = -10
10 * 20 = 200
```

A.24 模板模式

模板模式定义一个算法的骨架，可将一些具体的步骤延迟到子类中。模板方法使子类可以不改变一个算法的结构即可重定义该算法的某些特定步骤。模板模式示意图如图 A-26 所示。

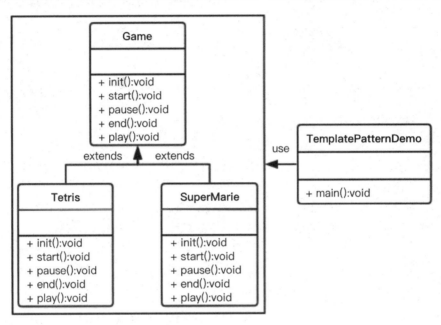

图 A-26 模板模式示意图

通过以下案例说明模板设计模式的使用。

（1）创建 Game 抽象类，代码如下：

```java
/**
 * @Author: zhouguanya
 * @Date: 2019/02/08
 * @Description: 游戏类
 */
public abstract class Game {
    abstract void init();
    abstract void start();
    abstract void pause();
    abstract void end();

    void play() {
        init();
        start();
        pause();
```

```
            end();
        }
}
```

（2）创建 Game 子类 Tetris，代码如下：

```java
/**
 * @Author: zhouguanya
 * @Date: 2019/02/08
 * @Description: 俄罗斯方块
 */
public class Tetris extends Game {
    @Override
    void init() {
        System.out.println("Init Tetris Game");
    }

    @Override
    void start() {
        System.out.println("Start Tetris Game");
    }

    @Override
    void pause() {
        System.out.println("Pause Tetris Game");
    }

    @Override
    void end() {
        System.out.println("End Tetris Game");
    }
}
```

（3）创建 Game 子类 SuperMarie，代码如下：

```java
/**
 * @Author: zhouguanya
 * @Date: 2019/02/08
 * @Description: 超级玛丽
 */
public class SuperMarie extends Game {
    @Override
    void init() {
        System.out.println("Init SuperMarie Game");
    }

    @Override
```

```java
    void start() {
        System.out.println("Start SuperMarie Game");
    }

    @Override
    void pause() {
        System.out.println("Pause SuperMarie Game");
    }

    @Override
    void end() {
        System.out.println("End SuperMarie Game");
    }
}
```

(4)创建模板模式测试类,代码如下:

```java
/**
 * @Author: zhouguanya
 * @Date: 2019/02/08
 * @Description: 模板模式测试类
 */
public class TemplatePatternDemo {
    public static void main(String[] args) {
        Game tetris = new Tetris();
        tetris.play();
        System.out.println("-----------分割线-----------");
        Game superMarie = new SuperMarie();
        superMarie.play();
    }
}
```

运行模板模式测试类,执行结果如下:

```
Init Tetris Game
Start Tetris Game
Pause Tetris Game
End Tetris Game
-----------分割线-----------
Init SuperMarie Game
Start SuperMarie Game
Pause SuperMarie Game
End SuperMarie Game
```

A.25 拦截过滤器模式

拦截过滤器模式用于对请求或响应做一系列处理。过滤器可以做认证/授权/记录日志，或者跟踪请求，然后把请求传给相应的处理程序。拦截过滤器模式示意图如图 A-27 所示。

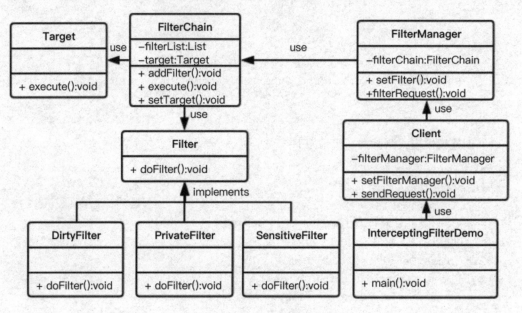

图 A-27 拦截过滤器模式示意图

通过以下案例说明拦截过滤器设计模式的使用。

（1）创建目标对象 Target，代码如下：

```
/**
 * @Author: zhouguanya
 * @Date: 2019/02/08
 * @Description: 目标对象
 */
public class Target {
    public void execute(){
        System.out.println("The Final Target Object");
    }
}
```

（2）创建过滤器接口 Filter，代码如下：

```
/**
 * @Author: zhouguanya
 * @Date: 2019/02/08
 * @Description: 过滤器
```

```
 */
public interface Filter {
    /**
     * 过滤方法
     */
    void doFilter();
}
```

(3)创建 DirtyFilter 过滤器,代码如下:

```
/**
 * @Author: zhouguanya
 * @Date: 2019/02/08
 * @Description: 脏话过滤器
 */
public class DirtyFilter implements Filter {
    /**
     * 过滤方法
     */
    @Override
    public void doFilter() {
        System.out.println("Execute Dirty Words Filter");
    }
}
```

(4)创建 PrivateFilter 过滤器,代码如下:

```
/**
 * @Author: zhouguanya
 * @Date: 2019/02/08
 * @Description: 隐私过滤器
 */
public class PrivateFilter implements Filter {
    /**
     * 过滤方法
     */
    @Override
    public void doFilter() {
        System.out.println("Execute Private Words Filter");
    }
}
```

(5)创建 SensitiveFilter 过滤器,代码如下:

```
/**
 * @Author: zhouguanya
 * @Date: 2019/02/08
 * @Description: 敏感词过滤器
```

```
 */
public class SensitiveFilter implements Filter {
    /**
     * 过滤方法
     */
    @Override
    public void doFilter() {
        System.out.println("Execute Sensitive Words Filter");
    }
}
```

（6）创建过滤器链 FilterChain，代码如下：

```
/**
 * @Author: zhouguanya
 * @Date: 2019/02/08
 * @Description: 过滤器链
 */
public class FilterChain {
    //过滤器集合
    List<Filter> filterList = new ArrayList<>();
    private Target target;

    /**
     * 添加过滤器
     */
    public void addFilter(Filter filter){
        filterList.add(filter);
    }

    /**
     * 执行过滤器
     */
    public void execute(){
        //前置拦截
        for (Filter filter : filterList) {
            filter.doFilter();
        }
        //执行目标对象
        target.execute();
    }

    /**
     * 设置目标对象
     */
```

```java
    public void setTarget(Target target){
        this.target = target;
    }
}
```

（7）创建 FilterManager 管理过滤器链，代码如下：

```java
/**
 * @Author: zhouguanya
 * @Date: 2019/02/08
 * @Description: 过滤器管理员
 */
public class FilterManager {
    private FilterChain filterChain;

    public FilterManager(Target target){
        filterChain = new FilterChain();
        filterChain.setTarget(target);
    }
    public void setFilter(Filter filter){
        filterChain.addFilter(filter);
    }

    public void filterRequest(){
        filterChain.execute();
    }
}
```

（8）创建客户端，代码如下：

```java
/**
 * @Author: zhouguanya
 * @Date: 2019/02/08
 * @Description: 客户端
 */
public class Client {
    FilterManager filterManager;

    public void setFilterManager(FilterManager filterManager){
        this.filterManager = filterManager;
    }

    public void sendRequest(){
        filterManager.filterRequest();
    }
}
```

（9）创建拦截过滤器模式测试类，代码如下：

```java
/**
 * @Author: zhouguanya
 * @Date: 2019/02/08
 * @Description: 拦截过滤器模式测试类
 */
public class InterceptingFilterDemo {
    public static void main(String[] args) {

        FilterManager filterManager = new FilterManager(new Target());
        //装配各种过滤器
        filterManager.setFilter(new DirtyFilter());
        filterManager.setFilter(new PrivateFilter());
        filterManager.setFilter(new SensitiveFilter());

        Client client = new Client();
        client.setFilterManager(filterManager);
        client.sendRequest();
    }
}
```

运行拦截过滤器模式测试类，执行结果如下：

```
Execute Dirty Words Filter
Execute Private Words Filter
Execute Sensitive Words Filter
The Final Target Object
```

参 考 文 献

[1] https://baike.baidu.com/item/IntelliJ%20IDEA/9548353?fr=aladdin

[2] https://baike.baidu.com/item/jdk/1011?fr=aladdin

[3] https://baike.baidu.com/item/tomcat/255751?fr=aladdin

[4] https://baike.baidu.com/item/Maven/6094909?fr=aladdin

[5] https://projects.spring.io/spring-framework/

[6] 埃克尔.Java 编程思想（第 4 版）[M].北京：机械工业出版社，2007

[7] 郝佳.Spring 代码深度解析[M].北京：人民邮电出版社，2013

[8] 黄健宏.Redis 设计与实现[M].北京：机械工业出版社，2014

[9] 倪炜.分布式消息中间件实践[M].北京：电子工业出版社，2011

[10] 周继锋，冯钻优，陈胜尊，左越宗.分布式数据库架构及企业实践——基于 Mycat 中间件[M].北京：电子工业出版社，2016

[11] 刘伟.设计模式[M].北京：清华大学出版社，2011

[12] 徐郡明.MyBatis 技术内幕[M].北京：电子工业出版社，2017

[13] 郝佳.Spring 代码深度解析[M].北京：人民邮电出版社，2013

[14] 倪超.从 Paxos 到 Zookeeper [M].北京：电子工业出版社，2015

[15] http://dubbo.io/

[16] https://www.oschina.net/p/sharding-jdbc

[17] http://www.mybatis.org/mybatis-3/

[18] https://baike.baidu.com/item/junit/1211849?fr=aladdin

[19] https://logging.apache.org/log4j/2.x/

[20] 鸟哥.鸟哥的 Linux 私房菜[M].北京：人民邮电出版社，2010

[21] 李林锋.Netty 权威指南第 2 版[M].北京：电子工业出版社，2015